Minnesota Studies in the Philosophy of Science

Minnesota Studies in the Philosophy of Science

Alan C. Love, General Editor
Herbert Feigl, Founding Editor

ALSO IN THIS SERIES

Scientific Pluralism
Stephen H. Kellert, Helen E. Longino, and C. Kenneth Waters, Editors
VOLUME 19

Logical Empiricism in North America
Gary L. Hardcastle and Alan W. Richardson, Editors
VOLUME 18

Quantum Measurement: Beyond Paradox
Richard A. Healey and Geoffrey Hellman, Editors
VOLUME 17

Origins of Logical Empiricism
Ronald N. Giere and Alan W. Richardson, Editors
VOLUME 16

Cognitive Models of Science
Ronald N. Giere, Editor
VOLUME 15

THE LANGUAGE OF NATURE

Reassessing the Mathematization of Natural Philosophy in the Seventeenth Century

GEOFFREY GORHAM, BENJAMIN HILL,
EDWARD SLOWIK, AND C. KENNETH WATERS
EDITORS

Minnesota Studies in the Philosophy of Science 20

University of Minnesota Press
Minneapolis
London

Published by the University of Minnesota Press
111 Third Avenue South, Suite 290
Minneapolis, MN 55401-2520
http://www.upress.umn.edu

Printed in the United States of America on acid-free paper

23 22 21 20 19 18 17 16 10 9 8 7 6 5 4 3 2 1

Library of Congress Cataloging-in-Publication Data
Names: Gorham, Geoffrey, editor.
Title: The language of nature : reassessing the mathematization of natural philosophy in the 17th century / Geoffrey Gorham [and three others], editors.
Description: Minneapolis : University of Minnesota Press, [2016] | Series: Minnesota studies in the philosophy of science ; volume 20 | Includes bibliographical references and index.
Identifiers: LCCN 2015036898| ISBN 978-0-8166-9950-6 (hc) | ISBN 978-0-8166-9989-6 (pb)
Subjects: LCSH: Physics—Philosophy—History—17th century. | Physics—Philosophy. | Mathematics—Philosophy—History—17th century. | Mathematics—Philosophy.
Classification: LCC QC7 .L224 2016 | DDC 530.15—dc23
LC record available at http://lccn.loc.gov/2015036898

CONTENTS

INTRODUCTION

GEOFFREY GORHAM, BENJAMIN HILL,
AND EDWARD SLOWIK

CONCEPTUAL BACKGROUND

No other episode in the history of Western science has been as consequential as the rise of the mathematical approach to the natural world, both in terms of its impact on the development of science during the scientific revolution but also in regard to the debates that it has generated among scholars who have striven to understand the history and nature of science. In his recent summary of this "mathematization thesis," Michael Mahoney recounts the stunningly quick ascendancy of the mathematization of nature, a mere two-hundred-year span that witnessed the overthrow of the Aristotle-inspired Scholastic approach to the relationship between mathematics and natural philosophy that had held sway up through the first half of the Renaissance: "For although astronomy had always been deemed a mathematical science, few in the early sixteenth century would have envisioned a reduction of physics—that is, of nature as motion and change—to mathematics" (1998, 702). Yet, by the end of the seventeenth century this radical change in approach had become dominant. In this introduction, we first summarize and explore some of the main conceptual issues crucial to the mathematization of nature during the scientific revolution. The mathematization thesis signifies above all the transformation of scientific concepts and methods, especially those concerning the nature of matter, space, and time, through the introduction of mathematical (or geometrical) techniques and ideas (Yoder 1989). We next analyze the prominence of mathematization as a historiographical framework within scholarship of the scientific revolution, especially in the twentieth century. Finally, we explain how the contributions to this volume explore, challenge, and reshape these conceptual and historiographical perspectives.

The ideal of mathematization has ancient roots (Bochner 1966). Indeed, as we will see in the next section, modern historiography has emphasized the revival of Platonism in the seventeenth century's drive to mathematize. The remnants of Plato's own Pythagoreanism are evident in the *Republic*, where he advocates an a priori astronomy insofar as the visible motions in the sky "fall short of the true ones—motions that are really fast or slow as measured in true numbers, that trace out true geometrical figures, that are all in relation to one another" (529d1-5; 1997, 1145–46). So Socrates urges: "let's study astronomy by means of problems, as we do geometry, and leave the things in the sky alone" (530b6-c1; 1146). And in the *Timaeus* Plato develops an elaborate geometrical cosmology and matter theory, guided by the conviction that the creator, in order to produce the best and most intelligible world, would produce a "symphony of proportion" (32c2; 1237). After Plato, Archimedes's program of mathematization in the sciences of hydrostatics and mechanics provided a model for Galileo and others (Clagett 1964).

Controversy about the value and limits of mathematization also goes back to the beginnings of philosophy. In Aristotle's view, Pythagoras and Plato excessively conflated the abstract realm of mathematics with the concrete realm of nature: "the minute accuracy of mathematics is not to be demanded in all cases, but only in the case of things which have no matter. Hence its method is not that of natural science" (995a15–18; 1984, 2:1572). So Aristotle concludes that the student of nature should not simply assume that matter and motion will conform to mathematical principles. Nevertheless, in his own *Physics*, he acknowledges the importance of "the more physical of the branches of mathematics, such as optics, harmonics, and astronomy" (194a8; 1984, 1:311). And in the methodological treatise *Posterior Analytics* he indicates that such sciences are subject to geometrical (e.g., mechanics and optics) or arithmetical (e.g., harmonics) demonstration (76a1, 21–25; 1984, 1:123) even though their subject matter is empirical: "it is for the empirical scientist to know the fact, and for the mathematicians to know the reason why" (78b32-3; 1984, 1:128). Aristotle assumed that the theorems of such sciences must be "subordinate" to the theorems of their corresponding mathematical sciences, since he prohibited demonstrations that crossed subject-genera (75b3-20; 1984, 1:122). This way of conceiving the "mixed sciences," as they came to be known, gained additional influence through the pseudo-Aristotelian treatise on mechanics, whose problems involving wheels, pulleys, and levers were routinely treated geometrically by philosophers through the sixteenth century, including Galileo (Bertoloni Meli 2006). Indeed, arithmetic, geometry, astronomy, and music—already identified as peculiarly mathematical by Plato (*Republic* Bk 7; 525a–31d; 1997, 1141–47)—were formally and pedagogically grouped together in the classical "quadrivium." Consequently, the idea that mathematics could be used to directly represent physical phenomena remained an open and contested question through the ancient and medieval periods. In the seventeenth century, the main foci of the ongoing debate can be grouped under three broad conceptual categories: instrumentalism versus realism, types of mathematization, and social context.

Instrumentalism versus Realism

Two important sources of skepticism about mathematization can be traced to the Aristotelian strictures mentioned previously, one metaphysical and one methodological. First, it was claimed that matter did not conform to the exactness of mathematics, and second, that the deductive structure of mathematical demonstration was inadequate to capture the causal relationships among natural bodies. Hence, outside of the classical "mixed sciences" of optics, mechanics, and astronomy, the utility of mathematics for understanding nature was severely limited. Based on these concerns, an instrumentalist tradition arose that provided a negative answer to the question, do mathematical objects and their relationships correspond to natural objects and their relationships? Instrumentalism regards the mathematical component of physical theories, for example, the epicycle-deferent system of Ptolemaic astronomy, as a mere calculating device for predicting phenomena (Machamer 1976). And this outlook remained influential through the beginning of the early modern period. It is expressed in Osiander's preface to Copernicus's *De Revolutionibus* (1543), which stipulates that since the astronomer "cannot in any way attain to the true causes, he will adopt whatever suppositions enable the motions to be computed correctly from the principles of geometry . . . these hypotheses need not be true nor even probable" (1978, xvi).

Yet, during the sixteenth and seventeenth centuries the mathematical constructions employed in the new Copernican theory of astronomy began to be accepted by many as providing knowledge of the actual relationships among celestial bodies. Thus, Kepler and Galileo urged that the aim of astronomy was physical truth, not merely to "save the phenomena" via mathematical models (Jardine 1979). And the same realist attitude was extended to the new mathematical work in optics and mechanics. Besides the increase in successful mathematically based approaches, such as Simon Stevin's work on statics and Galileo's account of free fall, the main catalyst for the increasing popularity of a realist conception of the link between mathematics and the physical world was almost certainly the rise of the mechanical conception of natural philosophy. By proposing that natural phenomena could be explained by means of machine models, the mathematical relationships that characterize the operation and part-whole relationships of the models offered an obvious and intuitive basis for positing those same mathematical relationships in the natural phenomena themselves. The growing appreciation of the success of mathematical techniques in explaining natural phenomena, combined with the rise of the mechanical philosophy and its realist conception of a hidden world of interacting material particles that have geometrical shapes and volumes, thus encouraged a realist conception of the relation between mathematics and physical reality. As John Henry put it, the "Scientific Revolution saw the replacement of a predominantly instrumentalist attitude to scientific analysis with a more realist outlook" (2008, 8). Galileo's famous declaration that the book of nature "is written in the language of mathematics, and its characters are triangles, circles, and other geometric figures" (1957, 238–39)—thus turned Osiander's preface on its head: it was precisely because nature itself was geometrical that mathematical physics had to be true.

Types of Mathematization

Galileo's "book of nature" comment also reveals the type of mathematics that informed much of his work on natural philosophy: geometry, the same approach used by ancient and medieval natural philosophers in the mixed mathematics tradition. At the start of the seventeenth century, geometry could thus lay claim as the most important branch of mathematics for investigating the physical world, especially given the historical precedent of the parallel structure between the synthetic or axiomatic conception of geometry developed in Euclid's *Elements* and the deductive methodology of Aristotelian-based Scholastic science. That is to say, axiomatic geometry derives theorems and other elaborate geometric results from a starting point consisting of basic definitions and concepts, and is thus a process that strongly resembles the logical structure of Aristotelian/Scholastic science whose explanatory methodology includes basic metaphysical postulates—"first principles"—as premises, and then goes on to produce specific scientific explanations of various phenomena from that basis (*Posterior Analytics*; 71b9-78a28; 1984, 1:115–25). Descartes, for example, declared in a 1638 letter to Mersenne that "all my physics are nothing but geometry" (AT 2 268), while Spinoza extended the *more geometrico* into metaphysics and ethics. Newton's *Principia* would constitute one of the last significant examples of this geometrical treatment of physics. A host of mathematical tools would be developed in the seventeenth century that would ultimately transform the acceptable standards of mathematization. While the trigonometric relations embodied in the Snell-Descartes law of refraction and Huygens's work on harmonic oscillators can be seen as the beginning of this change, the infinitesimal analysis that lay at the heart of the new calculus's treatment of transcendental curves would mark the most important challenge to the hegemony of the older geometric approach (Mahoney 1998).

The rapid development and increasing usefulness of analytic techniques in mechanics provided a powerful justification for their introduction, no matter how unintuitive or problematic these techniques may have seemed in comparison with the methods derived from classical geometry (Gaukroger 2010). Specifically, the debate about whether geometry should be supplemented by novel algebraic formalisms and techniques reverberated throughout the seventeenth century (Jesseph 1999), culminating in the opposition between Newton's fluxional version of the differential calculus and the analytical formulation of Leibniz. In this sense, the geometrical character of mathematics that had helped to usher in the scientific revolution, which itself was inspired by the newer mechanical philosophy and ancient geometry, could be seen as limiting the development of mathematization. Defenders of what we might call "geometric fundamentalism," including Hobbes and Barrow, pointed to the superior intelligibility of geometric proof and to the manifest applicability of geometry to space, time, and matter. Defenders of the new algebraic methods, the so-called mathematical pluralists, pointed to their flexibility and power and to their utility in representing continuous magnitudes, irregular curves, the infinite and infinitesimals, instantaneous velocity, and so on (Mancosu 1996). This is just one example of the way detailed historical scrutiny has recently complicated and enriched the grand narrative of mathematization.

Wholesale mathematization included the aim of extending the mathematical, specifically the geometrical, model of *demonstration* or method throughout the sciences. In mechanics and other areas of natural philosophy, the aspirational link with mathematics is evident in the many laws of nature, collision rules, and other quantified relationships among material phenomena that were posited by natural philosophers, such as Descartes' groundbreaking law for the conservation of "quantity of motion" (the product of size and speed; AT 8A 61-2). Given a mathematical formulation, these natural laws could thus be seen as acquiring the same level of necessity and certainty accorded to mathematics as a whole. But the attempts to extend the mathematization of various hypotheses in natural philosophy to the method of natural philosophy itself was never fully realized, despite the obvious success of the former venture by the end of the century. Galileo entitled his last major treatise the *Discourses and Mathematical Demonstrations Concerning the Two New Sciences* (1638). But Feyerabend and others have noted that this work and the *Dialogues* rely on a mix of strict demonstration, probabilistic arguments, and rhetoric (Feyerabend 2010; Jardine 1979). Hobbes modeled strictly philosophical method on geometry, but acknowledged the necessity of a hypothetical approach in physics (EW 1 387-8). Descartes hoped his physical principles would be accorded the "absolute certainty" of mathematics but seemed to concede they may possess only "moral certainty" (AT 8A 327). Moreover, although Descartes' physics aims for mathematical certainty, its content is remarkably free of mathematics, even granting the precedent set by his laws of nature. This split between the mathematization of *nature* vs. *method* is perhaps most evident in Spinoza: he hews to the *more geometrico* in the service of a metaphysical program that is quite unfriendly to mathematization (Schliesser 2014).

Furthermore, while certainty and demonstration were widely heralded, there was also considerable variation among the standards of proof and evidence. Even if mathematical demonstrations delivered certainty, many commentators, especially Aristotelians, denied that they provided substantive (i.e., causal) knowledge of natural processes (Mancosu 1996). The mathematical model of demonstration was not practiced by Bacon and his followers for a slightly different reason, however: they urged the investigation of nature through detailed, immediate "experiments" and the systematic collection of facts or "natural histories" (however, see chapter 2 in this volume).

A revival of 'atomistic' conceptions of nature in the early modern period also encouraged the mathematization trend. Conceiving of nature as discrete, rather than continuous as the Scholastics had typically done, atomism rendered mathematical methods that do not rely on geometric continuity more palatable to natural philosophers. Likewise, the mechanical conception of nature could be seen as helping to sanction the methods of analysis because, just as a machine can be viewed as the sum of its parts so the problem posed by an algebraic equation can be similarly resolved by examination and manipulation of its constituent components. Consequently, at the beginning of the eighteenth century, the art of analysis, exemplified by Pierre Varignon's refashioning of Newton's mechanics into the symbolic algebra of Leibniz's calculus, pointed to the future of nature's mathematization. Still, the mathematization of nature proceeded on many fronts, prompted by, and in turn stimulating, developments in fields besides mechanics and optics. Astronomy, for example, benefited from the introduction of logarithms, while Fermat and Pascal's investigations of gambling laid the foundations for probability theory, which would eventually have far-reaching applications in the sciences. Finally, the emphasis on experimentation and observation that gradually developed in the seventeenth century would usher in a growing reliance on quantification and measurement. Unlike the 'qualitative' approach to the sciences practiced by the Scholastics, the scientific revolution marked the transition to the quantitative outlook that underlies modern science (Roux 2010; Gingras 2001).

Social Context

The transformation of the content of mathematics in the seventeenth century mirrored the changing landscape of the social practices and institutions associated with mathematics. The growing power of Europe—in commerce, navigation, and technology among many other areas—was greatly facilitated by mathematical developments and applications. As the demand for mathematically proficient engineers and craftsmen rose, so their prestige and power in society increased. And with their increasing social and political influence, the authority and value of mathematics in society grew in proportion. The change was most evident in the universities, where mathematicians held an inferior status in comparison to natural philosophers at the beginning of the scientific revolution. Yet, by the end of the seventeenth century, many mathematicians and engineers had elevated their positions within the academy. As mathematical practitioners gained status, their knowledge claims garnered intellectual authority. At the same time, mathematics—previously denigrated by the intellectual elite as the purview of mere calculators, engineers, and merchants—won enhanced status through its increasing association with traditional physics and natural philosophy (Feingold 1984). As Biagioli has shown, Galileo himself was an early and rare example of a mathematician who was able to cross this disciplinary boundary and gain the title (and status) of "philosopher" (1989, 49; see also 1994).[1]

The efforts of the Medici court in promoting Galileo to the more esteemed rank of philosopher typifies this transition in the standing of mathematicians, a change in no small part explained by the progressively expanding importance placed on mathematical expertise for a host of social projects of wealthy patrons, courts, and institutions. The catalogue of these new or improving craft traditions is extensive; besides the more established mechanical arts of statics, hydrostatics, and kinematics, the practice of civil engineering (e.g., surveying, canal construction, and architecture), navigation, and military construction (e.g., artillery, fortifications) were greatly advanced by the development of mathematical techniques (Dear 1995). For instance, Galileo's use of geometric methods in his study of parabolic trajectories laid the groundwork for modern ballistics (Hall 1952). The arts were also deeply affected by the advance of mathematics, a trend begun in the Renaissance and exemplified in the use of perspective in painting. Leonardo da Vinci and Albrecht Dürer, whose work extended into the sixteenth century, were not only great artists but also skilled mathematicians and engineers. Many of the artists of the early modern period were inspired by the conviction that the essence of nature is mathematical; hence, the artistic content of their work, as well as the techniques used to produce those works, paralleled the rise of mathematics in the other craft traditions in society (Peterson 2011).

HISTORIOGRAPHY OF MATHEMATIZATION

The historiographical thesis that the scientific revolution, and by implication modern science as a whole, is guided by the project of mathematization has a long and controversial history of its own. In the preface to *Metaphysical Foundations of Natural Science* (1786), Kant asserts that "in any special doctrine of nature there is only as much proper science as there is mathematics therein" (2004, 6). It is the mathematical structure of science, as epitomized by Newtonian mechanics for Kant, that renders its fundamental laws necessary and a priori. The criterion of mathematization also explains why chemistry and psychology are not genuine sciences for Kant, the former because its principles are merely empirical and the latter because mathematics cannot be applied to the laws of inner sense.[2] So, already in Kant's influential reconstruction of modern science, mathematization serves the epistemic authority of certain sciences while marginalizing others. William Whewell, in *Philosophy of the Inductive Sciences* (1840), evinces a more ambivalent attitude. He acknowledges "how important an office in promoting the progress of the physical sciences belongs to mathematics," especially those sciences concerned with space, time, and motion (astronomy, optics, and mechanics). But Whewell also emphasizes "other ideas quite as necessary to the progress of exact and real knowledge as an acquaintance with arithmetic and geometry," especially cause, force, and substance (in sciences such as dynamics and chemistry) (1967, 1:156). More critically, in *The Crisis of European Sciences and Transcendental Phenomenology* (1936), Husserl claims that with Galileo we "observe the way in which geometry, taken over with the sort of naiveté that keeps every normal geometrical project in motion, determines Galileo's thinking and guides it to the idea of physics" (1970, 29). From a very different perspective, Joseph Needham opens the third volume of his monumental *Science and Civilization in China* by observing, "since mathematics and the mathematization of hypotheses has been the backbone of modern science, it seems proper that this subject should precede all others in our attempt to evaluate China's contributions" (1959, 1). Over several hundred years, and across a diverse range of scholarly perspectives, the assumption has been widely shared that, for better or worse, modern science and mathematics are inextricably linked.

Responsibility for the continuing prominence of mathematization historiography belongs to a trio of early twentieth-century historians, each born in 1892 but hailing from different countries. All three emphasized the increasingly mathematical treatment of problems that had long challenged Aristotelian natural philosophers of the middle ages: the planetary orbits, free fall, collision, and optical phenomena. Each emphasized the debt of modern mathematizers like Kepler and Galileo to ancient precursors, especially Plato, Aristarchus, and Archimedes. But their respective attitudes to mathematization are flavored by distinctive philosophical and normative presuppositions. In *The Metaphysical Foundations of Modern Physical Sciences* (1924), the American philosopher E. A. Burtt was particularly concerned with the implications of mathematization for "man's place in the world." The primary-secondary distinction of Galileo, as well as the rigid mind–body dualism of Descartes, are both portrayed as consequences of wholesale mathematization in science. The former denigrates human subjectivity: "in the course of translating this distinction of primary and secondary into terms suited to the new mathematical interpretation of nature, we have the first stage in the reading of man quite out of the real" (1924, 89). The latter relegates the mind to an obscure, isolated position within the brain: "the universe of the mind, including all experienced qualities that are not mathematically reducible comes to be pictured as locked up . . . away from the independent, extended realm in a petty series of locations inside of human bodies" (Burtt 1924, 123). For Burtt, then, mathematization is instrumental in the disenchantment of the world that was feared by some early, yet cautious, supporters of the new philosophy like the Cambridge Platonists. So we might say that Burtt was preoccupied with the *existential* implications of mathematization.

Alexandre Koyré is undoubtedly the historian most responsible for the twentieth-century embrace of the mathematization thesis. Koyré (1978) traced seventeenth-century mathematization to the influence of Plato (and Renaissance Platonism), especially in the case of Galileo who "argues for the superiority of Platonist mathematicism over abstract empiricism" (37). Galileo's Platonism is particularly evident for Koyré in his attitude toward the use of idealization and approximation in science. Not only was Galileo a skilled deviser of "thought experiments," even the experiments he actually carried out often required "mathematical license" or idealization about the hardness of surfaces, the sphericality of balls, and the parallel orientation of gravitational lines of force, and so on. In *Dialogues Concerning the Two Chief World Systems*, the Aristotelian Simplicio quips that "mathematical subtleties do very well in the abstract, but they do not work out when applied to sensible and physical matters" (Galileo 2001, 236). According to Koyré, Galileo's response was in effect that "the real and the material are homogeneous and that a geometrical figure can exist in a material form" (1978, 204).[3] Like Burtt, Koyré sees mathematization as key to a fundamental shift in man's conception of the universe. But unlike Burtt he is less concerned with man's place in the world than the world's place in the universe. In *From the Closed World to the Infinite Universe* (Koyré 1957), modern science, particularly through the work of Copernicus, Kepler, and Galileo, replaces the qualitative, closed, and finite cosmos of Christianized Aristotelianism with a purely quantitative, open, and infinite universe. Integral to this process is what Koyré calls the "geometrization of space": "the substitution of homogeneous and abstract—however now considered as real—dimension space of the Euclidean geometry for the concrete and differentiated place-continuum of pre-Galilean physics and astronomy" (1968, 6–7). Space and time are no longer framed in relation to privileged places (like the center of the earth) and events (like creation ex nihilo) but rather homogenized and extended infinitely in all directions. With the geometrization of cosmic space-time, matter and motion were also mathematized.[4] Matter, now assumed identical in the terrestrial and celestial spheres, becomes either pure res extensa, as in Descartes, or corpuscular, as in Boyle and Newton. Finally, motion and rest are conceived as equally real and are represented geometrically as curves and trajectories or symbolically as algebraic formulas. So we might say that Koyré is preoccupied with the *cosmological* side of mathematization.

Finally E. J. Dijksterhuis's 1961 *The Mechanization of the World Picture*, originally published in Dutch in 1950, offered in certain respects a corrective to the already influential writings of Koyré. For example, whereas Koyré held that revolution or "relative discontinuity" was typical of scientific change—a thesis later strengthened by Thomas Kuhn—Dijksterhuis held that "the" scientific revolution of the seventeenth century was unique in the severity of the conceptual rupture it involved. But he agreed that the revolutionary innovation of seventeenth-century science consisted precisely in "mathematization" (a near slogan in Dijksterhuis's writings): "the treatment of natural phenomena in words had to be abandoned in favor of mathematical formulation of the relation observed between them. In the present century, functional thinking with its essential mathematical mode of expression has not only been maintained, but has even come to dominate science" (1961, 501).

Dijksterhuis closely linked mathematization with the other major innovation commonly associated with the scientific revolution: mechanization. The pristine "mechanical philosophy" of Descartes conceived the functioning of natural processes by analogy with simple machines: levers, pulleys, and wheels. But such analogies could not provide an irreducible role for "force" and "attraction," which increasingly figured in analyses of impact, free fall, and projectile motion. "Even the most skilled mechanic," Dijksterhuis observed, "is unable to construct apparatuses in which material objects move in consequence of their mutual gravitation" (1961, 497). On Dijksterhuis's analysis, such concepts began to be associated with purely "functional" (i.e., mathematical) relationships rather than mechanical interactions in the traditional analogies.[5] Once mathematized, mechanics freed itself from intuitive metaphysical constraints, like no action at a distance, no velocity at an instant, and so on, that had confounded the early mechanical philosophy: "the science called mechanics had emancipated itself in the seventeenth century from its origin in the study of machines, and had developed an independent branch of mathematical physics dealing with the motion of material objects" (1961, 498). And so the heretofore "mixed" or "subordinate" science of mechanics, when finally given a rigorous mathematical formulation by Newton and his followers, became identified with the ancient Aristotelian science of physics itself. So we might say that Dijksterhuis is preoccupied with the *mechanical* side of mathematization.

The mathematization thesis was criticized for its emphasis on physics and astronomy at the expense of biology and medicine, and for its neglect of important natural philosophers like Bacon and Boyle. This led to a revised "two traditions" model of early modern science. According to Thomas Kuhn's influential version of the model, the "classical" tradition includes the familiar "mixed sciences" and, beginning in the fourteenth century, the science of local motion. This tradition—mathematical, rationalistic, and abstract—was practiced by Kepler, Galileo, and Descartes. The newer "Baconian" tradition, comprising sciences like chemistry, magnetism, and early electrical theory, was experimental, empiricist, and concrete. Given this methodological split, the Baconian approach had little impact on the rapid advance of sciences like astronomy and optics in the seventeenth century: "For a person schooled to find geometry in nature, a few relatively accessible and mostly qualitative observations were sufficient to confirm and elaborate theory" (Kuhn 1977, 38). Conversely, mathematization came to the Baconian sciences of chemistry and electricity much later than the classical sciences. Seen in this light, the "two traditions" historiography did not challenge, but rather confirmed, the dominant narrative of mathematization, privileging traditional physical sciences while reinforcing the conception of the chemical and life sciences as immature "fact-gathering." This is reflected in the historiography of the eminent scholar Richard Westfall. Emphasizing metaphysics, rather than methodology, his version of the two traditions thesis contrasts the "Pythagorean-Platonic" tradition, "which looked on nature in geometric terms, convinced that the cosmos was constructed according to principles of mathematical order," and the "mechanical philosophy," "which conceived of nature as a huge machine and sought to explain the hidden mechanism behind the phenomena" (Westfall 1978, 1). While acknowledging the "combined influence" of the two approaches, Westfall generally portrays the mechanical philosophy as "an obstacle to the full mathematization of nature" (42) that was eventually achieved by Newton (cf. Guicciardini 2009). In a relatively recent article, Westfall asserts that "the geometrization of nature is perhaps our most distinctive legacy from the scientific revolution," noting that it came first in physics but later spread to chemistry and molecular biology to such an extent that "to be a scientist today it is necessary to understand and do mathematics" (1990, 59).

In recent years, the mathematization thesis has been subject to varied criticism and analysis. From the perspective of the sociology of scientific knowledge, writers have explored the social and disciplinary implications of the increasing prestige of mathematics. Yves Gingras, for instance, documents how the mathematization of physics served to isolate emerging fields like "rational mechanics" and magnetism from public discussion (and hence criticism). Mathematization "had the effect of excluding actors from legitimately participating in the discourses on natural philosophy" (Gingras 2001, 385), thereby galvanizing the professional status of the new science. As Steven Shapin has discussed, such exclusion was one of Boyle's major concerns about excessive mathematization. While acknowledging the "usefulness" of arithmetic in experiment, and the elegance of geometrical proof, Boyle wrote in one of his own works on hydrostatics, "I had rather geometricians should not commend the shortness of my proofs than that those other readers, whom I chiefly designed to gratify, should not thoroughly apprehend the meaning of them" (Shapin 1994, 337). Whether because of, or despite, its exclusivity, geometry became part and parcel of physics in the wake of Newton's *Principia*. In the "Preface to the Reader," Newton explicitly uses mathematical precision to set the boundary between "rational" and merely "practical" mechanics: "rational mechanics will be the sense, expressed in exact proportions and demonstrations, of the motions that result from any forces whatever" (1999, 382).[6]

Even more recently, several philosophically oriented scholars have subjected the standard mathematization historiography to close scrutiny. In a searching critique of Koyré's "Platonist" Galileo, Gary Hatfield has argued that Galileo's mathematical approach (in contrast with Kepler's and Descartes') is not guided by any metaphysical presuppositions—Platonist, Aristotelian, or otherwise. Rather the application of geometry to nature by Galileo was vindicated simply by its successes, in a variety of theoretical contexts and experimental practices: "his achievement was to show how a mathematical approach to nature could be justified by its successes in practice, and specifically how it might be sufficiently justified by numerous local instances of application" (Hatfield 1990, 139). Lorraine Daston has taken a parallel line against Burtt, complaining that his fixation on metaphysics—especially the allegedly alienating metaphysics of the mechanical philosophy—and his disregard of social and political context, blinded him to the immense complexity of mathematization. He therefore failed to solve the central epistemological problem why mathematization was so compelling to so many: "Burtt's answer was that the new science required it, but this claim does not carry conviction: there were too many versions of the new science, with and without mathematics; too many versions of mathematized nature, with and without the mechanical philosophy; and too many versions of why nature should be mathematized to warrant any straightforward connection" (Daston 1991, 525; see also contributions to Garber and Roux 2012). Similarly, and even more recently, Sophie Roux has argued that if mathematization involves the application of mathematics to other fields of knowledge, there was little agreement in the seventeenth century about the meaning and aims of such cross-disciplinary application nor even about the nature of mathematics itself (2010, 324). Roux tentatively concludes, as Hatfield did, that "the grand narrative of mathematization has to be enriched by the dense spectrum of various mathematical practices" (327).

OVERVIEW OF THIS VOLUME

As a historiographical thesis, mathematization has many virtues. Its longevity attests to that. It is an elegant thesis, and highlights what seems to be a constitutive element of modern scientific practice, providing mathematically precise and rigorous explanations of natural phenomena. It also appears to neatly demarcate mature from immature sciences, thereby offering a framework for following the transformation of a discipline into a properly scientific one. And the thesis is strongly unifying: it unifies individual thinkers and disparate groups into a single movement; in other words, it prompts us to see how Cartesians and Newtonians, mechanists and iatrochymists, Galileo and Leibniz, all shared a philosophical outlook about the nature and aims of science. It also makes good sense of how the unification of phenomena was achieved during the period, such as the unification of celestial and terrestrial movements or the mechanistic unification of living and nonliving bodies, namely under a common mathematical formulation. Furthermore, the mathematization thesis respects the early moderns' own claims to be mathematizing nature. Finally, it is a very simple thesis. The idea of applying mathematics to nature in order to generate scientific understanding is as easy to formulate as it is to grasp. So it is thus not difficult to appreciate the perennial allure of mathematization as a historical thesis.

Recent historical work and historiographical trends, however, have put considerable pressure on the mathematization thesis, and in many cases have begun to undercut its power and plausibility as a narrative of the scientific revolution. It has always been recognized that there were outliers to the mathematization story: Gassendi and Charleton, Locke and Boyle, and Sydenham and La Mettrie should figure into the narrative of the scientific revolution, but they displayed little real interest in mathematics or quantitative explanations of natural phenomena. And it has always been recognized that within certain domains of natural philosophy (medicine, biology, and psychology, for example) very little mathematization was successful or even attempted. Although previously it may have been easier to view these as exceptions that prove the rule, they begin to look anomalous when taken in conjunction with more recent trends.

More detailed and careful historical work has begun to emphasize the importance of experience and experiment for the early stages of the scientific revolution as well as the central roles played by other conceptual moves and intellectual trends (Dobre and Nyden 2013). The rise of contextualist history of science and philosophy has also begun to highlight the many additional factors propelling the scientific revolution, such as the importance of scholarly societies, the impact of the discovery of the Americas, and sixteenth-century developments in economics and statecraft. The importance of developments within other, more "practical" disciplines, such as navigation and geography, art, anatomy, and pharmacology, are also being identified and explored. To say the least, the story of the emergence of modern science is much more complicated than the mathematization thesis generally suggests. Add to this the growing trend to deny that this emergence is revolutionary, as opposed to gradual or halting, and there seems to be no place for the mathematization thesis in current history of science.

Given these historiographical trends of the last thirty years, it is time to undertake a systematic reevaluation and potential reconceptualization of mathematization as a historical thesis. That is the aim of this volume. In calling for and offering some steps toward a reconceptualization, we are suggesting that the time is right to rethink (1) what mathematization does or should consist in; (2) how it squares with recent scholarship; and (3) its overall value as a historical framework for the emergence of science in the seventeenth century.

As one might expect, scholarly opinion varies on such questions, and the chapters presented here discuss a wide variety of issues and reconceptualizations. Possibilities range from the more extreme view that the mathematization thesis is not only historically useless but outright misleading to the very conservative position that it is historically accurate and requires little or no modification. Between these poles, there are positions that seek to rehabilitate the thesis by recharacterizing it in alternative terms,[7] limiting it to one of a variety of contributing factors,[8] or restricting it to particular individuals or sciences. If we could draw any generalization from this volume, and the current state of scholarship, it would be that the mathematization thesis as usually conceived is overly simplistic. While simplicity can be a virtue of historical explanation, in this case it depends on excessive vagueness in central concepts such as "mathematics," "nature," and "application." As they are made more determinate, the unity imposed by the mathematization thesis risks falling apart, and its usefulness as a grand narrative framework is compromised. As scholarship proceeds, we should expect the picture regarding mathematization to become even more nuanced and complicated.

In chapter 1, Carla Rita Palmerino ("Reading the Book of Nature: The Ontological and Epistemological Underpinnings of Galileo's Mathematical Realism") investigates Galileo's reputation as the "godfather" of mathematization. She investigates how Galileo's idealizations are part of a coherent and sophisticated realist ontology and epistemology of mathematics. According to Palmerino, for Galileo mathematical entities are mind-independent and rooted in the physical world, not unlike Barrow's geometrized notion of mathematics. But the mathematical structures of reality are too complex to be properly grasped on their own, thus necessitating the simplifying analyses in scientific models. Palmerino's essay closes the gap between Galileo's mathematics and his realist science by establishing how he conceived of mathematics in realist terms. This relieves the mathematization thesis of one of its persistent objections, which is that Galileo's mathematical science illicitly assumed the applicability of mathematics to nature. While it might be possible to see Palmerino's contribution as opening the door to a more sophisticated and fatal critique of the mathematization thesis, Palmerino herself does not invite this. A more natural understanding of her contribution seems to be to take it as part of a larger defense of the traditional mathematization thesis.

Dana Jalobeanu (chapter 2, "'The Marriage of Physics with Mathematics': Francis Bacon on Measurement, Mathematics, and the Construction of a Mathematical Physics") challenges the common view that Bacon is antagonistic to mathematization in natural philosophy. Jalobeanu argues, to the contrary, that there was an important quantitative aspect to Baconian natural histories. Other scholars have discussed the importance of careful and precise measurements within Baconian natural histories. But Jalobeanu is particularly interested in Bacon's claim that "a good marriage of Physics with Mathematics begets Practice," in which she perceives the seeds of a distinctively Baconian mathematization of nature. The linchpin of her position is an analysis of Bacon's "reductive experiments," which were aimed at reducing imperceptible powers or qualities to perceptible ones that could be precisely measured. These experiments were then to be expanded and multiplied until complete tables of the quantities, in every experimental configuration, are recorded. Jalobeanu's vision of a Baconian mathematization of physics suggests a relatively moderate revision of the mathematization thesis, making room for Bacon within the traditional narrative. A more critical reader, however, may view her analysis as further evidence against any univocal notion of mathematization.

Richard T. W. Arthur's chapter ("On the Mathematization of Free Fall: Galileo, Descartes, and a History of Misconstrual") illustrates how the mathematization historiography has distorted our understanding of how key figures of the scientific revolution struggled with one of its central problems. Arthur focuses on one of the supposed crowning achievements of mathematization, the law of free fall, and suggests that we have anachronistically mischaracterized and misunderstood the episode and its significance for mathematizing nature. Arthur's analysis centers on the concept of instantaneous velocity, which has been regarded as instrumental in both Galileo's and Descartes' accounts of free fall. But this was an impossible concept for them, Arthur argues, because they understood velocity as an affection of motion, which could never occur in an instant. By recognizing this anachronism, we can now explain why Galileo and Descartes can seem confused when they are not. On the nature of mathematization, Arthur indicates that "the process was nowhere near as smooth as it would appear" and only one of several drivers of philosophical development (which he terms "epistemic vectors") that simultaneously propel and constrain scientific thought. But Arthur's analysis may also suggest stronger conclusions. It undercuts the idea that physical space was geometrized in the way Koyré suggested, because motion was not yet conceived as a continuous magnitude. But it also suggests that there really was not a univocal conception of mathematics, motion, or nature extending from Galileo and Descartes through Leibniz and Newton and Clarke.

Roger Ariew (chapter 4, "The Mathematization of Nature in Descartes and the First Cartesians") provides the volume's most vigorous and sustained critique of the oversimplification of the mathematization thesis. Ariew concentrates on one special but important question—how the Cartesians viewed the relationship between mathematics and physics, especially when they thought they were articulating Descartes' own account. He shows that there was no consensus about the meaning and value of mathematization among the Cartesians. They variously held that mathematics was a tool for sharpening the mind and nothing more (Du Roure and Rohault); that mathematics employed the same faculty of the mind as physics but was not the basis for physical truth or method (Rohault); and that mathematics and physics are not even relevant to one another (Le Grand and Regis). But nowhere, Ariew emphasizes, did they embrace Descartes' notion of a *mathesis universalis*, which figures so prominently in the mathematization historiography. Indeed, why would they, Ariew asks, since Descartes' *Regulae* was unknown to them? Moreover, the relations between Descartes' own mathematics and physics were not due to the *mathesis universalis*, nor to the idea that physics was fundamentally mathematics, but to their both being rooted in the metaphysics of clear and distinct ideas. The disparate and often negative views of mathematization among the Cartesians undermines any wholesale or univocal version of the mathematization thesis. Ariew concludes that mathematization might well be a "twentieth-century invention, perhaps a construction we are forcing on the past."

Daniel Garber's chapter ("Laws of Nature and the Mathematics of Motion") proposes a more moderate reconceptualization of the mathematization thesis. His analysis of laws suggests that mathematization ought to be reconceived as one of many factors motivating and animating the scientific revolution. It may seem that mathematics and laws naturally coincide as mathematics captures the necessity and universality of laws. But Garber argues that for three key figures in the mathematization story—Galileo, Descartes, and Hobbes—the notions of law and the notions of mathematical representations of nature remained quite separate. In Descartes, the laws of nature are not formulated mathematically and there is little that is overtly mathematical in his natural philosophy as a whole. In Galileo, mathematics is everywhere, but the concept of law is missing from his physics and his overall scientific vision. And although Hobbes modeled philosophy and physics on geometry and offered important inertial principles, he was very careful not to label these statements "laws." Garber's analysis challenges the assumption that mathematization promoted scientific understanding via laws of nature, as well as the view that mathematization helped secure certainty and necessity in an otherwise voluntarist theological context. But Garber's analysis leaves open the possibility that the link between mathematization and laws was forged later, by Newton or Leibniz, for example, and that mathematization and the search for laws are parallel factors driving the early scientific revolution (perhaps among many others).

Yet another moderate strategy for reconceiving mathematization is developed by Douglas Jesseph (chapter 6, "Ratios, Quotients, and the Language of Nature"), which explores the debate between Wallis and Barrow about the conceptual foundations of mathematics. According to Jesseph, the disagreement between the arithmetically minded Wallis and the geometrically minded Barrow gives rise to different conceptions of the mechanistic project. Wallis embraced an arithmetic or numerical treatment by which ratios could be compounded and compared. Barrow, in contrast, adopted a geometrically grounded, relational analysis of ratios, which precluded compounding and numerical comparison. According to Jesseph, these divergent models of ratios influenced their respective conceptions of the relationship between mechanical motions and mathematics. For Wallis, mechanics was part of the mixed mathematical sciences in that it consisted of the application of general mathematics to motion. For Barrow, however, mechanical motion is geometrical because it formed the epistemic and ontological basis for geometry itself. One lesson of Jesseph's analysis is that historians of mathematization must attend to the complex seventeenth-century debates over the foundations of mathematics itself, particularly geometrical versus arithmetic constructions. On the whole, we can discern two parallel forms of mathematization emerging from the Wallis-Barrow dispute. The arithmetically based form fed into the Leibnizian calculus while the geometrically based form culminated in the fluxions of Newton's *Principia*.

Eileen Reeves's chapter ("Color by Numbers: The Harmonious Palette in Early Modern Painting") is yet another contribution suggesting that mathematization is not sufficiently unified to support a grand narrative of the scientific revolution. Reeves examines a very interesting case of the failure of mathematization—late sixteenth- and early seventeenth-century neo-Pythagorean efforts to explain colors as varying ratios of blackness and whiteness. Interestingly, their neo-Pythagoreanism itself seemed to function as an epistemological obstacle to the development of any coherent and plausible color theory. It was not until the differentiation of primary and secondary colors in the work of François Aguilon, who was anxious to distance himself from neo-Pythagorean predecessors, that color theory was able to fully develop. But, Reeves emphasizes, it was not through Aguilon's theoretical presentation that the theory of primary and secondary colors was able to spread, but rather through the work of painters and artists who worked with pigments and hues. Reeves neatly presents two of Diego Valázquez's paintings as visual commentaries on Aguilon's color theory. The significance of Reeves's analysis for the historiographical thesis of mathematization is twofold. Although neo-Pythagoreanism is often seen as part of the story of mathematization, here it is presented as an epistemological obstacle inhibiting the science of color theory. Furthermore, it underscores the diversity among domains of mathematization, since Pythagorean harmonics have little to do with the geometrization of space or the Euclidean model of scientific demonstration. There seems to be, at best, only a family resemblance among practices of seventeenth-century mathematization.[9]

Lesley B. Cormack (chapter 8, "The Role of Mathematical Practitioners and Mathematical Practice in Developing Mathematics as the Language of Nature") presents additional serious problems for a unified mathematization doctrine. As noted earlier, a key plank in Koyré's influential version of the mathematization thesis is the geometrization of space that supposedly began in the fourteenth century. Cormack emphasizes that, owing to the prominence of humanism, natural philosophy during the fifteenth and sixteenth centuries was for the most part not mathematically oriented. This suggests that even if physical space was geometrized in the fourteenth century, this played little role in the (alleged) mathematization of nature in the seventeenth century.[10] Rather than being in the hands of the natural philosophers or the universities, Cormack argues that mathematics was in the hands of "mathematical practitioners": painters, mapmakers, instrument designers, military consultants, navigators, and merchants. While it might seem this analysis demands only a corrective about the timing and transmission of the geometrization of space, a stronger conclusion more damaging to the mathematization thesis also seems possible—there was no geometrization of space prior to the rise of mathematical natural philosophy. Like the Egyptians and Mesopotamians who originally developed geometry and arithmetic for building monuments and conducting state business, the sixteenth-century mathematical practitioners were driven by the practical necessities of trade, war, and finance. With lives and fortunes on the line, they had no need for a philosophical breakthrough like the geometrization of physical space. Whether one draws the weaker or the stronger conclusion from Cormack's essay, at the very least she shows how the mathematization thesis needs to be augmented by including mathematical practitioners in the story.

The next three chapters examine the influential and complex role of the great polymath Leibniz in the process of mathematization. Kurt Smith (chapter 9, "Leibniz on Order, Harmony, and the Notion of Substance: Mathematizing the Science of Metaphysics and Physics") supports a reentrenchment of the traditional mathematization thesis, advocating its extension into the realm of metaphysics, particularly within the Leibnizian tradition. Smith shows how Leibniz used the idea of a mathematical procedure to make sense of his central metaphysical notions of order and harmony. The primary procedure in question is Leibniz's early version of what would become known as Cramer's Rule. He argues that the fact that there is a mathematical solution to a set of equations describing the various forces at work in a substance suggests that Leibniz conceived of harmony and order as the convergence on a determinate. So Leibniz's version of mathematization was not limited to his mechanics or natural philosophy, it affected even his ontology itself.

Justin E. H. Smith (chapter 10, "Leibniz's Harlequinade: Nature, Infinity, and the Limits of Mathematization") supports a similar restriction on the mathematization thesis as urged by Jesseph, but from a very different perspective. He focuses on Leibniz's reception of the iatromechanical tradition, and his biology and physiology more broadly, rather than Leibnizian mechanics. The iatromechanists, Smith emphasizes, conceived their project as fundamentally mathematical. Leibniz embraced this conception and supplemented it with his own theory of actual infinities. The notion of infinite structures within natural bodies allowed Leibniz to extend his solution to the problem of the continuum to the physiology of living things. This reflected Leibniz's ambitious hope to account for all of nature in a unified, broadly mathematical way. Although this mathematizing project was unsuccessful in the end (perhaps was even doomed from the start), Smith argues that it should be seen as an important, albeit unorthodox, part of the mathematization project of the seventeenth century. Like Jesseph, Smith can be seen as localizing the mathematization thesis, in this case to iatromechanism, while calling attention to mathematization programs far from mechanics and astronomy. The vast differences between traditional, geometrical forms of mathematization and Leibniz's infinity-based approach suggest moreover that Leibniz's approach represented a distinct and unique form of mathematization.

Ursula Goldenbaum (chapter 11, "The Geometrical Method as a New Standard of Truth, Based on the Mathematization of Nature") suggests that within the rationalist tradition mathematization extends into the realm of philosophical methodology. For her, the key is the geometrical method, which she views as an extension of the mathematization project into philosophy as a whole. This method was applied globally—even, as in Spinoza, to human emotions and ethics itself. The primary attraction of this method was that it promised to secure the certainty of philosophical knowledge the same way that mathematization promised to ground the certainty of scientific knowledge. So as for Kurt Smith, mathematization is not so much critiqued but rather extended into the rationalist philosophy of the seventeenth century.

Finally, Christopher Smeenk (chapter 12, "Philosophical Geometers and Geometrical Philosophers") explores the use Newton made of Barrow's geometrized conception of mathematics to bridge the gap between the mathematically ideal and the physically real. Smeenk shows how Newton used Barrow's geometrical conception of mathematics to overcome a prima facie barrier to the adoption of a mathematical methodology in physics: the tension between the universality and certainty of mathematics versus the particularity and uncertainty of real systems. According to Smeenk, Newton rejected the idea that mathematics is especially abstract or ideal but held instead that its proper domain of application is material objects rather than ideal entities. However, he did not promulgate a crude empiricist epistemology of mathematics in nature. Rather, he held that quantitative descriptions are rationally reconstructed on the basis of observations taken from within an appropriate conceptual framework. Like Jesseph, Smeenk's analysis of Newton's geometrically based mathematics suggests a way of reconceiving the mathematization thesis where different models of mathematization in Newton, Barrow, and Wallis develop independently or even in competition. But Smeenk's and Jesseph's points can also be taken in the more restrictive sense, whereby the mathematization of nature is seen as a feature only of the later, Newtonian and Leibnizian stage of the scientific revolution. In any case, taking their work in either of these directions requires a revision of the mathematization thesis, dethroning it as an overarching narrative of the scientific revolution.

This volume presents a variety of possible reconceptualizations of the mathematization historiography. One possibility is to reject it completely as oversimplified and misleading, both for historical and pedagogical purposes. More moderate revisions are also possible: mathematization might be one factor of many necessary to explain the scientific revolution; or it may be of limited historical utility helpful for explaining (1) the contributions of specific individuals or schools; (2) the unfolding of particular fields within natural philosophy; (3) the progress in certain subperiods of the scientific revolution. Finally, certain conservative positions seek to retain the substance of the mathematization thesis and even extend it beyond its traditional scope of natural philosophy to philosophical concerns more broadly. Regardless of what the scholarly consensus turns out to be, there is no doubt that the mathematization thesis has played an important role in shaping our conceptions of the scientific revolution and it deserves to be carefully reexamined in this new era of historical methodologies and frameworks.

ABBREVIATIONS

AT Descartes, R. 1976. *Oeuvres de Descartes* (citation by volume and page number)

EW Hobbes, T. 1839–45. *English Works* (citation by volume and page number)

NOTES

1. See also Henry (2008), chapter 2, for an overview of recent work. Westman (1980) has urged a similar point about the emerging authority of astronomy in the sixteenth century, one of the traditional mixed sciences.

2. See further Friedman (1992) and Massimi (2010).

3. The Platonist origin of Galileo's geometrical approach has been forcefully challenged by some historians. For example, Wallace (1984) has explored the influence of the Jesuit mathematicians of the Collegio Romano, especially Christopher Clavius, on Galileo's early philosophy. Nevertheless Koyré's version of the mathematization thesis gained wide acceptance within mainstream historiography of science in the twentieth century. The British historian Rupert Hall, while underscoring the complexity of the scientific revolution, and the dramatic transformations within mathematics itself, strongly reaffirms Koyré's thesis in numerous

works, including the late *Revolution in Science 1500–1750*: "The most eloquent and full defense of this process [the mathematization of nature] was given by Galileo whose mathematization of the science of the motions of real bodies furnished a model for physical science general during the following century" (1983, 12).

4. On the geometrization of space, time, matter, and motion, see more recently McGuire (1983), Jalobeanu (2007), and Palmerino (2011).

5. Koyré puts the same point about gravitation in terms of Galileo's famous saying: "a mathematical stricture that lays down the rule of syntax in God's book of nature" (1968, 13).

6. See further the recent articles by Domski (2013) and Dunlop (2013).

7. Sophie Roux suggests that it can be recast by focusing on the kinds of mathematical practices used during the period (2010, 319–37) or in terms of their polemical stances toward the Schoolmen (2013). Craig Martin (2014) develops this approach.

8. As H. Floris Cohen has done (2010).

9. For an incisive critique of a "family resemblance" approach to reconceptualizing mathematization, see Roux (2013, 57–58).

10. This can be related to suggestions by Noel Swerdlow (1993) and H. Floris Cohen (1994; 2010) that an ancient mathematical form of natural philosophy was rediscovered in the fifteenth and sixteenth centuries rather than being transmitted to them via a fourteenth-century discovery.

REFERENCES

Aristotle. 1984. *Complete Works of Aristotle*. Ed. Jonathan Barnes. 2 vols. Oxford: Oxford University Press.

Bertoloni Meli, E. 2006. *Thinking with Objects: The Transformation of the Science of Mechanics in the Seventeenth Century*. Baltimore: John Hopkins University Press.

Biagioli, M. 1989. "The Social Status of Italian Mathematicians 1450–1600." *History of Science* 27: 41–95.

———. 1994. *Galileo, Courtier: The Practice of Science in the Culture of Absolutism*. Chicago: University of Chicago Press.

Bochner, S. 1966. *The Role of Mathematics in the Rise of Science*. Princeton, N.J.: Princeton University Press.

Burtt, E. A. 1924. *The Metaphysical Foundations of Modern Physical Sciences*. London: Kegan Paul.

Clagett, M. 1964. *Archimedes in the Middle Ages*. 5 vols. Madison: Wisconsin University Press.

Cohen, H. F. 1994. *The Scientific Revolution: A Historiographical Inquiry*. Chicago: University of Chicago Press.

———. 2010. *How Modern Science Came into the World: Four Civilizations, One 17th Century Breakthrough*. Amsterdam: Amsterdam University Press.

Copernicus, N. 1978. *De Revolutionibus*. Translated by E. Rosen. Baltimore: John Hopkins University Press.

Daston, L. 1991. "History of Science in an Elegiac Mode: E. A. Burtt's *Metaphysical Foundations of Modern Physical Science* Revisited." *Isis* 82: 522–31.

Dear, P. 1995. *Discipline and Experience: The Mathematical Way in the Scientific Revolution*. Chicago: University of Chicago Press.

Descartes, R. 1976. *Oeuvres de Descartes*. Ed. D. C. Adam and P. Tannery. 11 Vols. Paris: J. Vrin.

Dijksterhuis, E. J. 1961. *The Mechanization of the World Picture*. Oxford: Oxford University Press.

Dobre, M., and T. Nyden. 2013. *Cartesian Empiricisms*. New York: Springer.

Domski, M. 2013. "Locke's Qualified Embrace of Newton's Principia." In *Interpreting Newton*, ed. A. Janiak and E. Schliesser, 48–68. New York: Cambridge University Press.

Dunlop, K. 2013. "What Geometry Postulates: Newton and Barrow on the Relationship of Mathematics to Nature." In *Interpreting Newton*, ed. A. Janiak and E. Schliesser, 69–102. New York: Cambridge University Press.

Feingold, M. 1984. *The Mathematician's Apprenticeship: Science, Universities and Society in England 1560–1640*. Cambridge: Cambridge University Press.

Feyerabend, P. 2010. *Against Method*. New York: Verso.

Friedman, M. 1992. *Kant and the Exact Sciences*. Cambridge, Mass.: Harvard University Press.

Galileo, Galilei. 1957. *Discoveries and Opinions of Galileo*. Ed. S. Drake. London: Anchor Books.

———. 2001. *Dialogues Concerning the Two Chief World Systems*. Ed. S. Drake. New York: Modern Library.

Garber, D., and S. Roux, eds. 2012. *The Mechanization of Natural Philosophy*. Dordrecht: Springer.

Gaukroger, S. 2010. *The Collapse of Mechanism and the Rise of Sensibility: Science and the Shaping of Modernity, 1680–1760*. Oxford: Oxford University Press.

Gingras, Y. 2001. "What Mathematics Did to Physics." *History of Science* 39: 383–416.

Guicciardini, N. 2009. *Isaac Newton on Mathematical Method and Certainty*. Cambridge, Mass.: MIT Press.

Hall, A. R. 1952. *Ballistics in the Seventeenth Century: A Study in the Relations of*

Science and War with Reference Principally to England. Cambridge: Cambridge University Press.

———. 1983. *The Revolution in Science 1500–1750*. London: Longman's Press.

Hatfield, G. 1990. "Metaphysics and the New Science." In *Reappraisals of the Scientific Revolution*, ed. D. Lindberg and R. Westman, 93–166. Cambridge: Cambridge University Press.

Henry, J. 2008. *The Scientific Revolution and the Origins of Modern Science*. 3rd ed. New York: Palgrave Macmillan.

Hobbes, T. 1839–45. *English Works*. Ed. W. Molesworth. 11 vols. London.

Husserl, E. 1970. *The Crisis of European Sciences and Transcendental Phenomenology*. Ed. D. Carr. Evanston, Ill.: Northwestern University Press.

Jalobeanu, D. 2007. "Space, Bodies and Geometry: Some Sources of Newton's Metaphysics." In *Notions of Space and Time*, ed. F. Linhardt. *Zeitsprunge, Forschungen zur Fruher Neuzeit* 11: 81–113.

Jardine, N. 1979. "The Forging of Modern Realism: Clavius and Kepler against the Skeptics." *Studies in the History and Philosophy of Science* 10: 141–73.

Jesseph, D. 1999. *Squaring the Circle: The War between Hobbes and Wallis*. Chicago: University of Chicago Press.

Kant, I. 2004. *Metaphysical Foundations of Natural Science*. Ed. M. Friedman. Cambridge: Cambridge University Press.

Koyré, A. 1957. *From the Closed World to the Infinite Universe*. Baltimore: Johns Hopkins University Press.

———. 1968. *Newtonian Studies*. Chicago: University of Chicago Press.

———. 1978. *Galileo Studies*. Trans. J. Mepham. Atlantic Highlands, N.J.: Humanities Press.

Kuhn, T. 1977. "Mathematical vs. Experimental Traditions in the History of Science." In *The Essential Tension*, 31–65. Chicago: University of Chicago Press.

Machamer, P. 1976. "Fictionalism and Realism in 16th Century Astronomy." In *The Copernican Achievement*, ed. R. S. Westman, 346–53. Berkeley: University of California Press.

Mahoney, M. 1998. "The Mathematical Realm of Nature." In *The Cambridge History of Seventeenth Century Philosophy*, ed. D. Garber and M. Ayers, 702–55. Cambridge: Cambridge University Press.

Mancosu, P. 1996. *Philosophy of Mathematics and Mathematical Practice in the Seventeenth Century*. Oxford: Oxford University Press.

Martin, C. 2014. *Subverting Aristotle: Religion, History, and Philosophy in Early Modern Science*. Baltimore: Johns Hopkins University Press.

Massimi, M. 2010. "Galileo's Mathematization of Nature at the Crossroad between the Empiricist and the Kantian Tradition." *Perspectives on Science* 18: 152–88.

McGuire, J. E. 1983. "Space, Geometrical Objects and Infinity: Newton and Descartes on Extension." In *Nature Mathematized: Historical and Philosophical Case Studies in Classical Modern Natural Philosophy*, ed. W. R. Shea, 69–112. Dordrecht: D. Reidel.

Needham, J. 1959. *Science and Civilization in China*. Vol. 3. Cambridge: Cambridge University Press.

Newton, I. 1999. *The Principia: Mathematical Principles of Natural Philosophy*. Ed. I. B. Cohen and A. Whitman. Berkeley: University of California Press.

Palmerino, C. L. 2011. "The Geometrization of Motion: Galileo's Triangle of Speed and Its Various Transformations." In "Forms of Mathematization," special issue, *Early Science and Medicine* 15: 410–47.

Peterson, M. 2011. *Galileo's Muse: Renaissance Mathematics and the Arts*. Cambridge, Mass.: Harvard University Press.

Plato. 1997. *Complete Works*. Ed. J. Cooper. Indianapolis, Ind.: Hackett Publishing.

Roux, S. 2010. "Forms of Mathematization." *Early Science and Medicine* 15: 319–37.

———. 2013. "An Empire Divided: French Natural Philosophy (1670–1690)." In *The Mechanization of Natural Philosophy*, ed. D. Garber and S. Roux, 55–95. Dordrecht: Springer.

Schliesser. E. 2014. "Spinoza and the Philosophy of Science: Mathematics, Motion and Being." In *The Oxford Handbook of Spinoza*, ed. M. Della Rocca. Oxford: Oxford University Press.

Swerdlow, N. 1993. "The Recovery of the Exact Sciences of Antiquity: Mathematics, Astronomy, Geography." In *Rome Reborn: The Vatican Library and Renaissance Culture*, ed. A. Grafton, 125–67. New Haven, Conn.: Yale University Press.

Shapin, S. 1994. *A Social History of Truth*. Chicago: University of Chicago Press.

Wallace, W. A. 1984. *Galileo and His Sources: Heritage of the Collegio Romano in Galileo's Science*. Princeton, N.J.: Princeton University Press.

Westfall, R. S. 1978. *The Construction of Modern Science*. Cambridge: Cambridge University Press.

———. 1990. "Making a World of Precision: Newton and the Construction of a

Quantitative Physics." In *Some Truer Method: Reflections on the Heritage of Newton*, ed. F. Durham, 59–87. New York: Columbia University Press.

Westman, R. 1980. "The Astronomer's Role in the Sixteenth Century: A Preliminary Study." *History of Science* 18: 105–47.

Whewell, W. 1967. *Philosophy of the Inductive Sciences*. 2 vols. New York: Johnson Reprints.

Yoder, J. G. 1989. *Unrolling Time: Christian Huygens and the Mathematization of Nature*. Cambridge: Cambridge University Press.

1

READING THE BOOK OF NATURE

The Ontological and Epistemological Underpinnings of Galileo's Mathematical Realism

CARLA RITA PALMERINO

On May 7, 1610, Galileo Galilei wrote to Belisario Vinta, Secretary of State of the Grand Duchy of Tuscany, about the terms of his future position as a court mathematician. In his letter Galileo expressed the wish that "His Majesty add the name of Philosopher to that of Mathematician," motivating his request with the fact that he had "spent more years studying philosophy than months studying pure mathematics" (Galilei 1890–1909, 10:353).

The fact that Galileo spoke of "pure mathematics," and not just of "mathematics," is highly significant. In his view, to be a philosopher meant to be a mathematician, but one who was interested in discovering the real constitution of the physical world. In the Third Day of the *Two New Sciences* (1638) Galileo describes his approach to the study of accelerated motion, and explains in which sense the philosopher's mathematical method differs from that of a pure mathematician:

> Not that there is anything wrong with inventing at pleasure some kind of motion and theorizing about its consequent properties, in the way that some men have derived spiral and conchoidal lines from certain motions, though nature makes no use of these paths. . . . But since nature does employ a certain kind of acceleration for descending heavy things, we decided to look into their properties so that we might be sure that the definition of accelerated motion which we are about to adduce agrees with the essence of naturally accelerated motion (Galilei 1974, 153; 1890–1909, 8:197).[1]

Similarly, in his *Dialogue* (1632), Galileo contrasts the ambition of the mathematical astronomer, who contents himself with saving the phenomena, to that of the "astronomer philosopher," who seeks "to investigate the true constitution of the universe—the most admirable problem that there is" (Galilei 1967, 341). For although whatever we read in the book of nature—a book that, as Galileo famously argues in the *Assayer* (1623), is written in the language of mathematics—"is the creation of the omnipotent Craftsman, . . . nevertheless that part is most suitable and most worthy which makes His works and His craftsmanship most evident to our view" (Galilei 1957, 3).[2]

But how is it possible for the natural philosopher to get access to the mathematical language of the book of nature and to find out which geometrical line or which mathematical formula corresponds to the essence of a particular physical phenomenon? As I shall try to show in this chapter, Galileo's answer to this question was much more sophisticated than scholars generally assume. In addressing the question concerning the relation between mathematical and physical truths, Galileo carefully kept ontological considerations distinct from epistemological considerations. In his view, the fact that physical phenomena cannot always be translated into simple mathematical laws was not an argument against mathematical realism, but was only a sign that the mathematical order of nature is often too complex to be grasped by the human mind.

IS THE BOOK OF NATURE REALLY WRITTEN IN THE LANGUAGE OF MATHEMATICS?

In *The Crisis of European Sciences and Transcendental Phenomenology* (1970), Edmund Husserl devoted considerable attention to Galileo's mathematization of nature. According to Husserl, Galileo took it for granted that geometry, which "produces a self-sufficient, absolute truth," could be applied to nature "without further ado," without reflecting "on how the free, imaginative variation of this world and its shapes results only in possible empirically intuitable shapes and not in exact shapes" (Husserl 1970, 49). In other words, Galileo overlooked the fact that concrete bodies are not "the geometrically pure shapes which can be drawn in ideal space," but "are thinkable only in gradations: the more or less straight, flat, circular, etc." (Husserl 1970, 25).

Whereas in Husserl's eyes Galileo was a Pythagorean Platonist mathematician who performed a "surreptitious substitution" of the ideal objects of mathematics for the real objects of the physical world (see Soffer 1990; De Gandt 2004; Moran 2012), many scholars believe that Galileo cannot be considered a Platonist precisely because he regarded mathematical laws as idealizations that do not have a counterpart in the actual world. In a passage of the *Two New Sciences*, for example, Galileo explicitly admits that: "No firm science can be given of such accidents of heaviness, speed, and shape, which are variable in infinitely many ways. Hence to deal with such matters scientifically, it is necessary to abstract from them. We must find and demonstrate conclusions abstracted from the impediments, in order to make use of them in practice under those limitations that experience will teach us" (1974, 225).

According to Noretta Koertge's influential interpretation, Galileo's concern with the "problem of accidents"—a problem for which he developed increasingly sophisticated solutions in his writings—shows that the *Assayer*'s image of the universe as a book written in the language of geometry should not be "taken as a significant or careful statement of Galileo's philosophical views" (Koertge 1977, 402). In her view, Galileo endorsed neither a Pythagorean ontology nor a Platonist epistemology. Likewise, Robert Butts perceived an inconsistency between Galileo's practice of science and his professed mathematical realism, arguing that, "Galileo's argument that mathematics applies to the world was more a metaphysical faith than a philosophically established conclusion. He seems to have concluded that if the world does not conform to the truths of mathematics, so much the worse for the world" (Butts 1978, 81). Similarly, in an article published in 1985, Ernan McMullin argued that Galileo "is not entirely single-minded" in maintaining the view that the effects of the physical impediments can be calculated. "He sometimes lapses back into a Platonic pessimism about the 'imperfections of matter, which is subject to many variations and defects.' . . . But if this were the case, the Book of Nature would not really be written in the language of mathematics, or would, at least, be poorly written" (McMullin 1985, 251). Finally, Maurice Finocchiaro has recently claimed that Galileo's remark in the *Assayer* that the book of nature is written in mathematical language was "more a plea for independent-mindedness" than a statement of mathematical realism. Moreover, "if and to the extent that the remark on the book of nature . . . can be taken as an expression of mathematical realism or Platonism, it should be noted that the remark is an epistemological reflection, not an instance of concrete scientific practice, and one can raise the question whether Galileo's words and deeds correspond" (Finocchiaro 2010, 115–16). In Finocchiaro's view, they often don't correspond.

In this chapter I shall try to show that Galileo's claim that nature is written in the language of mathematics, far from being a rhetorical statement or an unwarranted metaphysical conviction, is grounded in coherent ontological and epistemological arguments. In his works Galileo repeatedly argues that mathematical entities are ontologically independent from us and that the physical world has a mathematical structure. This structure is, however, too complex to be fully grasped by our finite intellect, which is why we need to simplify physical phenomena in order to be able to deal with them mathematically. What scholars have regarded as an opposition between the abstract and the concrete, the mathematical and the physical, was intended by Galileo as a distinction between what is mathematically simple, and hence easy for our intellect to grasp, and what is mathematically complex and hence unknowable.

In the following pages I shall first focus on Galileo's use of the metaphor of the book of nature, paying particular attention to his views concerning the different properties of verbal and mathematical language. Then I shall turn my attention to Galileo's reflections on the relation between mathematical and physical truths which, as I shall try to demonstrate, are fully in accordance with his scientific practice.

GALILEO ON VERBAL AND MATHEMATICAL LANGUAGE

In his book on *Nominalism and Constructivism in Seventeenth-Century Mathematical Philosophy*, David Sepkoski observes that "the epistemology of mathematization is fundamentally linked to the epistemology of language" (Sepkoski 2007, 2). Indeed, early modern authors such as Gassendi, Hobbes, Locke, and Berkeley adopted a nominalistic theory of language that also influenced their views concerning the relation between mathematics and the physical world. While Kepler and Galileo conceived of mathematics as the "language of nature," these authors regarded it "as a 'language' for describing nature that was subject to the same epistemological conventions that govern the structure, objects, and claims to knowledge of natural languages" (125).

As I shall try to document in the following pages, Galileo's works contain very important, albeit unsystematic, considerations concerning the conventional nature of verbal language, which display interesting analogies with those of his nominalist contemporaries. Contrary to the latter, however, Galileo was not willing to extend his conclusions to mathematical language. Rather, the chief function of Galileo's use of the metaphor of the book of nature is precisely that of contrasting the exact and "obligatory" character of mathematical language to the imprecise and arbitrary character of verbal language.[3]

In the famous passage of the *Assayer* to which we have already referred, Galileo reminds his Jesuit opponent that philosophy is not

> a book of fiction created by some man, like the *Iliad* or *Orlando Furioso*—books in which the least important thing is whether what is written in them is true. Well, Sarsi, that is not the way matters stand. Philosophy is written in this grand book—I mean the universe—which stands continually open to our gaze, but it cannot be understood unless one first learns to comprehend the language and interpret the characters in which it is written. It is written in the language of mathematics, and its characters are triangles, circles, and other geometrical figures, without which it is humanly impossible to comprehend a single word of it. (1957, 3)

In other places Galileo uses the same *topos* to convey the originality of his approach to philosophy. While his Scholastic opponents spend their time commenting on Aristotle's books and disputing *ad utranque partem*, he prefers to study "the book of nature, where things are written in one way only" (1890–1909, 248). Finally in the *Copernican Letters*, which represent the manifesto of his ideas concerning the relationship between revealed and physical truths, Galileo compares the book of nature, which God wrote at the moment of creation, to the book of Scripture, which he dictated to evangelists and prophets:

> Holy Scripture and nature derive equally from the Godhead, the former as the dictation of the Holy Spirit, and the latter as the most obedient executrix of God's orders; moreover, to accommodate the understanding of the common people it is appropriate for Scripture to say many things that are different (in appearance and in regard to the literal meaning of the words) from the absolute truth; on the other hand, nature is inexorable and immutable, never violates the terms of the laws imposed upon her, and does not care whether or not her recondite reasons and ways of operating are disclosed to human understanding. (Galilei 1989, 93)

As has been observed by Giorgio Stabile, in these lines Galileo relies on the medieval image of God's two books, but attributes a greater binding force to the natural law (*lex naturae*) than to the divine law (*lex divina*). While the

Bible must follow the logic of ordinary language, which is conventional and hence negotiable, the book of nature, being the reification of God's word, is unmediated by human conventions, fixed and inviolable (Stabile 1994, 55–56). It is in this regard interesting to see that Galileo's distinction between the respective status of verbal and mathematical languages also influences his judgment concerning the accessibility of God's two books. Augustine, an author Galileo often quotes, claimed in his *Enarrationes in Psalmos* (XLV, 7), that while the pages of the Bible could only be enjoyed by those who know how to read, the book of the universe is accessible to everyone. Galileo believes, on the contrary, that the book of nature is more difficult to decipher than Scripture, because the latter is adjusted to the intellectual capacities of common people, whereas the former is not. As he explains in a letter to Fortunio Liceti in January 1641: "the book of philosophy is that which stands perpetually open before our eyes. But being written in characters different from those of our alphabet, it cannot be read by everybody; and the characters of this book are triangles, squares, circles, spheres, cones, pyramids and other mathematical figures, the most suited for this sort of reading" (1890–1909, 18:295).

As we have seen, Galileo contrasts the book of nature with works of fiction, which do not pretend to tell the truth, with Scholastic books, which tell questionable truths, and with scriptural books, which do tell the truth, but a truth that is often merely "adumbrated."[4] What these books have in common is the fact that they are written in a language that is ambiguous by its very nature.

Galileo's considerations concerning natural language, which are scattered throughout his works, touch upon three main themes, notably the arbitrary character of names, the inconstancy of meanings, and the link between words and appearances. In fact, Galileo's ideas on these subjects show an interesting resemblance with those contained in Locke's *Essay Concerning Human Understanding*.

In the Third Book of the *Essay*, where he explains that words are made "arbitrarily the mark of an idea" by a "voluntary imposition," Locke (1690) criticizes Scholastic philosophers for coining names such as "saxietas, metallietas, lignietas and the like . . . , which should pretend to signify the real essences of those substances whereof they knew they had no ideas" (3.2.1, 187; 3.8.2, 230).

The conventional character of verbal language was already emphasized by Galileo, often in the context of a critique of the essentialist definitions put forward by his Aristotelian opponents. In answering Ludovico delle Colombe, Galileo notices, for example, that "the explications of terms are free" (1890–1909, 4:632) and that the attribution of a name can hence never be mistaken. However, precisely because of their arbitrary character, words cannot reveal the essences of things. A similar point is made in the *Letters on the Sunspots* and in the *Assayer*, where Galileo takes issue with the scientific nomenclature used by his Aristotelian opponents. Welser is free to call the sunspots "stars" and Grassi may well refer to a comet as a "planet" provided they don't pretend that their word choice can solve a dispute concerning the nature of celestial bodies. In the *Third Letter on the Sunspots* to Mark Welser, we read: "In truth, I am not insisting on nomenclature, for I know that everyone is free to adopt it as he sees fit. As long as people did not believe that this name conferred on them certain intrinsic and essential conditions . . . one might also call solar spots 'stars,' but they have, in essence, characteristics that differ considerably from those of actual stars" (Galilei and Scheiner 2010, 289).

And in the *Assayer* Galileo observes: "I am not so sure that in order to make a comet a quasi-planet, and as such to deck it out in the attributes of other planets, it is sufficient for Sarsi and his teacher to regard it as one and so name it. If their opinions and their choices have the power of calling into existence the things they name, then I beg them to do me the favor of naming a lot of old hardware I have about my house, 'gold'" (1957, 253).

In the *Dialogue* Galileo also points out that verbal language is full of misleading synonymies, which can sometimes hinder the process of knowledge (1967, 403). When for example Simplicio voices his skepticism concerning Gilbert's geomagnetic theory, Salviati asks himself whether it is not only because of "a single and arbitrary name" that his Aristotelian interlocutor is reluctant to accept the idea that the earth is a big lodestone. If our planet had not been called "earth," a term which also signifies "that material which we plow and sow," but rather "stone," then "saying that its primary substance was stone would surely not have met resistance or contradiction from anybody."

In the *Essay Concerning Human Understanding*, Locke was to regard the unsteady application of names as one of the great abuses of language. For although words are "intended for signs of my ideas, to make them known to others, not by any natural signification, but by a voluntary imposition, it is plain cheat and abuse, when I make them stand sometimes for one thing and sometimes for another" (1690, 3.10, 240–51). Galileo made a somewhat similar point when he observed that misunderstandings and errors do not originate from the "first definition" of a name, which being conventional can never be mistaken, but from the fact that "one doesn't stick to the terms originally included in the definition, or forms different concepts of the defined thing" (1890–1909, 4:632). Galileo repeatedly accuses Scholastic authors of being incoherent in the application of terms that they themselves have coined. If, following Aristotle, one defines the term "place" as the "surface of the surrounding body," then it makes no sense to inquire, as Ludovico delle Colombe does, whether the outermost heaven is in a place. Similarly, if one agrees with Aristotle that humid bodies are those that are not confined within limits of their own, but adapt to the form of their container (De gen. 2.2, 329b), then one must reach the conclusion that fire is humid (Galilei 1890–1909, 4:632–33). Disputes concerning the imposition of names are the business of grammarians, not of philosophers. The latter must however make sure that terms are not "first defined in one way, and then applied to scientific demonstrations in another" (Galilei 1890–1909, 4:698–700).

Another issue that is dear to Galileo and Locke alike is the relation between names and appearances. In the third book of his *Essay*, Locke introduces a famous distinction between the real essence and the nominal essence of things, which he anchors in the distinction, made in the second book, between the primary qualities of bodies (i.e., solidity, extension, motion or rest, number or figure) and their secondary qualities, which are the sensations produced in us by the primary qualities. Due to our "ignorance of the primary qualities of the insensible parts of bodies," the real essence of a substance, which Locke identifies with its internal constitution, remains unknown to us, and we have "no name that is the sign of it" (1690, 4.3.12, 271; 3.3.18, 196). The abstract general ideas we form of substances "with names annexed to them, as patterns, or forms," refer to their nominal essences, which is the collection of particular qualities that one observes together in a substance (3.3.13, 193). As Locke explicitly acknowledges, "the ideas that our complex ones of substances are made up of, and about which our knowledge concerning substances is most employed, are those of their secondary qualities." (4.3.11, 271). Given "that there is no discoverable connexion between any secondary quality and those primary qualities," the gap between the real and the nominal essence of substances remains unbridgeable (4.3.12, 271). According to Locke, the only category of objects for which nominal and real essences coincide is constituted by geometrical figures, the definition of which is "not only the abstract idea to which the general name is annexed, but the very *essentia* or being of the thing itself; that foundation from which all its properties flow, and to which they are all inseparably annexed" (3.3.18, 195).

The relation between names and things, attributes and essences is also addressed, though in an unsystematic way, in Galileo's writings. In the *Dialogue*, Simplicio states with conviction that the cause of free fall is called "gravity." Salviati reminds him that to know the name of a thing is not the same as to know its essence:

> You are wrong, Simplicio; what you ought to say is that everyone knows that it is called "gravity." What I am asking you for is not the name of the thing, but its essence, of which essence you know a bit more than you know about the essence of whatever moves the stars around. I except the name which has been attached to it and which has been made a familiar household word by the continual experience that we have of it daily. But we do not really understand what principle or what force it is that moves stones downward, any more than we understand what moves them upward after they leave the thrower's hand, or what moves the moon around. We have merely, as I said, assigned to the first the more specific and definite name "gravity," whereas to the second we assign the more general term "impressed force," and to the last-named we give "spirits," either "assisting" or "abiding"; and as the cause of infinite other motions we give "nature." (1967, 234–35)

In the *Letters on Sunspots* (Galilei and Scheiner 2010, 91) Galileo claims that "names and attributes must accommodate themselves to the essence of the things, and not the essence to the names, because things come first and names afterwards." It is however vain "to try and penetrate the true and intrinsic essence of natural substances," which means that we have to content ourselves with definitions dependent on the perceived qualities of bodies:

> If upon inquiring into the substance of clouds, I am told that it is a moist vapor, I will then wish to know what vapor is. Perhaps I will be informed that it is water, attenuated by virtue of warmth and thus dissolved into vapor, but being equally uncertain of what water is, I will in asking about this finally hear that it is that fluid body flowing in rivers that we constantly handle and use. But such information about water is merely closer and dependent on more [of our] senses, but not more intrinsic than [the information] I had earlier about clouds. (Galilei and Scheiner 2010, 254)

That information derived from our senses is not able to reveal the intrinsic essences of things, is something Galileo also claims in a famous passage of the *Assayer* that anticipates Locke's distinction between primary and secondary qualities. When explaining that tastes, odors, and colors do not reside in the perceived objects, but only in the perceiving subject, Galileo declares that they are "mere names." The fact that we have imposed upon sensory qualities "special names, distinct from those of the other and real qualities mentioned previously" (i.e., size, figure, quantity, motion, and rest) makes us wrongly believe that "they really exist as actually different from those" (1957, 274). What in fact happens is that our senses transliterate the mathematical language of the book of nature into the language of experience, which is riddled by synonymies and homonymies. The famous fable of sounds, told in the *Assayer*, is nothing other than a way of proving that "the bounty of nature in producing her effects" is such that "our senses and experience" sometimes judge as identical phenomena that are in fact produced by utterly different causes (258–59).

Although our definitions of natural substances are inevitably dependent on the senses, it is not a vain enterprise to try and provide an accurate analysis of some of their properties. In the case of sunspots, Galileo notices, for example, that attributes such as their "location, motion, shape, size, opacity, mutability, appearance, and disappearance" can "be learned by us and then serve as our means better to speculate upon other more controversial conditions of natural substances" (Galilei and Scheiner 2010, 255). Natural philosophers must hence draw their attention to the quantitative properties of bodies, which contrary to sensorial qualities are not "mere names," but have an independent ontological status. While names have been arbitrarily imposed by men onto things, numbers and geometrical figures have been inscribed by God into things, and hence have the power of disclosing their essences.

On the basis of what is just said, it might appear strange to see Galileo argue, in the *Two New Sciences*, that mathematical definitions are "the mere imposition of names, or we might say abbreviations of speech," which is exactly what many years before he had declared about verbal language.[5] The arbitrary character of mathematical language resides, however, only in the choice of the signifier, not in that of the signified. As Galileo repeatedly explains, the language of mathematics is strict and unambiguous. Each sign carries one and only one meaning, and the propositions we can construct out of these signs can but be true or false. While verbal language is the language of persuasion, mathematical language is the language of certainty: "Geometrical things cannot be affected by cavils and paralogisms, as they are true in one way only, can be explained in one way only and intended in one way only" (1890–1909, 458).

The extent to which Galileo thinks that the language of the book of nature is accessible to the mathematical philosopher is a question that is addressed in the following section.

DECODING THE BOOK OF NATURE

Galileo explicitly acknowledges that mathematical laws are idealizations that do not exactly correspond to the behavior of physical bodies. This has led some scholars to regard the claim that the book of nature is written in the language of mathematics as a rhetorical statement, which is neither representative of Galileo's philosophical views nor in accordance with his scientific practice.

As I have argued elsewhere (Palmerino 2006, 39), the fault with this interpretation is that it attributes an ontological meaning to considerations that are in fact epistemological. When Galileo claims, for example, that in the study of motion one must abstract from the "accidents of heaviness, speed, and shape, which are variable in infinitely many ways," he just means that their random variations are too complex to be translated into a simple mathematical law. As Galileo explains in the *Assayer*

> Regular lines are called those that, having a single, firm and determinate description, can be defined and whose accidents and properties can be demonstrated . . . But the irregular lines are those that, not having any determination whatsoever, are infinite and casual, and thus indefinable, and of which therefore no property can be demonstrated and nothing, in sum, can be known. To say that "this *accident* happens according to an irregular line" is for that reason the same as to say "I do not know why it happens." (1890–1909, 458)

In Galileo's eyes the problem is hence not that irregular lines (and physical accidents) are not mathematical, but rather that their mathematical structure is beyond the reach of our intellectual skills.

The issue of the relation between physical and mathematical truths is explicitly addressed by Galileo in the second day of the *Dialogue* (1967, 203). Having patiently listened to a long mathematical demonstration proposed by Salviati, the Aristotelian Simplicio objects that mathematical truths lose their validity when they are applied to physical matters. "A sphere touches a plane in one point" is, for example, a typical case of a proposition that is true in the abstract, but not in the concrete. Salviati offers a very articulated answer to this objection, the first step of which consists in providing a mathematical proof of the validity of the proposition at stake.[6] Such a proof can of course not satisfy Simplicio, who regards it as valid "for abstract spheres, but not for material ones" (206). Challenged by Salviati to explain why what is conclusive for immaterial and abstract spheres should not apply to material ones, Simplicio observes that "material spheres are subject to many accidents," like for example porosity and weight, which make it impossible to "achieve concretely what one imagines of them in the abstract" (206–7). Salviati is however not willing to accept the equation between "abstract" and "perfect," on the one hand, and "concrete" and "imperfect" on the other hand. "Even in the abstract, an immaterial sphere which is not a perfect sphere can touch an immaterial plane which is not perfectly flat in not one point, but over a part of its surface, so that what happens in the concrete up to this point happens the same way in the abstract" (207). Contrary to Simplicio, Salviati uses the attributes "perfect" and "imperfect" not to draw a boundary between mathematics and physics, but to distinguish what is regular, and hence mathematically simple, from what is irregular, and hence complex.[7] Galileo's spokesman is ready to admit that the behavior of physical bodies is not always translatable into a simple mathematical law, but he insists that "the philosopher geometer, when he wants to recognize in the concrete the effects which he has proved in the abstract, must deduct the material hindrances, and if he is able to do so, I assure you that things are in no less agreement than arithmetical computations. The errors, then, lie not in the abstractness or concreteness, nor in geometry or physics, but in a calculator who does not know how to make a true accounting" (207–8). In order to be able to "make a true accounting," the philosopher geometer must hence abstract from those material hindrances that render physical phenomena too complex to be grasped. As Salviati explicitly states in the first day of the *Dialogue* (1967, 103), the human intellect can in fact only understand a limited number of mathematical propositions, but "with regard to those few which the human intellect does understand . . . its knowledge equals the divine in objective certainty, for here it succeeds in understanding necessity, beyond which there can be no greater sureness." When Galileo argues in the *Two New Sciences* that it is not possible for the natural philosopher to reach a "firm science . . . of such accidents of heaviness, speed, and shape, which are variable in infinitely many ways" (1974, 225) he does not mean that these accidents are not mathematical, but just that we are not able "to deal with them scientifically," due to their complex mathematical character.

In commenting on Galileo's reflections on physical-mathematical reasoning, Maurice Finocchiaro has recently argued that "we have neither separation nor identification of the two domains, but rather correspondence. We can never know in advance that a particular mathematical truth corresponds to physical reality, or that a particular physical situation is representable by a particular mathematical entity, but we can claim in advance as a matter of methodological prescription that every physical situation is representable by *some* mathematical entity" (2010, 119). While I believe that these lines perfectly catch the meaning of Galileo's ideas concerning the relation between physical and mathematical truths, I don't fully understand why Finocchiaro claims that "when so interpreted," the *Assayer*'s remark on the book of nature being written in the language of mathematics, "is a long way from the extreme mathematical realism or Platonism sometimes attributed to Galileo" (119). If "mathematical realism or Platonism" means—as Finocchiaro claims—the "identification or conflation of mathematical and physical truth," then Galileo is indeed not a Platonist. As I have argued elsewhere (Palmerino 2006, 41–42), Galileo in fact believes that while what is true in physics must necessarily be true in mathematics, not all mathematical propositions must necessarily find an instantiation in the physical reality. However, if by "mathematical realism or Platonism" one intends the view that mathematical entities are ontologically independent from our mind, then Galileo is certainly a Platonist. In his writings he not only asserts that the structure of reality is intrinsically mathematical, but also claims that mathematical propositions are true independently of whether they are known by us. While God knows all mathematical propositions, we only understand a limited amount of them.

In his book, Finocchiaro takes issue not only with those scholars who explicitly talk of Galileo's Platonism, but also with those who implicitly attribute to him a conflation between mathematical and physical truths. In his view, "the most common instance of such implicit conflation is the interpretation that Galileo was certain about the truth of Copernicanism because of its mathematical simplicity" (Finocchiaro 2010, 115n21). Also in this case, I only partially agree with Finocchiaro's point of view. To be sure, Galileo did not think that the criterion of mathematical simplicity could yield a demonstration of the validity of the Copernican system, for otherwise he would not have put forward the physical proof of the earth's motion based on the phenomenon of the tides. By arguing that the flux and reflux of the sea could not be brought about by any other cause than the double motion of the earth, Galileo tried to transform what was just a probable hypothesis into a demonstrated scientific truth, the only truth capable of outrivaling the authority of the book of Scripture. If one looks at Galileo's scientific practice, one sees however that he did attribute an important heuristic function to the principle of simplicity of nature. Besides being invoked in the *Dialogue* to argue that the Copernican system is more probable than the Ptolemaic one (1967, 123–24), that principle is presented in the *Two New Sciences* as the guiding assumption in Galileo's search for the true law of free fall.

> Further, it is as though we have been led by the hand to the investigation of naturally accelerated motion by consideration of the custom and procedure of nature herself in all her other works, in the performance of which she habitually employs the first, simplest, and easiest means. And indeed, no one of judgment believes that swimming or flying can be accomplished in a simpler or easier way than that which fish and birds employ by natural instinct. Thus when I consider that a stone, falling from rest at some height, successively acquires new increments of speed, why should I not believe that those additions are made by the simplest and most evident rule? . . . And we can perceive the increase of swiftness to be made simply, conceiving mentally that this motion is uniformly and continually accelerated in the same way whenever, in any equal times, equal additions of swiftness are added on. (1974, 153–54)

Also in this case, however, Galileo regards mathematical simplicity as an indication not of the truth, but of the plausibility of his definition of natural accelerated motion. In order to conclusively demonstrate that the law of free fall that can be mathematically derived from this demonstration corresponds to the "acceleration employed by nature in the motion of her falling heavy bodies," he invokes the result of the famous experiment with a bronze ball rolled down a groove in an inclined plane (1974, 169–70). Mathematical reasoning can help the natural philosopher to draw the boundaries of what is physically possible, but only sensory experiences can establish whether a specific physical phenomenon obeys a specific mathematical law (see also Palmerino 2006, 43).

Galileo's repeated appeals to the principle of simplicity of nature seem to be at odds with other passages of his works, where he claims that nature does not conform to the human criteria of perfection and simplicity. In a letter of July 1611 to Gallanzone Gallanzoni, Galileo observes for example that if a man had been allotted the task of defining the relation among the respective movements of the celestial spheres, he would have chosen the "first and most rational proportions" (1890–1909, 149). God, however, "without consideration for our sense of symmetry, arranged those spheres according to proportions which are not only incommensurable and irrational, but totally inaccessible to our intellect." Galileo illustrates this point with an interesting mathematical example. A person having an insufficient understanding of geometry might wonder why the circumference of the circle "has not been made exactly three times as long as the diameter or corresponding to it in some better known proportion." The answer is that, if this had been the case, many "admirable" properties of the circle would have been lost: "the surface of a sphere would not have been four times as big as the maximum circle, nor would the volume of a cylinder have been 3/2 of that of a sphere, and in sum no other geometrical property would have been true as it is now" (1890–1909, 149–50). In order to appreciate the perfection and simplicity of the sphere, one should hence know all of its mathematical properties, which is something our finite intellect cannot achieve.

Two important conclusions can be drawn from the letter to Gallanzone. First, Galileo regards mathematical entities as divinely created objects that are ontologically independent from our mind. Second, he believes that God always operates in the most simple and rational manner, even when his acts look irrational to us.

This can help explain the presence in the *Dialogue* of two apparently contradictory statements. Galileo claims, on the one hand, that we should not "make human abilities the measure of what nature can do," as there is no single physical phenomenon of which we can achieve "a complete understanding" (1967, 101). At the same time, however, he insists that the mathematical philosopher, in his attempt to unveil the real constitution of the universe, should let himself be guided by the assumption that nature "does not act by means of many things when it can do so by means of few."[8]

The principle of simplicity of nature finds an ontological justification precisely in the passage of the *Dialogue* in which Galileo addresses the question of the relation between mathematical and physical truths. It might hence be useful to resume our analysis of this crucial text.

After having made the point that "abstract" and "concrete" are not necessarily synonymous with "perfect" and "imperfect," Salviati observes that "meeting in a single point is not at all a special privilege of the perfect sphere and a perfect plane." So, even if Simplicio were right in claiming that neither a perfect sphere nor a perfect plane can be found in nature, there are still good reasons to think that an imperfect material sphere and an imperfect material plane would touch each other at a single point. For in order to touch each other "with parts of their surfaces," these parts must either be both "exactly flat, or if one is convex, the other must be concave with a curvature which exactly corresponds to the convexity of the other." Such conditions are, however, "much more difficult to find, because of their too strict determinacy, than those others in which their random shapes are infinite in number" (1967, 208). Salviati's remark seems to have a double function. The first is to guarantee right of existence to the mathematical point, an entity that plays a crucial role in Galileo's physics. In the *Two New Sciences*, in fact, Galileo advocates the composition of space, time, and matter out of nonextended physical atoms.[9] Second, by observing that random shapes, being infinite in number, are more likely to be found in nature than regular shapes, Salviati provides an explanation of the fact that material bodies, with their variable accidents, don't behave according to simple mathematical laws. Salviati's remark, however, puzzles Sagredo, who does not see why it should be more difficult to obtain from a block of marble a perfect sphere or pyramid rather than a perfect horse or grasshopper. Salviati's reply reads as follows:

> I say that if any shape can be given to a solid, the spherical is the easiest of all, as it is the simplest, and holds that place among all solid figures which the circle holds among surfaces— . . . The formation of a sphere is so easy that if a circular hole is bored in a flat metal plate and a very roughly rounded solid is rotated at random within it, it will without any other artifice reduce itself to as perfect as spherical sphere as possible . . . But when it comes to forming a horse or, as you say, a grasshopper, I leave it to you to judge, for you know that few sculptors in the world are equipped to do that. (209)

Sagredo agrees with Salviati that "the great ease of forming a sphere stems from its absolute simplicity," whereas the production of complex figures is rendered difficult by their "extreme irregularity." This is for him a reason to wonder whether objects with a regular shape are really as rare as many believe: "If of the shapes which are irregular, and hence hard to obtain, there is an infinity which are nevertheless perfectly obtained, how can it be right to say that the simplest and therefore the easiest of all is impossible to obtain?" (210). Sagredo's curiosity remains unsatisfied as Salviati suddenly puts an end to the discussion to go back to "serious and important things." The right answer to Sagredo would of course be that, in purely probabilistic terms, it is equally difficult to find an object that exactly corresponds to any given irregular figure than to any given regular figure. But this is not the conclusion that Galileo wants to suggest. In his view, the fact that regular figures are "the simplest and easiest of all" increases the probability of their occurrence in nature. And, in turn, the fact that "nature . . . does not act by means of many things when it can do so by means of few" (117) increases the chance that the mathematical philosopher will be able to decode some chapters of the book of nature.

CONCLUSION

At this point we can return to the question raised in the introduction, as to whether Galileo's remark that the book of nature is written in the language of mathematics may be taken as a literal expression of his philosophical views. As I have tried to show in this chapter, this question must be answered in the affirmative, as there is no conflict between Galileo's professed mathematical realism and his scientific practice. When Galileo claims that the natural philosophers must abstract from accidents and impediments in order to be able to mathematize physical phenomena, he does not mean that these phenomena are intrinsically nonmathematical, but just that their mathematical setup is too complex to be grasped by our finite intellect. Moreover, the very fact that the mathematical philosopher is able, simply by "eliminating the material hindrances," to translate physical phenomena into exact mathematical laws, is in Galileo's eyes a sign of the fact that nature is organized according to the criteria of mathematical order and simplicity.

In his above-mentioned book on seventeenth-century mathematical philosophy, David Sepkoski (2007, 83) analyzes Isaac Barrow's considerations on the ontological and epistemological basis of mathematics, which he regards as being "reminiscent of Gassendi's." In Sepkoski's view, Barrow's constructivist view of mathematics, according to which "mathematical objects are created by the mathematician and do not necessarily represent real objects in physical nature," finds "a broad and general epistemological justification" in Gassendi's nominalism (124–25). By claiming that "a mathematical number has no existence proper to itself," Barrow destabilizes the ontological foundations of arithmetic (Barrow 1734, 41, 103).[10] Sepkoski admits, however, that Barrow "is more hesitant when it comes to geometry, since he believes that on some level geometrical demonstrations do correspond with physical realities" (Sepkoski 2007, 103).

Barrow's considerations concerning the ontological status of geometrical objects show, in my opinion, that his philosophy of mathematics is far closer to Galileo's than to Gassendi's. Like the former, and contrary to the latter, Barrow seems in fact willing to admit that geometrical figures have a real existence outside of our intellect.[11]

In a passage of *The Usefulness of Mathematical Learning*, Barrow (1734) explicitly takes issue with the Jesuit Giuseppe Biancani, according to whom mathematical figures have "no other existence in the nature of things than in the mind alone." In reaction to this claim, Barrow observes that,

> if the Hand of an Angel (at least the Power of God) should think fit to polish any Particle of Matter without Vacuity, a Spherical Superfice would appear to the Eyes of a figure exactly round; not as created anew, but as unveiled and laid open from the Disguises and Covers of its circumjacent Matter. Nay I will go farther and affirm that whatsoever we perceive with any sense is really a mathematical Figure, though for the most part irregular; for there is no reason why irregular figures should exist everywhere, and regular ones can exist nowhere. Moreover if it be supposed that Mathematical things cannot exist, there will also be an end of those ideas or types formed in the mind, which will be no more than mere Dreams or the Idols of Things no where existing. (77)

The similarity between Barrow's (1683–86) and Galileo's (1890–1909, 4:52) views concerning the relation between mathematical and physical truths is truly striking. In the lines just quoted, Barrow claims, exactly like Galileo in the *Dialogue*, that all physical objects possess a geometrical shape, and although these are "for the most part irregular," there is no reason to exclude a priori that regular figures can be found in nature. Moreover, even if a mathematical entity does not find instantiation in the physical world, it still exists in the mind of God, who, as both Galileo and Barrow claim, quoting the Platonizing Book of Wisdom, "arranged all things by number, weight and measure."

NOTES

I wish to thank the participants of the Work-in-Progress Seminar of the Center for the History of Philosophy and Science of Radboud University Nijmegen (Delphine Bellis, Hiro Hirai, Klaas Landsman, Christoph Lüthy, Elena Nicoli, Kuni Sakamoto, Francesca Vidotto, and Rienk Vermij) for their comments on an earlier draft of this paper, as well as Antonio Cimino for his elucidation of Husserl's interpretation of Galileo.

1. Galilei 1974, 153 (= 1890–1909, 8:197). As Ofer Gal and Raz Chen-Morris have recently observed there is a striking resemblance between this passage and a statement found in the dedicatory letter of Kepler's *Ad Vitellionem Paralipomena*: "And I have not satisfied my soul with speculations of abstract Geometry, namely with pictures, 'of what there is and what is not' to which the most famous geometers of today devote almost their entire time. But I have investigated the geometry that, by itself, expresses the body of the world following the traces of the Creator with sweat and heavy breath" (Dedication to the Emperor, in Kepler [1937], quoted in Gal and Chen-Morris [2012]).

2. A similar point is made by Galileo in the "First Letter on Sunspots": "The latter [= the philosophical astronomers], besides the task of saving the appearances in whatever way necessary, try to investigate, as the greatest and most marvelous problem, the true constitution of the universe, because this constitution exists, and it exists in a way that is unique, true, real and impossible to be otherwise" (Galilei and Scheiner 2010, 95).

3. I have dealt with Galileo's use of the *topos* of the book of nature in Palmerino (2006). On the same topic see, among others, Stabile (1994), Howell (2002), and Biagioli (2003).

4. For Galileo's theory of "adumbratio," see Stabile (1994, 54–56).

5. Compare Galilei (1974, 36) with Galileo's answer to Di Grazia: "Everybody has free authority in the imposition of names and in the definition of terms, as similar definitions are nothing else than abbreviations of speech" (1890–1909, 4:697).

6. I cannot agree with Rivka Feldhay (1998), according to whom Salvia-

ti's answer to Simplicio is "surprisingly poor" if analyzed against the background of the Renaissance debate *de certitudine mathematicarum*.

7. For a critique of the Aristotelian notion of perfection, see also Galilei (1890–1909, 4:446; 6:319–20; 7:35; 11:149–50).

8. Galilei (1967, 117). See also page 60 ("natura nihil frustra facit"), page 117 ("nature . . . does not act by means of many things when it can do so by means of few"), and page 123 ("frustra fit per plura quod potest fieri per pauciora").

9. For Galileo's views on the composition of space, time, and matter see Palmerino (2011).

10. Barrow's claim that "a mathematical number has no existence proper to itself" is quoted in Sepkoski (2007, 100).

11. This is what Gassendi denies in his *Disquisitio metaphysica, seu dubitationes et instantiae adversus Renati Cartesii metaphysicam*: "at non est dicendum . . . triangulum esse reale quid, veramque naturam praeter intellectum" (Gassendi 1658). This passage is quoted and discussed in Osler (1995).

REFERENCES

Aristotle. 1941. *On the Generation of Animals*. Trans. A. L. Peck. Loeb Classical Library. Cambridge, Mass.: Harvard University Press.

Augustine. 1865. *Enarrationes in Psalmos*. In *S. Aurelii Augustini Opera Omnia: Patrologiae Latinae Elenchus*, ed. J.-P. Migne. Paris, vols. 36–37.

Barrow, I. 1734. *The Usefulness of Mathematical Learning Explained and Demonstrated*. London: S. Austen.

Biagioli, M. 2003. "Stress in the Book of Nature: The Supplemental Logic of Galileo's Realism." *Modern Language Notes* 118: 557–85.

Butts, R. 1978. "Some Tactics in Galileo's Propaganda for the Mathematization of Scientific Experience." In *New Perspectives on Galileo*, ed. R. Butts and J. C. Pitt, 59–85. Dordrecht: Reidel.

De Gandt, F. 2004. *Husserl et Galilée. Sur la crise des sciences européennes*. Paris: Vrin.

Feldhay, R. 1998. "The Use and Abuse of Mathematical Entities: Galileo and the Jesuits Revisited." In *Cambridge Companion to Galileo*, ed. P. Machamer, 80–146. Cambridge: Cambridge University Press.

Finocchiaro, M. A. 2010. *Defending Copernicus and Galileo: Critical Reasoning in the Two Affairs*. Dordrecht: Springer.

Gal, O., and R. Chen-Morris. 2012. *Baroque Science*. Chicago: University of Chicago Press.

Galilei, G. 1890–1909. *Le Opere di Galileo Galilei*. Ed. A. Favaro. 20 vols. Florence: Barbera. (All translations are the author's.)

———. 1957. *Excerpts from The Assayer*. In *Discoveries and Opinions of Galileo*. Trans. S. Drake, 229–80. New York: Anchor.

———. 1967. *Dialogue Concerning the Two Chief World Systems*. Trans. S. Drake. 2nd ed. Berkeley: University of California Press.

———. 1974. *Two New Sciences*. Trans. S. Drake. Madison: University of Wisconsin Press.

———. 1989. "Letter to Christina." In *The Galileo Affair: A Documentary History*, ed. M. Finocchiaro, 87–118. Berkeley: University of California Press.

Galilei, G., and C. Scheiner. 2010. *On Sunspots*. Trans. E. Reeves and A. Van Helden. Chicago: University of Chicago Press.

Gassendi, P. 1658. *Opera Omnia in sex tomos divisa*. Lyon: Anisson & Devenet.

Howell, K. J. 2002. *God's Two Books. Copernican Cosmology and Biblical Interpretation in Early Modern Science*. Notre Dame, Ind.: University of Notre Dame Press.

Husserl, E. 1970. *The Crisis of European Sciences and Transcendental Phenomenology*. Trans. D. Carr. Evanston, Ill.: Northwestern University Press.

Kepler, J. 1937. *Gesammelte Werke*. Ed. Walther von Dyck and Max Caspar. Munich: C. H. Beck.

Koertge, N. 1977. "Galileo and the Problem of Accidents." *Journal of the History of Ideas* 38: 389–408.

Locke, J. 1690. *Essay Concerning Human Understanding*. London: The Bassett.

McMullin, E. 1985. "Galilean Idealization." *Studies in History and Philosophy of Science* 16: 247–73.

Moran, D. 2012. *Husserl's Crisis of the European Sciences and Transcendental Phenomenology: An Introduction*. Cambridge: Cambridge University Press.

Osler, M. 1995. "Divine Will and Mathematical Truth: The Conflict between Descartes and Gassendi on the Status of Eternal Truths." In *Descartes and His Contemporaries: Meditations, Objections and Replies*, ed. R. Ariew and M. Grene, 145–58. Chicago: University of Chicago Press.

Palmerino, C. R. 2006. "Galileo and the Mathematical Characters of the Book of Nature." In *The Book of Nature in Modern Times*, ed. K. van Berkel and A. J. Vanderjagt, 27–45. Leuven: Peeters.

———. 2011. "The Isomorphism of Space, Time and Matter in Seventeenth-Century Natural Philosophy." *Early Science and Medicine* 16: 296–330.

Sepkoski, D. 2007. *Nominalism and Constructivism in Seventeenth-Century Mathematical Philosophy*. New York: Routledge.

Soffer, G. 1990. "Phenomenology and Scientific Realism: Husserl's Critique of Galileo." *The Review of Metaphysics* 44: 67–94.

Stabile, G. 1994. "Linguaggio della natura e linguaggio della scrittura in Galilei. Dalla *Istoria* sulle macchie solari alle lettere copernicane." *Nuncius* 9 (1): 37–64.

2

"THE MARRIAGE OF PHYSICS WITH MATHEMATICS"

Francis Bacon on Measurement, Mathematics, and the Construction of a Mathematical Physics

DANA JALOBEANU

CONSIDERATIONS ON THE NATURE OF SCIENCE played a very important role in Francis Bacon's project for a Great Instauration. On a general level, his approach was foundational; knowledge was required to grow like a pyramid, on a solid basis of natural history, sustaining physics and metaphysics: "For knowledges are as pyramides, whereof history is the basis: so of Natural Philosophy the basis is Natural History; the stage next the basis is Physic; the stage next to the vertical point is Metaphysic."[1]

Furthermore, Bacon preserved the pyramidal model in his detailed investigations into the nature of particular sciences. His claim is that in order for something to become *scientia*, it has to be constructed on a properly organized natural and experimental history.[2] Given the importance and centrality of this concept, it is not surprising that Bacon wrote extensively about the nature, characteristics, and ways of composing such a well-organized and properly recorded natural and experimental history.[3] In one such methodological text on the subject, one can find the following precept:[4] "everything to do with natural phenomena, be they bodies or virtues, should (as far as possible) be set down, counted, weighed, measured and defined. For we are after works, not speculations, and, indeed, a good marriage of Physics with Mathematics begets practice [*Physica autem et Mathematica bene commistae, generant Practicam*]" (OFB, 6:465–66).

This is surely a puzzling set of requirements. First, there is a quantitative requirement; a proper natural and experimental history would contain quantitative descriptions of bodies and virtues. The resulting natural history will not be simply a selection of facts, but a well-ordered collection of measured quantities.[5] As Graham Rees and Cesare Pastorino have already shown, measurement is a very important feature of Bacon's natural history; it is what distinguishes it from its humanist predecessors (Rees 1985; 1986; Pastorino 2011a). In theory, at least, Bacon envisaged natural and experimental histories composed of numerical results, or tables, obtained through the careful "weighing" of experimental and instrumental results.[6] In practice, however, Bacon's natural histories are rarely openly quantitative; and with few notable exceptions they do not contain numerical tables. This makes even more puzzling Bacon's reference to the "good marriage," the good mixture of mathematics and physics necessary in order to generate practice. The producing of "works" belongs, for Bacon, to the realm of operative science; and it is only possible once the "science" itself has taken off the ground. A well-measured and ordered natural history becomes, in this case, a *prerequisite* for, and not the *result* of the "marriage" in question. Can this striking claim refer, therefore, to a peculiar, Baconian form of mathematical physics? What does Bacon mean here by "physics," "mathematics," and their "marriage"?

In a slightly later methodological text, dealing this time with the place of mathematics in a general tree of science, one can find a similar phrase: "Physics and Mathematics produce Practice or Mechanics" (SEH, 4:369, 1:576 [*Physicam et Mathematicam generare Practicam sive Mechanicam*]). This time, however, the phrase is attributed to Aristotle and given as a quote. In fact, Bacon refers here to a key passage from a well-known and highly debated treatise, the pseudo-Aristotelian *Mechanica*.[7] The passage itself was subject to many interpretations (and different translations) in the second part of the sixteenth century, mainly because it addresses directly the possibility of using mathematics to treat problems of physics, such as "the moving of heavy bodies by art, for human benefit" (Berryman 2009, 106).[8] The passage refers, more precisely, to the intermediary status of mechanical problems that "share in both physical and mathematical speculations,"[9] and invites reflection upon the status of mechanics as a mixed-mathematical science. It portrays mechanics as resulting from a "consent" or "collaboration" of physics and mathematics. The precise nature of such a consent and collaboration was the subject of heated debates involving mathematicians, natural philosophers, and practitioners of mechanical arts.

Thus, when Bacon announced a discussion on the "marriage of Physics and Mathematics" he was most probably taking a stand in an ongoing debate, relative to what we call today the "mathematization of nature." The purpose of this chapter is to reconstruct Bacon's position: his original and somewhat idiosyncratic views on "physics," "mathematics," and "mechanics," and his particular way of describing the consent and collaboration between the two disciplines. My claims are the following: (1) Bacon did take a position in the wider debate over the status of mixed-mathematical sciences and the possibility of a mathematical physics; (2) his position was characterized by an instrumental and practical understanding of mathematics; and (3) he saw the "marriage of physics and mathematics" as a prerequisite to the emergence of a quantitative science of nature. In other words, I claim that there is a Baconian version of "mathematical-physics," which becomes evident if we reconstruct more carefully what Bacon and his contemporaries meant by "physics" and "mathematics" and by the various forms of collaboration and consent between the two disciplines.

BACON ON MATHEMATICS, MEASUREMENT, AND EXPERIMENTAL PRACTICE: THE FOUR IDOLS OF BACONIAN SCHOLARSHIP

Any attempt to write about Francis Bacon's views on the "marriage" of physics and mathematics faces serious historical and historiographical problems. For a long time, the history of early modern thought has been dominated by the famous Kuhnian divide between proper, "mathematical sciences" and the "Baconian sciences." Presumably originating with Bacon, Baconian science was supposed to have a qualitative, preparadigmatic, and essentially nonmathematical character. Although this divide has been repeatedly refuted in the last two decades, its lingering echoes pose numerous historiographical and terminological problems. Both "physics"[10] and "mathematics"[11] had a fluid and changeable meaning until the mid-seventeenth century; the landscape of late sixteenth-century theoretical and practical mathematics is full of subtle complexities of which the historian of early modern thought should be aware.[12] It is only recently that the subject has reached outside the borders of the history of mathematics itself, into the wider community of people interested in early modern science.[13] On the other hand, Bacon's thoughts on mathematics and his project to use physics and mathematics conjointly in order to explore the labyrinth of nature are, perhaps, the least explored of his many unfinished projects. This is only partly surprising. After all, the subject has long been obliterated by the persistence of what I have called elsewhere the four idols of Baconian scholarship (Jalobeanu 2013a; 2013b; 2015), namely recurrent general judgments of obscure origin and remarkable persistence, impermeable to refutation.[14]

One of the oldest and most entrenched idols of Baconian scholarship can be exemplified by the repeated claims that Bacon disliked and distrusted mathematics.[15] Such claims have many features of an idol of the tribe:[16] they are useful and widespread simplifications, based on common received views and essentialist historiographical presuppositions. They implicitly attribute to "mathematics" some atemporal essentialist nature, disregarding the historical character and the evolution of mathematical knowledge and mathematical disciplines in the sixteenth century. They have also displayed a remarkable persistence and resistance to refutation.[17] They are partly responsible for the fact that so little work has been done, to date, to unearth the sources of Francis Bacon's attempts to reform astronomy, astrology, and the mechanical arts. A second, related category of idols originates in a too literal interpretation of Bacon's metaphors, especially his celebrated metaphor of the "alphabet of nature." Since these idols relate to language and interpretation, they display common features with Bacon's idols of the market. This category of idols illustrates the power of words over interpretations. Indeed, Bacon repeatedly claims that nature is a labyrinth and the explorer of nature a hunter; that everything in nature results from endlessly active and invisible combinations of appetites and motions (i.e., letters of the alphabet) that one needs to "become like a child" and learn the *abecedarium* of nature in order to be able to perform the work of *interpretation*. He also claims that the results of investigation need to be written down, that "experience itself has to be taught how to read and write," that is, to become literate. It is extremely tempting to give such claims a quasiliteral interpretation, transforming Bacon's project of an experimental investigation of nature into a form of literary pursuit.

A third category of idols can be recognized in the repeated claims that Bacon's science is purely speculative; that Bacon never did experiments but only mimicked the language of experimental practice in order to argue for what was fundamentally a purely speculative system (Jalobeanu 2013b, 7–8). Of the same kind are the claims according to which Bacon's natural history was a collection of recipes and phenomena borrowed from others and never tested, or tried, in practice. There are many versions of this claim, which are recurrent and remarkably persistent to refutation. Their persistence explains why relatively little work has been done to explore Bacon's natural histories and to trace their sources and evaluate their originality.[18]

Last but not least, a fourth category of idols can be recognized in the repeated claims that Bacon rejected the physico-mathematics of Galileo and the mechanics of his contemporaries; that he was isolated among his more scientifically minded contemporaries, "writing philosophy like a lord Chancellor."[19]

What the four categories of idols have in common is a series of assumptions about the nature of mathematics, the nature of scientific enterprise, and Francis Bacon's "isolation" in the scientific and philosophical community of his day. Each of these assumptions has been individually refuted more than once. This chapter does not propose another refutation. I would rather use the framework of the four idols of Baconian scholarship as a background for a more precise reconstruction of Francis Bacon's peculiar form of mathematical physics.

FRANCIS BACON ON MATHEMATICS AND MEASUREMENT: AGAINST THE FIRST IDOL

Francis Bacon gave mathematics a very important role in his restoration of sciences. Mathematics are said to be "the great appendix of natural philosophy, both speculative and operative" (SEH, 4:369)[20] they are "of so much importance both in Physics and Metaphysics and Mechanics and Magic" (370). As appendices, arithmetic and geometry should function as "handmaids" of physics and metaphysics. They have an intermediate but essential role in the instauration of sciences.

Meanwhile, since arithmetic and geometry are also "sciences," and parts of philosophy, Bacon argues that they should be constructed according to the proper method, on a natural historical foundation. Hence it is not surprising that in the list of natural histories Bacon appended to the end of his *Instauratio Magna*, one can find two natural histories of "mathematics": a history of "the natures and powers of numbers" and a history of "the powers and natures of figures" (OFB, 11:485). What would the two histories consist of? One cannot infer much from their titles. However, taken together with the other things Bacon has to say about mathematics, a certain amount of reconstruction is possible. If, moreover, we read such claims in the wider context of the late sixteenth-century debates over the nature of mathematics, Bacon's position becomes clearer. Here is an important element in this reconstruction: in the posthumous *New Atlantis*, among the laboratories and houses of sciences of Solomon's House, Bacon lists a "mathematical house, where are represented all instruments, as well of geometry as astronomy, exquisitely made" (SEH, 3:164). Mathematics, therefore, deals with instruments; a collection of instruments is required both to construct mathematical histories and to provide tools for other sciences, particularly astronomy. Note that for Bacon instruments are not only the tools of mixed mathematical sciences, such as the science of astronomy, but also of pure mathematics, such as geometry. Such an instrumental and practical view on mathematics squares with its role as a "handmaid" for physics, and with Bacon's further insistence on the importance of mathematics for measurement and calculus. In *De augmentis scientiarum* (DA), Bacon mentions a couple of important and yet unsolved problems of his days. In arithmetic, they are the discovery of "formulas for the abridgment of computation sufficiently various and convenient, especially with regard to progressions." To these, Bacon claims, "there is no slight use in Physics" (SEH, 4:370–71). In geometry, he also lists a problem of measurement, computation, and calculus: "the doctrine of solids" (i.e., the problem of calculating areas and volume of solids).

Bacon's views on mathematics are by no means singular at the beginning of the seventeenth century. In fact, in late sixteenth-century England it was quite common to define mathematics in instrumental terms—as a science of measurement and calculation. Equally common was to argue for the universal propaedeutic value of mathematics and for its practical applications in every other science. Such views are widespread among the "mathematical practitioners"[21] of the late sixteenth century; but they can also be found in the works of more traditional mathematicians, especially in the case of vernacular mathematics.[22] The first English translation of Euclid produced by Henry Billingsley (with a preface and textual additions by John Dee) is well stocked with instructions for constructing geometrical figures, with explanations of geometrical instruments and their uses, and with strategies for calculating areas and volumes (sometimes in practice, and with approximation). This instrumental presentation of geometry receives full-length justification in John Dee's accompanying preface. Although Dee emphasizes the divine nature of arithmetic and geometry, his preface argues at length for their use in commerce, navigation, architecture, the art of assaying, mining and so on (Billingsley and Dee 1570). He also distinguishes between several uses of "vulgar geometry" for measuring, approximation, and, more generally, for describing the world of physics. A slightly later English *Elements of Geometry*, translating Ramus's geometrical books from *Schola Mathematicae*, begins with the following definitions:

> 1. Geometrie is the Arte of measuring well
> 2. The thing propounded to the wel measured, is Magnitude
> 3. Magnitude is a continual quantitie
> 4. A terme is the end of a magnitude
>
> Therefore a magnitude is both infinitely made, continued, and divided by these things wherewith it is termed (Ramus and Hood 1590).[23]

Ramus's definitions sketch an instrumental, constructivist, and inductive approach to geometry; the book also offers constructional strategies for drawing basic geometrical figures, followed by instructions on how to compose, decompose, and measure various figures.[24] This attitude to geometry is reinforced in Thomas Hood's "advice to the auditors" of his mathematical lectures,[25] in which geometry is presented as an essential introduction in "mathematical sciences" (mixed mathematics) but also in "anie human knowledge" (Ramus and Hood 1590).

Similar views on mathematics can be found in the more complete and less idiosyncratic French edition of Euclid, translated by Pierre Forcadel and published in 1564.[26] Forcadel's preface argues for the importance of mathematics using the same strategy: on the one hand, the object of mathematics is seen as superior to the objects of any other science; on the other hand, Forcadel argues that mathematics is, in fact, astronomy: "Mathematics treats mainly of celestial business; the motions of the sky, the course of the Sun, the Moon and other planets."[27] Again, mathematics is said to have universal practical value for any theoretical and practical knowledge (including government).

This universal practical value makes mathematics a useful tool for the treatment of physical problems as well. This, however, does not mean that mechanics, astronomy, and optics become, in any way, mathematical *physics*. They are mixed-mathematical sciences, classified under the general heading of geometry.[28] There are, however, other ways of dealing with the "consent" between mathematics and physics, one step further along the road toward a "good marriage of physics with mathematics." One quite striking example of consent and collaboration between mathematics and natural philosophy can be found in a composite and peculiar treatise published in 1571 under the name *A Geometrical Practical Treatise Named Pantometria*. The treatise was published by Thomas Digges (1571) and dedicated to Francis Bacon's father, Nicholas. In the dedicatory letter, the treatise is said to contain: "mathematical demonstrations, and some such other rare experiments and practical conclusions." The said "experiments" cover the construction of instruments and technologies for the production of geometrical and topographical instruments, devices for calculating lengths, surfaces, and volumes, etc. For example, the third book, *Stereometria*, contains "rules to measure the Superficies and Crasitude of Solide bodies, whereof, although an infinite rote of different kindes might be imagines, yet shall I only entreate such as are both usually requisite to be moten, and also many sufficiently induce the ingenious to the mensuration of all other Solides, what forme or figure so ever they beare" (Digges and Digges 1571, 81).

The book is a weird mixture of classical Euclidean geometry with practical problems of measurement; it is also a composite work of two English mathematicians, father and son.[29] The "practical" part relating to measurement belongs to Leonard Digges and was left unpublished for the young and talented Thomas Digges to complete, among other things, with a "doctrine of . . . the five Platonicall solids."[30] Thomas Digges's preface to the reader emphasizes not only the practical utility of geometry but also its experimental character: geometry is a science of measurement. Due to geometry: "man notwithstanding be here imprisoned in a mortal carkasse, and thereby detained in this most inferior and vilest portion of the universall world, fardest distant from that passing pleasant and beautifull frame of celestial orgbes, yet his divine minde ayded with this science of Gemoetrical mensurations, founde out the Quantities, Distances, courses, and strange intricate miraculous motions of these responded heavenly Globes of the Sunne, Moone, Planets and Starres fixed, leaving the precept hereof to his posteritie."[31]

This science of measurement is seen as having universal applications: it can be used to create instruments, to measure the land, to navigate, to perfect ballistics and fortification, to discuss architecture, or the storage of goods (Digges and Digges 1571).[32] What is more important, however, is that this science is seen as leading to the development of more sophisticated measuring instruments and techniques, which, in turn, can perfect the art itself. Last but not least, the science of measurement extends beyond the traditional list of mixed-mathematical sciences, into natural philosophy itself. Thomas Digges's preface claims that one cannot understand Aristotle without being a good mathematician, because "in sundrie of his works also of naturall Philosophie, as the Physickes, Meteores, de Caelo & Mundo, &c. yee shall finde sundry Demonstratins, that without Geometrie may not possibly be understanded." In other words, this kind of mathematics would advance the sciences traditionally considered to belong to "mixed mathematics," but also other sciences, whose mathematical character has not been apparent so far.

Francis Bacon's arguments for the importance of mixed mathematics are similar. He begins with a general definition: "Mixed Mathematic has for its subject some axioms [*Axiomata*] and parts of natural philosophy [*portiones physica*], and considers quantity in so far as it assists to explain, demonstrate, and actuate these." Following the definition, Bacon draws a list of mixed-mathematical sciences, presenting them as "parts of nature" that cannot be comprehended without the aid of mathematics: "For many parts of nature can neither be invented [comprehended] with sufficient subtlety [*nec satis subtiliter comprehendi*], not demonstrated with sufficient perspicuity, nor accommodated to use with sufficient dexterity, without the aid and the intervention of Mathematic: of which sort are Perspective, Music, Astronomy, Cosmography, Architecture, Machinery and some others" (SEH, 4:371).

If Bacon's list of mixed-mathematical sciences is quite traditional, his view on the role of mathematics in physics is not quite so, although it is strikingly similar to Digges's views on the matter. Bacon claims that there should be a dynamic interplay between physics and mathematics; that without this no progress can be made. Moreover, he envisages the situation in which new parts of physics will require fresh assistance from mathematics. "In Mixed Mathematics I do not find any entire parts now deficient, but I predict that *hereafter there will be more kinds of them*, if men be not idle. For as Physics advances farther and farther every day and develops new axioms, it will require fresh assistance from Mathematic in many things, and so the parts of mathematics will be more numerous [*eo Mathematicae opera nova in multis indigebit, et plures demum fient Mathematicae Mixtae*]" (SEH, 4:371).

In what way does physics require the assistance of mathematics? Bacon's general answer to this question is "measurement." In order to construct a natural history on which one can build a sound physics, one has to measure, count, and weigh instances. This involves instruments and experimental techniques; it also involves mathematical instruments and techniques of calculus. All these are used in "measuring, weighing and counting"; observations and experiments further recorded in natural history and useful for the construction of physics. But this "measuring of nature" is, for Bacon, a more general concern. For example, he states that "in every inquiry into nature we must note the *Quantity* or, as it were, the dose of body needed to produce a given effect, and add a dash of guidance concerning *Too Much* or *Too Little*" (OFB, 11:383).

It is clear from such statements that Bacon also has methodological and theoretical concerns associated with the very process of measurement. This is supplemented, in the second book of the *Novum Organum*, by a selection of instances destined to provide "general and catholic observations" (OFB, 11:419) and more general guidance for the investigation of nature. Such are, for example, what he calls *mathematical instances* or *instances of measure* (OFB, 11:367). They are explicitly introduced in order to solve the problem of "inaccurate determination and measurement of the powers and actions of bodies." They underline a general theory of measurement: "Now the powers and actions of bodies are circumscribed and measured either by point in space, moment of time, concentration of quantity, or ascendency of virtue, and unless these four have been well and carefully weighed up, the sciences will perhaps be pretty as speculation, but fall flat in practice" (OFB, 11:367).

There are four such instances, and they are extremely diverse: some are operations or operational procedures governing the investigation of a given nature; some are experiments destined to circumscribe and measure the range of a given phenomenon; others are attempts to set down instruments or techniques for measuring distances, time, or the range of a certain virtue or quality. Among the mathematical instances, some are even called "instances of quantity" or "doses of nature" (OFB, 11:381–83). All these instances provide techniques and examples of measuring; together, they seem to delineate a general theory of measurement of the "powers and actions of bodies." What they measure, however, is a bit more complicated, and relates to the peculiarities of Bacon's physics. Disentangling the meaning and structure of Bacon's physics is a prerequisite to understanding what he meant by the "marriage of physics and mathematics." This is the subject of the next section.

PHYSICS AND THE "ALPHABET OF NATURE": BRINGING PHYSICS CLOSER TO MATHEMATICS AND THE SECOND IDOL OF BACONIAN SCHOLARSHIP

For Bacon, physics deals with three large domains of reality: "the principles of things," "the structure of the universe," and the realm of phenomena, that is, "all the varieties and lesser sums of things" (SEH, 4:347). His investigation focuses especially on this third domain, called "diffused physics"

[*Physicam Sparsam, sive de Varietate Rerum*] (SEH, 1:551) and said to be "a gloss or paraphrase attending upon the text of natural history" (SEH, 4:347; see also OFB, 4:83). This "gloss" has two parts: the concrete and the abstract physics.[33] The first is barely distinguishable from natural history. Indeed, Bacon's concrete physics has the same objects and the same structure as his natural history.[34] It deals with "the heavens or meteors, or the globe of earth and sea, or the greater colleges, which they call the elements, or the lesser colleges or species, as also with pretergeneration and mechanics" (SEH, 4:347).

Concrete physics is a "gloss" on natural history because it simply adds causal explanations to the "facts" of natural history. In some cases, such as mechanics, concrete physics is already a mixed-mathematical science. What about the other parts of concrete physics? Interestingly, Bacon illustrates the requirements of concrete physics with his own project of reforming the "science" of the heavens. He claims that: "Among these parts of Physic that which inquires concerning the heavenly bodies is altogether imperfect and defective," and has been ill handled in more than one way. On the one hand, astronomy—although built on phenomena—has transformed the science of heavens into a study of an abstract and simplified "system of machinery arbitrarily devised and arranged to produce [motions]" (SEH, 4:349).[35] In addition, astronomers have done a "lax and careless job" (OFB, 6:167)[36] when they made observations and supplemented the lack of data with ad hoc assumptions, dogmas, and theories (ibid.). By contrast, Bacon's proposal is to reform both astronomy and astrology in order to build a proper *physica coelestis*. A concrete physics of heavens would have to study the "substance, motion and influence of the heavenly bodies as they really are," and also their "physical reasons" (SEH, 4:348).[37] It would be built on a proper natural history of the heavens, carefully compiled and properly measured.[38] Such a natural history would not limit its inquiries to the "exterior" of the heavenly bodies, but would also inquire into their "interiors" [*viscera*]. Bacon recommends a general inquisition, covering "physical causes, as well of the substance of the heavens both stellar and interstellar, as of the relative velocity and slowness of the heavenly bodies; of the different velocity of motion in the same planet; . . . of their progressions, stationary positions, and retrogressions; of the elevation and fall of motions in apogee and perigee; of the obliquity of motions, either by spirals . . . or by the curves which they call Dragons; of the poles of rotation" (SEH, 4:348).

Bacon claims that such a general inquiry has never yet been performed, mainly because the received astronomy has replaced the real problems of a celestial physics with simplified problems of calculus and prediction. On the other hand, he seems to be aware of some of the difficulties involved in such an investigation; for example, how difficult it is to get precise astronomical observations.[39] He recommends "estimates" [*aestimativas*] (OFB, 11:466) and "comparative measurements"[40] when "precise proportions are not available" (467; see also 6:167, 169). He mentions the methods of distance measurement currently in use in astronomy and the need for "other aids to be devised for this matter, which human industry may contrive" (OFB, 6:169). Such difficulties of measurement become even greater if one takes into consideration the part of celestial physics that deals with the nature and composition of heavenly matter, namely the "substance of the heavenly bodies and every sort of quality, power and influx . . . what is found in the bowels of nature and is actually and really true" (OFB, 6:111).[41] What is found in the "bowels" and "viscera" of nature are the appetites and motions of matter and what Bacon calls "configurations," or "schematisms" of matter; and these are precisely the elements of his abstract physics.[42] Bacon calls them the letters of the alphabet of nature,[43] "by which all that variety of effects and changes which we see in the works of nature and art is made up and brought about" (SEH, 5:425).

How can such primordial, constitutive elements of the universe be subject to measurement and quantitative natural history? Bacon clearly states that these "letters" of the alphabet of nature take place in the "recesses of nature" (SEH, 4:356); that they are "imperceptible" and "intangible" (OFB, 11:351); and that all information about this level of reality "comes via reduction" [*per Deductionem procedit*] (ibid.). On the other hand, the purpose of abstract physics is also precise measurement. The explorer of nature should find ways to "call upon nature to render her account" (SEH, 5:427). The inquisition into simple motions, sums of motions, configurations, and the other elements of the abstract physics should be pursued in such a way that "they are not diffuse [*vagae*], lacking in rigour, and in manner intellectually satisfying, but useless in practice. This is why *we must get closer to the mathematics, namely to measures and scales of motions* [Quamobrem accedendum propius ad mathematica, sive mensuras & scalam motuum], *without which, well counted and weighed and defined the doctrine of motions may falter and not be reliable translated into practice*" (OFB, 13:210–11).[44]

How is the measurement possible if the appetites, motions, and configurations of nature are not accessible to the senses? This is where Bacon's program seems to break down, a wide gap separating his ideal of "getting closer to mathematics" from his metaphysics of simple and compound motions. Furthermore, Bacon gives no composition rules for his language. Even if we know the letters, we do not have a grammar to put the words together.

This is precisely the point at which the influence of the second idol of Baconian scholarship was persuasive and long lasting. To date, there is a wide gap between scholars interested in Bacon's speculative metaphysics,[45] those focusing on Bacon's method, and those mainly intrigued by the structure and composition of Baconian natural histories. The three groups tend to focus on different texts, exploiting the corresponding apparent divide in Bacon's writings between speculative metaphysics of matter and seemingly experimental natural history. However, a less literal reading of Bacon's metaphorical language about the alphabet and the language of nature might show us that the break results in a far smaller gap than it had appeared to so far.

In fact, Bacon does have an answer to the problem of bringing abstract physics closer to mathematics. The answer is in the adoption of a mixed strategy: on the one hand, he proposes an experimental approach and particular kinds of "reductive experiments" [*experimenta deductoria*] (OFB, 13:215) that are able (in principle) to establish connections between the invisible causes and the visible aspects of the phenomena. On the other hand, he elaborates a reductionist strategy based on a minimal list of simple motions, or appetites. The list is provisional and subject to further corrections.[46] There can be other simple motions, or there can be less simple motions; or some simple motions might be different than initially stated. This clearly happens due to some form of empirical and experimental input. Bacon's reductionist strategy is quite sketchy; but we have enough of it to see that for him the problem of inferring *per Deductionem* and the theory of measurement were closely intertwined; and that they have a strong experimental component. It remains to explain what these "reductive experiments" [*experimenta deductoria*] are and in what way can they serve the project of bringing physics "closer" to mathematics.

MEASUREMENT IN PRACTICE: EXPERIMENTS, INSTRUMENTS, AND LEVELS OF PRECISION

A characteristic feature of Baconian experimentation is the way in which Bacon borrows observations and experiments of more traditional natural histories and turns them into experimental series aiming at precise quantitative measurements.[47] He can begin, for example, with Pliny's observation that sailors once obtained freshwater on a ship from fleeces of wool hung around the sides of the ship at night. He "tries" this by simply putting a "woolen fleece" on the ground "for a long while" and observing that it gains weight "which could not happen unless something pneumatic had condensed into something with weight" (OFB, 13:141). However, such a trial is, for Bacon, just the beginning of an entire series of experiments one can find scattered through his writings, all involving the study of the same phenomena: the unusual capacity of the porous fibers of wool to condense air into water. One such experiment is re-creating the conditions on Pliny's ship by hanging a pack of wool in a deep well, just above the level of the water. Here is how Bacon recorded the result: "I have found that in the course of one night the wool increased to five ounces and one dram; and the evident drops of water clung to the outside of the wool, so that one could as it were wash or moisten one's hand. Now I tried this time and time again and, although the weight varied, it always increased mildly" (OFB, 13:141).

We have, therefore, a measurement and a quantitative result. Mark that Bacon seemed to have been aware of several practical problems of measurement, such as the slight variations of the results in repeated trials and the importance of giving an estimate as the mean value and the quantitative result of the experimental series. He also seemed to have been particularly interested by what happens if one varies the experimental conditions of one's experiment. Bacon records repeated trials with the wool placed on the ground, hung in a well at various distances from the water, and placed on the top of a closed wooden vessel containing vinegar. In each case, wool is instrumental in producing (or perhaps facilitating) a process of condensation of air (or vapors) into water (or liquid). In other words, Bacon has transformed an ancient, natural historical observation into a technology. Furthermore, by measuring different quantitative results according to different external conditions, Bacon points out how such a technology can be further developed into an instrument. One can use the capacity of wool to condense vapors in order to create an instrument measuring the properties of the surrounding air. Air is more prone to condensation in closed spaces, in cold, near the water, under certain influences of the stars, and so on. The "woolen instrument" can measure the dispositions of air to condensate by translating them into units of weight. In Century IX of the posthumous *Sylva sylvarum*, such experiments are used in a large-scale program of measuring the qualities of the air in a given spatial region. In order to find out, for example, which part of a property has the healthier air, one has to place weather glasses and packs of wool in various locations throughout the property, record simultaneous results, and draw tables of these properties of air. It is important to note that this experimental research program is developed before one knows exactly what one is measuring—weather glasses and packs of wool are simply used conjointly to find out more about properties of air in a given region without knowing precisely what these properties are (whether what we measure is the temperature and humidity or whether we measure the way air captures and transmits the influences and radiations of the stars).

Such experiments are precisely what Bacon calls *experimenta deductoria* or *instantias deductoria* (OFB, 11:350): they figure, for example, under the heading *Summonsing Instances* in the second part of *Novum Organum*. Summonsing Instances are experiments capable of reducing the imperceptible to the perceptible [*deducunt Non-Sensibile ad Sensibile*]. A particular kind of such reductions occurs precisely in situations such as those described earlier: "It is evident that air, spirit and suchlike things which are fine and subtle in their entirety cannot be seen or felt so that reductions are absolutely necessary when inquiring into them."[48] Similarly, "subtler textures and configurations of things . . . are imperceptible and intangible. The consequence is that information about these also come via reduction [*per Deductionem*]" (OFB, 11:350–51). The way in which such invisible effects are made visible is through instruments that reduce the variations of measured properties to variations of visible properties: *weight* (in the case of the pack of wool) or *length* (in the case of the weather glass).[49] Repeated experiments under different external conditions and simultaneous experiments amount to calibrating the instrument. One can argue that this is exactly what Bacon is doing when placing the pack of wool into the deep well, into a situation of maximum humidity, and recording the increase in weight in this situation. He is thus determining the limits, or the range of variation, of a given property or phenomenon. Furthermore, by requiring the performance of simultaneous measurements with packs of wools and weather glasses, Bacon seems to go one step further into the creation and calibration of his instruments.[50]

The practical process of calibrating instruments is doubled, in Bacon's case, by theoretical concerns regarding the accuracy and precision of such instruments. We have seen already that Bacon distinguishes (in the case of astronomical observations) between precise measurements and rough estimates or comparative measurements: "where precise proportions are not available to us we must for sure fall back on rough estimates and comparisons, as, for instance (if we happen to distrust the astronomers' calculations of distances) that the Moon stands within the Earth's shadow; that Mercury is above the Moon, and the like" (OFB, 11:467). A third category is introduced for practical purposes: the "setting down the extremes . . . where average proportions are not available, let us set down the extremes; for instance that a weaker loadstone will lift so much weight relative to the weight of the stone itself, whereas one with the greatest virtue will lift sixty times its own weight" (OFB, 11:467). What we have here is an attempt to determine the range of the virtue under observation. In the case of the woolen instrument, this will determine the maximum and minimum of a scale. In the case of magnetic virtue, setting down the extremes is equivalent with determining the orb of virtue of a particular magnet and the proportion between magnetic virtue and size (weight). More generally, Bacon indicates as a major mode of operation [*modus operandi*] for every experiment, the determination of "how much or *dose* in nature: what of distance, which is not unfitly called the orb of virtue or activity; what of rapidity or slowness; what of short or long delay; what of the force or dullness of the thing; what of the stimulus of surrounding things" (SEH, 4:357). When moving from the "main effect" to the "stimulus of surrounding things" we are already in deep waters in terms of theory of measurement, because this involves distinguishing between the variation of what is currently called the major parameter and the additional influences of the external conditions. In some cases, these are small or can be minimized. But Bacon is concerned with all sorts of situations in which the supplementary influences coming from external conditions are present and cannot be neglected. Much remains to be done before Bacon's experimental research programs will be seriously investigated. Substantial work in this area has been constantly jeopardized by the third idol of Baconian scholarship, that is, the strong entrenched belief in the qualitative nature of Bacon's physics. As the examples discussed so far clearly indicate, there is little truth attached to such claims. Quite on the contrary: Bacon's experiments show a permanent preoccupation for obtaining measurements of increasing precision and tables of quantitative results.

A THEORY OF MEASUREMENT: MEAGER TABLES, COMPLEX TABLES, AND SUCCESSFUL APPROXIMATIONS

In what way does all this process of experimentation and measuring bring physics closer to mathematics? Is Bacon's "marriage of mathematics with physics" a mere attempt to obtain precise, numerical results? The purpose of the final section of this chapter is to show that Bacon's theory of measurement is more complex than that; that it has techniques for transforming limits into estimates and estimates into precise measurements. In order to do this, I will turn to another example of "reductive experiment," Bacon's much discussed table of densities.[51]

The experiment is designed in the following manner: two identical silver boxes of a cubic shape are filled respectively with equal volumes of various substances. They are subsequently weighed with a balance to determine their relative densities. The etalon in the experiment is gold—so, the first measurement amounts to measuring, with a balance, the relative weight of a given substance with respect to the same volume of gold. By filling the second cube with every possible substance able to fill a cubic space, Bacon is able to draw a table of densities expressed numerically, in grains. In principle, as has been said, this table of densities is not only a quantitative (or, rather, numerical) experiment, but it might also be given as an example of a good natural history; that is, facts carefully weighed, measured, and recorded (Pastorino 2011b). However, this is not what Bacon claims. He claims that the table reveals "many unexpected things" and shows the limits of our knowledge (OFB, 11:353). He also claims the table is "meager" [*indigestissima*]. It is incomplete (lacking entries for numerous substances that cannot be reduced to the cubical volume). It is also full of unexplained gaps (results are not ranged in a progressive series, etc.). Bacon also seems to have recognized that such a table is highly unreliable. For example, the substances entered in the table on a particular position can exist in more than one state. Metals, for example, can be solid blocks, but can also be grinded in powder, can rust and produce a special kind of powder through oxidation, can be melted and mixed with other metals, and so on. In addition, all tangible bodies might be, in themselves, subject to condensation under certain conditions: will their place in the table change in this case? Bacon lists in *Sylva sylvarum* and *Historia densi et rari* (HDR) numerous examples in which condensation seems to take place (rusting metals, swollen leaden statues, condensations by fire, etc.). He comes up with another table that compares the weight of similar substances occupying the same volume in two cases: when substances are in their "natural state" and when they are reduced to powder, rust, or distilled solutions. By contrast with the first table, the second is much more complex and introduces into discussion at least one more parameter; namely, the way in which powders and solutions were produced, in the first place. Obtaining equal volumes of crude powder by grinding, powder obtained through the rusting of a metal, or powder obtained through more complex chemical procedures will change the numerical values in the final table. So the second table is a very complex object, merely sketched, and never fully developed. Furthermore, Bacon claims that both tables (the first, simpler one, and the second, more complex, one) are "pretty meager":

> The only precise table of bodies and their openings [dilatation] would be one which displayed the weight of the individual bodies whole first, then of their crude powders, next of their ashes, limes and rusts; next of their amalgamations, then of their vitrification (in those capable of vitrification) then of their distillations (once the weight of the water they are dissolved in was taken away) and of all other alterations of the same bodies; so that in this manner a judgment might be formed of the openings of bodies and very close-knot connections of the nature in its whole state. (OFB, 13:59)

What Bacon seems to be saying is that both tables are rough estimates; neither is fine-grained enough to allow an ordering of substances according to the relative density in such a way that the results are "precise enough" to answer the questions raised thus far. Putting them together, however, might result in a finer grained table/ordering of substances according to their densities. The resulting table will have more entries, filling some of the gaps found in the first table. So, for the purpose of representing with the help of a table the wide range of densities in nature, the two tables taken together can be considered as more "precise" (fine-grained) than the first one alone. This degree of precision can also be increased. Indeed, Bacon sketches a full experimental program that results in a multiplication of tables: the experiment will amount to measuring the relative weights of equal volumes of substances in all the above-mentioned states of powder and distilled liquors, obtained through consistently applying the same experimental procedures. The envisaged result will be a precise (fine-grained enough) table of bodies and their openings (their various states of aggregation) and a more fine-grained representation of the general scale of densities in the universe.

CONCLUSION

The purpose of this chapter was to reconstruct Bacon's original and somewhat idiosyncratic views on the consent and collaboration between physics and mathematics. I have shown first that his instrumental and practical conception of mathematics was not singular at the beginning of the seventeenth century, and that his attempts to extend the collaboration of mathematics and physics beyond the received realm of mechanical practices, into more general questions of a celestial physics, for example, were shared by many of his contemporaries. I also have shown that by defining mathematics as a science of quantity and defending the preeminence of arithmetic as a science of quantity Bacon was also in good company. Perhaps we can even see his views on mathematics as a way of taking a stand in a wider contemporary debate over the nature of physics, mathematics, and mechanics. For this, however, a wider contextual reconstruction would have been required. I have shown in the third and fourth sections of this chapter that Bacon saw the "marriage of physics and mathematics" as a prerequisite to the emergence of a quantitative science of nature. He was not simply interested in measuring, computing, and registering properties of visible phenomena. His experimental program shows awareness of the role of estimates, approximations, and, more generally, of the need for a theory of measurement in physics. I have shown how, for Bacon, experiments were used to devise technologies, instruments, and experimental research programs destined to bring natural phenomena into a form that will make the marriage of mathematics and physics possible. Bacon devised a methodical way of getting "closer to mathematics" through experimental procedures destined to discover new instruments and "reductive experiments." Such experiments are used, in turn, to devise a complex program of measurement of increasing precision: from "setting down de extremes" to comparative measurement and eventually to precise proportion. At each step along the way, new experiments are added to the series in order to compare and refine the previous results. The ideal, final result of these procedures would be a complex table with many columns, each column containing numerical results. To the untrained eye, the complete tables resulting from the example discussed here would be indistinguishable from astronomical tables, for example. In other words, the results of such experimental research programs look very much like mixed mathematics. Without being Galilean science, we have a mathematical inquiry into nature in which physics has not been transformed into mathematics, but it has been put under the form of mixed mathematics. This, I claim, is what Bacon means by the marriage of mathematics and physics. This is also, I think, the very purpose of Baconian science, the driving force behind his painstaking efforts to measure, weigh, and experiment with nature.

ABBREVIATIONS

AL	*The Advancement of Learning* (OFB 4)
ANN	*Abecedarium novum naturae* (OFB 13)
DA	*De augmentis scientiarum* (SEH 1), English translation (SEH 4 and 5)
DGI	*Descriptio globi intellectualis* (OFB 6)
HDR	*Historia densi et rari* (OFB 13)
HNE	*Historia naturalis et experimentalis* (OFB 12)
HVM	*Historia vitae et mortis* (OFB 12)
OFB	Rees, G., and L. Jardine, eds. 1996–2006. *The Oxford Francis Bacon.*
PHU	*Phenomena universi* (OFB 6)
SEH	Spedding, J., R. L. Ellis, and D. D. Heath, eds. 1857–74. *The Works of Francis Bacon, Baron of Verulam, Viscount St. Alban, and Lord High Chancellor of England*

NOTES

1. (AL, 85). In the later DA Bacon revised the passage in a significant manner: he placed at the basis "history and experience" and claimed that "sciences" in general should follow this pyramidal pattern (SEH, 4:361; SEH, 1:561 [*Sunt enim Scientiae instar pyramidum, quibus Historia et Experientia tanquam basis unica substernuntur; ac proinde basis Naturalis Philosophiae est Historia Naturalis. Tabulatum primum a basi est Physica; vertici proximum Metaphysica*]).

2. Bacon uses the term *scientia* in a fairly traditional manner, to mean demonstrative, universal knowledge, as opposed to *historia*, which treats of "individuals." He claims that natural history should serve as the basic material or "prime matter" of philosophy (see SEH, 1:501–2; OFB, 11:37, 39; SEH, 5:510–11; and SEH, 1:494ff).

3. The nature, structure, and principles of organization of Baconian natural history are currently the subject of contention and debate among scholars. It is perhaps fair to say that after a long period of neglect, in which it was considered a mere collection of facts, Bacon's natural history has become again the focus of scholarly debates. For a recent survey of the field, see the articles in Corneanu, Giglioni, and Jalobeanu (2012). See also Jalobeanu (2010; 2012) and Manzo (2009).

4. Bacon has a number of methodological texts containing precepts on how to write natural history. Among them, the most relevant are *Parasceve* (OFB, 11), the introductory preface to HNE (OFB, 12) and DGI (SEH, 5). See Jalobeanu (2012).

5. Bacon's instruments of ordering natural histories are lists and tables as it will become clear from the third and fourth sections of this chapter.

6. According to Rees, this principle of quantification was at least in part motivated by the interest in generating a practical and productive philosophy, and it was not fully substantiated by Bacon's ways of recording natural histories, which were often lacking proper quantitative data (Rees 1985, 33). Rees also claims that only for two of Bacon's natural histories, HVM and HDR, are "quantitative measurements . . . fundamental" and that Bacon's posthumously published *Sylva sylvarum* "contains very little research conducted on quantitative lines." I think that this is not correct; on the contrary, attempts to measure are fundamental for Bacon's natural historical approach. What can vary is the degree of precision.

7. *Mechanica* is the earliest surviving text of mechanics found in the Aristotelian corpus, written, most probably, by a member of the early Peripatetic school. It has a number of striking features that have puzzled many scholars since its recovery. The text was included in the Aristotelian corpus published by Aldus in 1495–98 and had a very interesting sixteenth-century circulation and posterity among humanists, natural philosophers, and mathematicians. It was still extremely popular at the end of the sixteenth century and became part of the university curriculum in more than one Italian university. On the reception of *Mechanica* in the sixteenth century see Rose and Drake (1971). Both the question of authorship and the nature and philosophical relevance of "mechanics" in this text are still subject to discussion among historians of mathematics (see Berryman 2009). For a recent survey of the question of authorship see Coxhead (2012).

8. Duhem has already remarked that *Mechanica* represents a quite striking attempt to unify the action of a number of different devices under a single analysis, and to offer a mathematical account of their action. Berryman's characterization of the text is that "it seems to be an attempt to make philosophical sense of the 'law of the lever' and its operation in various situations" (114).

9. The text refers to the status of mechanical problems, which "are not altogether identical with physical problems, not entirely separated from them, but they have a share in both mathematical and physical speculations" (847a; Aristotle and Hett 1936, 332).

10. In his 1605 *The Advancement of Learning* Bacon introduces "physicke" with the following specification: "taking it according to the derivation, & not according to our Idiome, for Medicine" (OFB, 4:82).

11. On the multiple meanings and traditions of renaissance mathematics see Goulding (2010), Cifoletti (1990), and Axworthy (2009).

12. This is the reason why, until very recently, historians of philosophy could speak of "mathematizing nature" as if geometry had a unique and essentialist meaning throughout the sixteenth and seventeenth centuries. In fact, geometry and arithmetic were evolving subjects with multiple meanings and often belonging to very different traditions.

13. For example, through the interesting works on expertise and experts in early modern Europe, or through more general discussions on mechanics and mechanical practitioners in early modern Europe (see Ash 2004; Long 2001).

14. For Bacon, the idols are "the deepest fallacies of the human mind," originating in a "corrupt . . . predisposition of the mind" and able to "infect all the anticipations of the intellect" (SEH, 4:431).

15. Although fully articulated only in the twentieth century, in the works of historians and philosophers of science, this idol actually originated in the seventeenth century and can often be identified in the works of Bacon's followers. See the discussion in Giglioni (2013). In a seminal article, Kuhn made this evaluative judgment on Bacon's dislike of mathematics the very basis of a general classification of early modern sciences into "mathematical" and "Baconian" (see Kuhn 1977).

16. Bacon's idols of the tribe originate "from the evenness of the substance of the human spirit, or from its preconceptions, its narrowness, its restlessness, contamination by the affections, the inadequacy of the senses, or mode of impression" (OFB, 11:89). They are "rooted in human nature itself" or in the "race" or "tribe" of men (79–80). Under this label Bacon discusses various forms of simplifications and generalizations. Of relevance for us now are the mind's tendency to accept a limited number of simplifications and affirmatives and then to "pull everything into line and agreement with them," or to suppose that "there is more order and equality in things than it actually finds" (OFB, 11:83).

17. Repeated refutations came from two directions in the past decades. On the one hand, some Bacon students have shown the importance of mathematics (particularly arithmetic) for Bacon's natural historical enterprise. On the other hand, important work has been done to disclose the peculiar character of sixteenth-century "mathematics" and the various traditions within the large field of mathematics (see for example Feingold 1984; Goulding 2010; Pumfrey 2011; Ash 2004). On Bacon and mathematics see Rees (1986).

18. Although this situation is about to change due to the efforts of Graham Rees and his Oxford Francis Bacon team, it is still easy to see how little attention and research has been devoted to Bacon's natural historical writings by contrast with his philosophical or literary output. A number of relatively recent works

have tried to refute the particular claim that Bacon's natural histories consist of large collections of random data about nature (see for example Corneanu, Giglioni, and Jalobeanu 2012; see also Anstey and Hunter 2008; Hunter 2007).

19. I have discussed the implication of this idol for the field of Baconian studies in Jalobeanu (2013b).

20. It is worth noting that in the earlier AL, Bacon classified mathematics as a branch of metaphysics. In DA he moves it from metaphysics to this intermediate, auxiliary, and very important place as an appendix of both physics and metaphysics (and their corresponding operative sciences, mechanics and magic).

21. The term "mathematical practitioner" is admittedly vague and was subject to numerous debates and attempts of refinement and replacement. I am using it here in a general sense, to designate a practical approach to mathematical problems.

22. Giovanna Cifoletti has explored the "vernacular scientific project" of late sixteenth-century France. A similar project of vernacular geometry can be found in England toward the end of the century, in the works of John Dee, and Leonard and Thomas Digges. On vernacular mathematics and vernacular geometry see Cifoletti (1990) and Taylor (2011).

23. For a description of this translation and its context see Johnston (1991, 335).

24. On the peculiarities of Ramus's views on Euclid, his edition of Euclid's *Geometry* and associated texts, see Goulding (2010).

25. As the dedicatory letter emphasizes, the book is the result of Hood's two years of teaching as a "mathematical lecturer of the city of London." Hood held this position from its establishment in 1588 until 1592.

26. This is followed by a second volume, containing books seven to nine, published in 1565. On the context of this translation, see Cifoletti (1990).

27. "les Mathematiques traitent principalement les negoces celestes, les mouvements des cieux, les cours du Soleil & de la Lune & des autres planettes" (Forcadel 1564).

28. A slightly more complicated case is that of John Dee's conception of mathematics. In his *Mathematical Preface* he is certainly arguing for a form of putting together mathematics and physics. However, in the classification, all the ensuing new mathematical sciences are still classified as mathematics (see Rampling 2011; Johnston 2012).

29. Historians tend to say that *Pantometria* was largely written by Leonard Digges and published posthumously by Thomas, with minimal additions. It is

however fair to say that these additions were fully investigated. Indeed, the whole treatise has never been thoroughly explored and to date there is no modern edition of it (Bennett 1991; Taylor 1967; Taylor 2011).

30. It is worth mentioning that in the preface Thomas Digges mentions explicitly that the doctrine of the five solid bodies is added "not to discourse of their secrete or mystical appearances to the Elementall regions and frame of Coelestial Spheres, as things remote and far distant from the method, nature and certaintie of Geometrical demonstration" but to determine their properties, including superficies and volumes (see Digges and Digges 1571, 97).

31. Preface to the reader.

32. On Digges's views on mathematics see also Pumfrey (2011).

33. "Physic diffused, which touches on the variety and particularity of things, I will again divide into two parts: Physic concerning the things Concrete, and Physic concerning things Abstract; or Physics concerning Creatures, and Physic concerning Natures . . . But as all Physics lies in a middle term between Natural History and Metaphysic, the former part (if you observe it rightly) comes nearer to Natural history, the latter to Metaphysic. Concrete Physics is subject to the same division as Natural history; being conversant either with the heavens or meteors, or the globe of earth and sea, or the greater colleges, which they call the elements, or the lesser colleges or species, as also with pretergenerations and mechanics. For in all these Natural History investigates and relates the fact, whereas Physic likewise examines the causes" (SEH, 4:347).

34. Bacon's concrete physics has a part dealing with celestial bodies, one dealing with meteors, a third one dealing with the "greater masses" or the elements, another one dealing with "lesser masses," one dealing with pretergenerations; to these, Bacon adds mechanics. This addition of mechanics is highly relevant to the kind of mixed-mathematical inquiry Bacon had in mind for his physics. For a slightly different interpretation on philosophical mechanics see Weeks (2008).

35. See also SEH (4:348) with the claim that astronomy is solely interested in "mathematical observations and demonstrations," without paying attention to what happens in the "interiors" [*viscera*] of the heavens. What Bacon describes here is the *theorica planetarum*—a two-sphere system and the calculus necessary to compute relative motions and positions of objects on the two spheres. For a more general discussion see Barker and Goldstein (1998) and Westman (1980).

36. In more detailed criticisms of astronomy Bacon delineates three different reasons for why traditional astronomy is not only false but also idolatrous.

One is that by assuming more order in the universe than there really is, astronomy instantiates one of the most common idols of the tribe. The second is that astronomers patched up their lack of data and inability of calculus with ad hoc assumptions and hypotheses. The third is that such procedures stand in the way of collecting true and accurate data.

37. Bacon compares astronomy with the stuffed ox offered by Prometheus as a sacrificial victim, instead of a real ox; similarly, astronomy is an empty science, "stuffed" with seemingly complex mathematical calculations in order to seem real. His own, reformed science will be, by contrast, a *Living Astronomy*. Bacon was seriously interested in building up this project. The posthumous *Sylva sylvarum* mentions twice a special section on heavenly bodies. Such a section was never published, but can be found in the draft manuscript of *Sylva sylvarum* and it has been discussed by Graham Rees (1981). On Bacon's cosmology see also Rees (1975, 101, 161–73).

38. In his projected natural history of the heavens, Bacon argues for an instrumental and experimental approach. He quotes approvingly "the industry of mechanics and the eagerness and enthusiasm of certain learned men, that by means as it were of the skiffs and boats of optical instruments have begun . . . to do new trade with the celestial phenomena" (OFB, 6:115). DGI also quotes Galileo's astronomical discoveries and proves Bacon's familiarity with both *Sidereus nuncius* and the discovery of sunspots.

39. In DGI, Bacon mentions the distortions of the atmosphere and the differences between naked eye observations and telescopic observations (OFB, 6:155, 157); he discusses apparent magnitudes and real magnitudes (167), and emphasizes the need of a reformation of "optical calculations" (ibid.). He enumerates the standard methods of measurement in astronomy, including parallax calculations (169). In NO, further problems of astronomical measurement are enumerated, such as, for example, the need for reliable clocks. Bacon even mentions the possibility of a limited speed of light, which would imply the introduction of an "apparent time" beside the "real time" (see OFB, 11:376–77).

40. On comparative measurements see also NO (OFB, 11:379–80).

41. The passage belongs to a longer excerpt of the same project, Bacon's description of a proper natural history of the heavens, from the unfinished *Descriptio globi intellectualis*.

42. The abstract physics has two parts: a doctrine concerning the "configurations of matter" and a doctrine concerning the "appetites and motions." The latter is subsequently divided into simple motions and compound motions, or processes (see SEH, 4:355–56).

43. Surely as the words or terms of all languages, in an immense variety, are composed of few simple letters, so all the actions and powers of things are formed by a few natures and original elements of simple motions (SEH, 5:426). The metaphor of the alphabet is persistent in Bacon's writings and one of his later works bears the very title *Abecedarium novum naturae.*

44. My emphasis and with a slightly amended translation; what we have here is the same equivalence of mathematics with measurement we have seen in the previous section of this chapter.

45. On Bacon's appetitive metaphysics see Giglioni (2010; 2011; 2013). See also Weeks (2007a; 2007b).

46. The list of simple motions from *Novum organum* ends with the following specification: "I do not deny that other species could perhaps be added, or that the divisions set out could be shifted the better to match the truer veins of things, or that, lastly, that their number could be reduced" (OFB, 11:413).

47. I have given more examples of this strategy in my 2016 publication.

48. Patet quod Aer & Spiritus, & huiusmodi res, quae sunt toto corpore tenues & subtiles, nec cerni nec tangi possint. Quare in Inquisitione circa huiusmodi corpora, Deductionibus omnino est opus (OFB, 11:346–47).

49. The weather glass is also on Bacon's list of Instantia Deductoria; it makes the invisible "degrees of Heat or Cold" visible. In the weather glass "air expanded pushes the water down and contracted draws it up, and in that way the reduction to what can be seen takes place, and not before or in any other way" (OFB, 11:355).

50. Simultaneous measurements of this kind require not only identical instruments, but also good clocks and many researchers to do the job; this gives us some hints as to the complexity of Bacon's project for measuring the properties of the air.

51. Bacon's table of densities appears, with slight variations in PHU, NO, and HDR. It is said to be a summonsing instance (OFB, 1:353). In the subsequent reconstruction I am indebted to the following conceptual and contextual reconstructions: Pastorino (2011a; 2011b) Jalobeanu (2011; 2013c; 2015).

REFERENCES

Anstey, P., and M. Hunter. 2008. "Robert Boyle's 'Designe About Natural History.'" *Early Science and Medicine* 13, no. 2: 83–126.

Aristotle. 1936. *Minor Works.* Trans. W. S. Hett. *Loeb Classical Library.* Cambridge, Mass.: Harvard University Press.

Ash, E. H. 2004. *Power, Knowledge, and Expertise in Elizabethan England*. Baltimore: Johns Hopkins University Press.

Axworthy, A. 2009. "The Epistemological Foundations of the Propaedeutic Status of Mathematics According to the Epistolary and Prefatory Writings of Oronce Fine." In *The Worlds of Oronce Fine. Mathematics, Instruments and Print in Renaissance France*, ed. Alexander Marr, 31–52. Donington: Shaun Tyas.

Barker, P., and B. R. Goldstein. 1998. "Realism and Instrumentalism in Sixteenth-Century Astronomy: A Reappraisal." *Perspectives on Science* 6: 235–58.

Bennett, J. A. 1991. "Geometry and Surveying in Early-Seventeenth-Century England." *Annals of Science* 48, no. 4: 345–54.

Berryman, S. 2009. *The Mechanical Hypothesis in Ancient Greek Natural Philosophy*. Cambridge: Cambridge University Press.

Billingsley, Sir Henry, and J. Dee. 1570. *The Elements of Geometrie of . . . Euclide* [Books 1–15] . . . Now First Translated into the Englishe Toung, by H. Billingsley . . . Whereunto Are Annexed Certaine . . . Annotations . . . of the Best Mathematiciens . . . with A . . . Preface . . . By M. J. Dee, Specifying the Chiefe Mathematicale Sciences, What They are, etc. ff. 464. London: John Daye.

Cifoletti, G. 1990. "Oronce Fine's Legacy in the French Algebraic Tradition: Peletier, Ramus and Gosselin." In *The Worlds of Oronce Fine. Mathematics, Instruments and Print in Renaissance France*, ed. A. Marr, 172–91. Donington: Shaun Tyas.

Corneanu, S., G. Giglioni, and D. Jalobeanu. 2012. "Introduction: The Place of Natural History in Francis Bacon's Philosophy." *Early Science and Medicine* 17: 1–10.

Coxhead, M. A. 2012. "A Close Examination of the Pseudo-Aristotelian Mechanical Problems: The Homology between Mechanics and Poetry as Technē." *Studies in History and Philosophy of Science* 43, no. 2: 300–306.

Digges, L., the Elder, and T. Digges. 1571. *A Geometrical Practise, Named Pantometria, Diuided into Three Bookes, Longimetra, Planimetra, and Stereometria, Containing Rules Manifolde for Mensuration of All Lines, Superficies and Solides*. London: Henrie Bynneman.

Feingold, M. 1984. *The Mathematicians' Apprenticeship: Science, Universities and Society in England, 1560–1640*. Cambridge: Cambridge University Press.

Forcadel, P. 1564. *Les Six Premiers Livres Des Elements D'euclide Traduicts Et Commentez Par P. Forcadel De Bezies*. ff. 190. Paris: H. de Marny & G. Cavellat.

Giglioni, G. 2010. "Mastering the Appetites of Matter: Francis Bacon's Sylva Sylvarum." In *The Body as Object and Instrument of Knowledge: Embodied*

Empiricism in Early Modern Science, ed. C. T. Wolfe and O. Gal, 149–67. Dordrecht: Springer.

———. 2011. *Francesco Bacone. Pensatori.* Rome: Carocci.

———. 2013. "How Bacon Became Baconian." In *The Mechanization of Natural Philosophy*, ed. D. Garber and S. Roux, 27–54. Dordrecht: Springer Netherlands.

Goulding, R. 2010. *Defending Hypatia: Ramus, Savile and the Renaissance Rediscovery of Mathematical History.* Dordrecht: Springer.

Hunter, M. 2007. "Robert Boyle and the Early Royal Society: A Reciprocal Exchange in the Making of Baconian Science." *British Journal for the History of Science* 40: 1–23.

Jalobeanu, D. 2010. "The Philosophy of Francis Bacon's Natural History: A Research Program." *Studii de stiinta si cultura* 4: 18–37.

———. 2011. "Core Experiments, Natural Histories and the Art of Experientia Literata: The Meaning of Baconian Experimentation." *Societate si Politica* 5: 88–104.

———. 2012. "Francis Bacon's Natural History and the Senecan Natural Histories of Early Modern Europe." *Early Science and Medicine* 17: 197–229.

———. 2013a. "The Four Idols of Baconian Scholarship." *Procedia—Social and Behavioural Sciences* 71: 123–30.

———. 2013b. "Francis Bacon, Early Modern Baconians and the Idols of Baconian Scholarship: Introductory Study." *Societate si Politica* 7: 5–28.

———. 2013c. "Learning from Experiment: Classification, Concept Formation and Modeling in Francis Bacon's Experimental Philosophy." *Revue Roumaine de Philosophie* 57, no. 2: 75–93.

———. 2015. *The Art of Natural and Experimental History. Francis Bacon in Context.* Bucuresti: Zeta Books.

———. Forthcoming. *The Hunt of Pan: Francis Bacon's Art of Experimentation and the Invention of Science.* Bucuresti: Zeta Books.

Johnston, S. 1991. "Mathematical Practitioners and Instruments in Elizabethan England." *Annals of Science* 48, no. 4: 319–44.

———. 2012. "John Dee on Geometry: Texts, Teaching and the Euclidean Tradition." *Studies in History and Philosophy of Science* 43, no. 3: 470–79.

Kuhn, T. 1977. "Mathematical Versus Experimental Traditions in the Development of Physical Science." In *The Essential Tension: Selected Studies in Scientific Tradition and Change*, 31–65. Chicago: University of Chicago Press.

Long, P. O. 2001. *Openness, Secrecy, Authorship: Technical Arts and the Culture of Knowledge from Antiquity to the Renaissance*. Baltimore: Johns Hopkins University Press.

Manzo, S. 2009. "Probability, Certainty, and Facts in Francis Bacon's Natural Histories: A Double Attitude against Skepticism." In *Skepticism in the Modern Age: Building on the Work of Richard Popkin*, ed. J. R. Maia Neto, G. Paganini, and J. C. Laursen, 123–37. Leiden: Brill.

Pastorino, C. 2011a. "Weighing Experience: Experimental Histories and Francis Bacon's Quantitative Program." *Early Science and Medicine* 16: 542–70.

———. 2011b. *Weighing Experience: Francis Bacon, the Inventions of the Mechanical Arts, and the Emergence of Modern Experiment*. Indianapolis: Indiana University.

Pumfrey, S. 2011. "'Your Astronomers and Ours Differ Exceedingly': The Controversy over the 'New Star' of 1572 in the Light of a Newly Discovered Text by Thomas Digges." *British Journal for the History of Science* 44: 29–60.

Rampling, J. M. 2011. "The Elizabethan Mathematics of Everything: John Dee's 'Mathematicall Praeface' to Euclid's Elements." *BSHM Bulletin: Journal of the British Society for the History of Mathematics* 26, no. 3: 135–46.

Ramus, P., and T. Hood. 1590. *The Elementes of Geometrie* . . . Trans. T. Hood. Fol. A1. London, n.p.

Rees, G. 1975. "Francis Bacon Semi-Paracelsian Cosmology." *Ambix* 22: 81–101, 161–73.

———. 1981. "An Unpublished Manuscript by Francis Bacon: Sylva Sylvarum Drafts and Other Working Notes." *Annals of Science* 38: 377–412.

———. 1985. "Quantitative Reasoning in Francis Bacon's Natural Philosophy." *Nouvelles de la republique de letttres* 1: 27–48.

———. 1986. "Mathematics and Francis Bacon's Natural Philosophy." *Revue internationale de philosophie* 40: 399–427.

Rees, G., and L. Jardine, eds. 1996–2006. *The Oxford Francis Bacon*. 15 vols. planned, 6 vols. to date. Oxford: Clarendon Press.

Rose, P. L., and S. Drake. 1971. "The Pseudo-Aristotelian Questions of Mechanics in Renaissance Culture." *Studies in the Renaissance* 18: 65–104.

Spedding, J., R. L. Ellis, and D. D. Heath, eds. 1857–74. *The Works of Francis Bacon, Baron of Verulam, Viscount St. Alban, and Lord High Chancellor of England*. 14 vols. London: Longman (facsimile reprint Stuttgart-Bad Cannstatt: Frommann-Holzboog, 1961–63).

Taylor, E. G. R. 1967. *Mathematical Practitioners of Tudor and Stuart England 1485–1714*. Cambridge: Cambridge University Press.

Taylor, K. 2011. "Vernacular Geometry: Between the Senses and Reason." *BSHM Bulletin: Journal of the British Society for the History of Mathematics* 26, no. 3: 147–59.

Weeks, S. 2007a. *Francis Bacon's Science of Magic*. PhD thesis, University of Leeds.

———. 2007b. "Francis Bacon and the Art-Nature Distinction." *Ambix* 54, no. 2: 101–29.

———. 2008. "The Role of Mechanics in Francis Bacon's Great Instauration." In *Philosophies of Technology: Francis Bacon and His Contemporaries*, ed. G. Engel Claus Zittel, R. Nani, and N. C. Karafilys, 133–96. Intersections: Yearbook for Early Modern Studies. Leiden: Brill Academic Publishers.

Westman, R. 1980. "The Astronomer's Role in the 16th Century: A Preliminary Study." *History of Science* 18: 105–47.

3

ON THE MATHEMATIZATION OF FREE FALL
Galileo, Descartes, and a History of Misconstrual

RICHARD T. W. ARTHUR

IN ANY ATTEMPT TO UNDERSTAND conceptual history, the cardinal rule is *not* to assume that thinkers in the past were trying to express what we now understand with perfect clarity. For it will almost always turn out both that their understanding was different from ours—that apparently innocuous details like using proportions instead of equations "shifts" their whole understanding with respect to ours—and that "we," in any case, do not understand the matter as perfectly as we would like to believe. Now, if we have learned this lesson of humility, it is partly from reading the classic studies of Alexandre Koyré, E. A. Burtt, Cornelis de Waard, Marshall Clagett, Thomas Kuhn, Stillman Drake, and other doyens of the history of science, and partly from finding them violating that rule themselves. For it is a principle that is impossible to apply in a total sense: there are always aspects of our understanding of a historical problem that we are unwittingly projecting onto past thinkers when we should not. So if in what follows I should appear to be hard on earlier historians of science, this must constantly be borne in mind; I do not pretend to be free of the vice myself.

The case I want to discuss here is the understanding of motion in the first half of the seventeenth century, specifically the case of the mathematization of free fall. Galileo famously established in his *Discorsi* that the distances traversed by a heavy body falling from rest in successive equal times are as the odd numbers 1, 3, 5, 7, . . . , or equivalently, that the total distances fallen are proportional to the squares of the times of fall (I shall refer to this as Galileo's "law of fall" or "t^2 law").[1] Given the impact that the Galilean law of fall had on the mathematization of physics in the hands of Huygens, Leibniz, and Newton, this mathematical model has come to be regarded as a constitutive element in the new paradigm of natural philosophy known as the mechanical philosophy. Kuhn's term "paradigm" has of course been almost voided of content by its initial ambiguity and its subsequent constant overuse and misapplication. But here, one might suppose, we are on safer ground: the mathematical modeling of free fall is an *exemplar* that featured large in Huygens's *Pendulum Clock*, in Leibniz's derivation of the measure of vis viva as proportional to v^2, and in Newton's derivation of his inverse square law of gravitation.

It is also standardly supposed that the way of modeling motion implicit in this exemplar may be unproblematically ascribed to Galileo (and to a lesser extent, Descartes). On this conception, the trajectory of a moving body is represented by a curve on a graph of space traversed against time elapsed, and the body has at every instant of its motion an instantaneous velocity that is a function of time elapsed, and whose magnitude is given by the slope of the tangent to the curve at that instant. Such an understanding is usually ascribed to the original efforts of these thinkers themselves. For it is to Galileo that we owe the geometric representation of the curved trajectory of a body in motion, and to Descartes the expression of the curve as an algebraic equation, with both thinkers resolving such motion into orthogonal components; while the concept of instantaneous velocity derives from the notion of a degree of speed used by Galileo in his analysis of uniform acceleration, as well as from Descartes' notion of conatus.

Nevertheless, I shall argue, Galileo, Descartes, and others did not yet have our modern understanding of motion as a function of instantaneous velocity, since velocity for them was an affection of motion and there is no motion in an instant.[2] The initial evidence for this is in the form of paradoxes and incongruities that arise from historians of science projecting this modern understanding back onto those authors, and the unconvincingness of their attempts to attribute the resulting confusions to the early natural philosophers themselves. This prompts the question, how was motion conceptualized prior to the modern account involving instantaneous velocity? As soon as this question is raised, the initial appearance of a clear exemplar of mathematization of motion begins to evaporate; that is not to say that such an exemplar does not eventually emerge, but that the process was nowhere near as smooth as it would appear from our projections of the modern understanding back onto its originators.

There were, I contend, several strands in early seventeenth-century thinking about motion that variously complemented or contradicted one another. There existed a strong presumption from Aristotelian philosophy that motion, like time, is continuous. But there was also a widespread conviction that changes in motion occurred discontinuously, with increases in the velocity of motion occurring by the addition of discrete increments of uniform motion—a discretist conception of acceleration that Stillman Drake refers to (rather unhappily) as "the quantum theory of speeds." There was also a second noncontinuist model of motion as consisting in an alternation of motions and periods of rest, with differences in speed accounted for in terms of different proportions of motions and rests, associated in particular with Arriaga (1632). Then there was Galileo's model of nonuniform motion, indebted to the Scholastic theory of intension and remission of forms, according to which acceleration occurs "continuously from moment to moment, and not interruptedly from one quantified part of time to another,"[3] as the moving body goes through an actual infinity of degrees of speed which, taken together, constitute its "overall velocity." This theory ran headlong into the notorious difficulties of the composition of the continuum, a fact that provided continued motivation for the two discretist models just mentioned.

One could characterize this state of affairs by saying that prior to the advent of the functional model of motion established in the eighteenth century, the understanding of motion was in a state of Kuhnian crisis: there was no universally accepted paradigm, and instead several competing theories or paradigms, none of which commanded universal assent. I am not convinced that that is the best way to conceive things, since it suggests that there were definite, well-formed theories or factions in competition with one another. Yet there was not, for example, an Aristotelian theory of nonuniform motion, nor was there a "quantum theory of speeds" in the sense of a clearly articulated theory. In the last section of this chapter I will revisit this issue, and suggest a different way of characterizing the situation. But before we can intelligently discuss such historiographical matters, we need to examine the specifics of the case before us, the mathematization of free fall by Galileo and his predecessors, and by Descartes.

GALILEO AND HIS PREDECESSORS

Let us begin with Galileo's criticism in the *Discorsi* of his predecessors' view that a falling body will move more swiftly as its distance from its point of origin increases. As is well known, Galileo had arrived at the correct law for

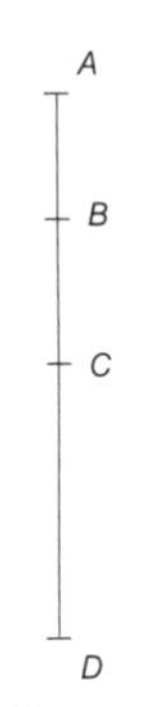

Figure 1. Galileo to Sarpi, 1604.

freely falling bodies in both its forms in 1604. In a letter to his friend Paolo Sarpi of October 16, 1604, he writes that "*spaces traversed in natural motion are in the double proportion of the times*"—the t^2 rule for the proportion of distance to time of fall—"and consequently that *the spaces traversed in equal times are as the odd numbers starting from 1*"[4]—the odd-number rule for successive distances covered in successive times. But in the same letter he announces the principle on which this law is based:

> And the principle is this, that the natural moving body goes by increasing its velocity in the proportion that it is distant from the beginning of its motion; as, for example, with a heavy body falling from the point *a* through the line *abcd*, I suppose that the degree of velocity which it has in *c* to the degree it has in *b* is as the distance *ca* is to the distance *ba*, and so consequently in *d* it will have a degree of velocity more than in *c* according as the distance *da* is more than *ca*.[5]

We tend automatically to interpret Galileo's "degree of velocity" as an instantaneous velocity, and therefore take his principle to be that the velocity of fall at a given instant is proportional to the distance through which the body has fallen from rest. This is how the principle was interpreted by Koyré, following accounts previously given by Paul Tannery and Ernst Mach. But then "the correct formula for the law '*the speed of the moving body is proportional to the distance covered*' would be an exponential function," writes Koyré, citing Tannery (1926, 441ff.). The argument for this claim is sketched "in our language of today" by Mach, and can be filled out as follows.[6] If the instantaneous velocity v is related to distance fallen s in a given time t by

$$v = \mathrm{d}s/\mathrm{d}t = as,$$

with a constant, then

$$\int \mathrm{d}s/s = \int a \; \mathrm{d}t = \int \mathrm{d}s/s \,,$$

and

$$\log_e s = at + c^7$$

Here there is a difficulty that if $s = 0$ at $t = 0$, we obtain $c = \log_e 0$. But $\log_e 0$ is undefined. This can be circumvented if instead we assume that the body is already in motion at $t = 0$. For if at $t = 0$ we have $s = A$ (and thus $v = aA$), with A another constant, we obtain $c = \log_e A$, so that

$$\log_e s - \log_e A = \log_e (s/A) = at$$

giving

$$s = A \exp(at)$$

and

$$v = \mathrm{d}s/\mathrm{d}t = aA \exp(at)$$

Thus both the distance and the velocity increase exponentially with time elapsed.

Galileo soon realizes his error (most probably by 1610, although we have no direct documentary evidence), and in the *Dialogo* of 1632 correctly characterizes the degrees of velocity (represented by transverse lines) as increasing in length in proportion to increasing time, not to distance fallen. As he expresses it in the *Discorsi*, "in equal time intervals, the body receives equal increments of velocity; . . . the acceleration continues to increase according to the time and duration of motion."[8] At this point in the dialogue Galileo has Sagredo say that "it seems to me that this could be defined with perhaps greater clarity without altering the conception as follows: a uniformly accelerated motion is that in which the velocity will have increased in proportion to the increase in the space that has been traversed."[9] This affords Galileo the opportunity to admit (through Salviati) that this is how he once conceived things, and to offer an argument to show why this proposition is not a clearer or even an equivalent way of describing the case, but is in fact "false." (We will come back to that argument presently.) He proposes instead the following principle: "We call that motion equably or uniformly accelerated which, starting from rest, acquires equal moments of velocity in equal times."[10] Now Mach and company interpret this principle as stating that the instantaneous velocity increases in proportion to the time,[11] so that $dv = g\,\mathrm{d}t$, where g is a constant, giving

$$v = \mathrm{d}s/\mathrm{d}t = gt$$

Now,

$$s = \int g\,t\,\mathrm{d}t$$

and, with $s = 0$ at $t = 0$, the t^2 law correctly follows:

$$s = 1/2 g t^2$$

The question now arises, how is it that Galileo could have identified the wrong principle in the first place? Here it must be admitted that the fallacy is a subtle one, which is hard to see without the benefit of the calculus. If no time has elapsed, the falling body has no speed and has fallen through zero distance; then as it falls, its speed increases as time elapses, but so does the distance it covers. It goes faster the farther it is from its starting point, and it does not seem to matter whether the remoteness from its starting point is thought of temporally or spatially. But this does not explain why Galileo chooses initially to enunciate the principle in terms of space covered.

One explanation offered by historians of science (Koyré cites Emil Wohlwill and Pierre Duhem) is that in so doing, Galileo was simply drawing on the tradition. Leonardo da Vinci had written that "The heavy body in free fall acquires a degree of motion with each degree of time, and a degree of speed with each degree of motion."[12] Yet he had still "asserted that the speed is proportional not to the time elapsed but to the distance covered," as had Giovanni Battista Benedetti and Michel Varron, and the former was certainly an influence on Galileo (Koyré 1978, 70). But Koyré rejects the sufficiency of this explanation in terms of tradition, since it only pushes the difficulty further back without resolving it (69). So why did Leonardo, Benedetti, and Varron, and also later Galileo and Descartes, prefer to express the relation in terms of a proportionality to distance covered?

"The reason," writes Koyré, "seems at once both simple and profound. It is entirely a matter of the role that geometry plays in modern science, of the relative intelligibility of spatial relations" (1978, 73). The mathematizing of the laws of nature "comes to the same thing" as the geometrization of space, "for how could something have been mathematised—before Descartes—except by geometrising it?"[13] Thus Galileo's error of using a representation that "is only valid for an increase which is uniform in relation to *time*" is attributed by Koyré to his "thorough-going geometrisation," which "*transfers to space that which is valid for time*" (78).

This is not the place to contest Koyré's grand narrative of the scientific revolution as "the geometrization of nature," which has been ably done by others.[14] But I cannot let his appeal to the notion of the "geometrization of time" go unchallenged. One finds similar Bergsonian considerations in Burtt, who contends that, as a result of Galileo's success in mathematizing motion by treating time as a straight line, "time as something *lived*" has been "banished from our metaphysics."[15] This ignores the fact that one finds the comparison of time with a line even in Aristotle.[16] But leaving such grander claims aside, Koyré makes two specific claims here, both of which seem doubtful, if not outright false. The first is that the mathematization of time necessarily involves its geometrization, and the second is that it is this geometrization that leads Galileo to substitute space for time in announcing his principle. Concerning the first, it is not the case that the only way to mathematize time prior to Descartes was through geometry. The idea that the successive moments of time might form an arithmetical and not a geometrical progression was far from being unavailable; indeed, as we will see in detail in the section below on Clagett and the Merton School, Leonardo da Vinci, Beeckman, Fabry, and Cazré all characterized the succession of moments as a discrete order, and consequently represented it in terms of an arithmetical progression. Concerning the second claim, if one instead conceives time as constituting a continuous ordering—as did Galileo in asserting a 1–1 correspondence between the instants of this time and degrees of speed—it is indeed natural to represent such a continuous ordering by analogy with the ordering of points on a line.[17] One may call the employment of this spatial model to represent temporal order a "spatialization." However, this is not in itself a fallacy: there may be (and indeed are) other features of time besides the ordering of its instants, but representing a temporal ordering by analogy with a spatial ordering is traditional and unproblematic. Galileo, of course, continues to represent time in this way in his published works. The root of the fallacy we are exploring does not lie in representing time by a line—as Galileo does in the *Dialogo* (see Figure 2), where he writes of "the infinite instants that there are in the time *DA* corresponding to the infinite points on the line *DA*,"[18] setting them in 1–1 correspondence with the infinite degrees of velocity of a uniformly accelerated motion—but in conceiving these same degrees of velocity to accrue uniformly in relation to the distance of the body from the beginning of its fall. If the degrees of velocity increase uniformly in time, then the degree of velocity, say, halfway through the time of the fall is not the same as the degree of velocity at a point halfway through the distance of the fall.

Figure 2. Galileo, *Dialogo, Second Day.*

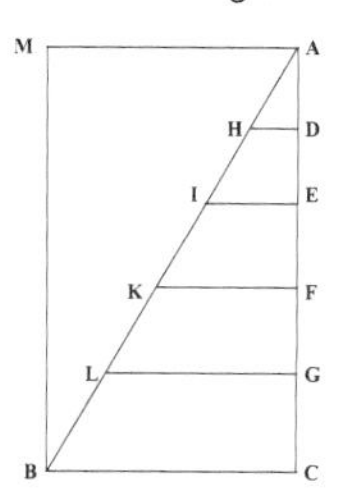

Of course, one might argue that a more charitable reading of Koyré's point is that it is the very representation of time by a line in space that led Galileo, Descartes, and others to take this false step of taking temporal distance from the beginning of the motion to be proportional to the spatial distance from it. Perhaps there is something to this. But what I wish to suggest here is that there is a much more compelling way of understanding why these authors did not see the fallacy of moving backward and forward between talk of equal intervals of a motion through space and of equal intervals of a motion through time. This has to do with their notion of *velocity*, and its concomitant modeling by the theory of proportions.

As Enrico Giusti has argued, the operant notion of velocity in Galileo's time was the Aristotelian one: velocity (or "celerity," swiftness) is *an affection of a whole motion understood as completed*. On this conception, the greater the velocity, the less time it will take for a given body to traverse a given distance, or the greater the distance it will cover in the same time. This corresponds to what Aristotle stated in his physics: "If one thing is faster than another, it will cover a greater distance in an equal amount of time, and it will take less time to traverse an equal distance, and it will take less time to traverse a greater distance. Some people take these properties to define 'faster'" (*Physics* vi.2, 232a 24–27; Aristotle 1996, 141). Such a conception of velocity was still current in the seventeenth century,[19] as can be seen by the definition of *velox* ("fast") given by the Jesuit physicist Honoré Fabry, writing in opposition to Galileo in 1646: "*Def. 2*: A fast [*velox*] motion is that by which more space is traversed in an equal time, or an equal space in less time; and a slow motion is defined contrariwise" (Fabry 1646, 1). Damerow, Freudenthal, McLaughlin, and Renn (1992) have made very much the same point in their book, a point endorsed and aptly summarized by Vincent Jullien and André Charrak in their study of Descartes' writings on free fall as follows: "Speed, such as it is in usage in the pre-classical tradition (before Galileo), which P. Souffrin has called 'holistic speed' ('global speed' might be preferable), is the measure of a movement accomplished in an elapsed time and/or a space traversed" (Jullien and Charrak 2002, 37–38).[20] Giusti (1990) calls this the *velocità complessive* ("overall velocity"), which is the term I shall use.

Now it is crucially important to realize that this is not just a term, but a concomitant of how motion was represented mathematically. Velocities are affections of motions, and these are compared quantitatively using proportions. It is true that Aristotle did not compare velocities directly, but as Clagett explains (1959, 217), the philosophers of the Merton School interpreted his definition to allow this. Thus, according to Thomas Bradwardine, "for every two local motions continued through the same or equal times, the velocities and spaces are proportional, so that one of the velocities is to the other as the space traversed by one velocity is to the space traversed by the other," and "for every two local motions over the same or equal spaces, the velocities and times are always inversely proportional, so that one of the velocities is to the other as the space traversed by one velocity is to the space traversed by the other" (Clagett 1959, 233). That is, if $T_1 = T_2$, then $V_1{:}V_2 = S_1{:}S_2$, and if $S_1 = S_2$, then $V_1{:}V_2 = T_2{:}T_1$. But these are the velocities with which those motions are accomplished, overall velocities, not velocities at a time, as in a contemporary understanding. Again, this concurs with Jullien and Charrak, in their summary of the conclusions of Damerow et al.: "The terms speed, *velocitas* or *celeritas*, employed by themselves, do not designate the speed in an instant, or in 'a point in the trajectory'" (2002, 37).

This contrasts with the reasonings of Mach, Tannery, and Koyré expounded above, who read back into the work of Galileo and his predecessors the conception that Newton's work has accustomed us to, where there is a velocity of motion at every single instant. To Galileo and his contemporaries, however, such a notion of instantaneous velocity would have appeared self-contradictory. There cannot be any motion in an instant, since a motion must take place over time. So, if velocity is an affection of a motion, there cannot be such a thing as an instantaneous velocity.

This, of course, is why Galileo (following in the tradition of the Merton School and Oresme) adopted and developed the notions of *degree of velocity* and *moment of velocity*: precisely in an attempt to explicate how an accelerated motion gets progressively faster without assuming a motion in an instant. He does not have the modern concept of an integral of instantaneous velocities with respect to time;[21] instead he appropriates the idea initiated by the Merton School that a motion has a certain *intensity* at any given instant. Each such intensity can be represented quantitatively by a transverse line, in such a way, Galileo asserts, that "all the degrees of velocity" can compose into the overall velocity of the motion in the same way that "all the lines" can be seen to add up to the whole corresponding area. Thus in Folio 128 of the *Fragments Connected with the Discorsi* Galileo wrote:

> Thus the degrees of velocity continually increase at all the points of the line *af* according to the increment of the parallels drawn from all these same points. Moreover, because the velocity with which the moving body has

> come from *a* to *d* is composed of all the degrees of velocity acquired in all the points of the line *ad*, and the velocity with which it has traversed the line *ac* is composed of all the degrees of velocity that is has acquired in all the points of the line *ac*, it follows that the velocity with which it has traversed the line *ad* has the same proportion to the velocity with which it has traversed *ac*, as all the parallel lines drawn from all the points of the line *ad* up to *ah* have to all the parallels drawn from all the points of the line *ac* up to *ag*.[22]

Here we see that the conception is that all the degrees of velocity add up to a velocity, but this velocity is the overall velocity, Giusti's *velocità complessive*, the swiftness with which the motion is accomplished. By this construction Galileo has proved this to be proportional to the area of the corresponding triangle, and thus to the square of the distance of fall. But now he concludes:

> Thus the velocity with which the line *ad* is traversed to the velocity with which the line *ac* is traversed has double proportion to that between *da* and *ca*. And since the ratio between the velocities is the inverse of the ratio of the times (for to increase the velocity is the same as to decrease the time) it follows that the time of the motion through *ad* is to the time of the motion through *ac* as the subduplicate proportion of the distance *ad* to the distance *ac*. Thus the distances from the point of departure are as the squares of the times, and, dividing, the spaces traversed in equal times are as the odd numbers from unity.[23]

Figure 3. Galileo, *Frammenti attententi ai Discorsi*.

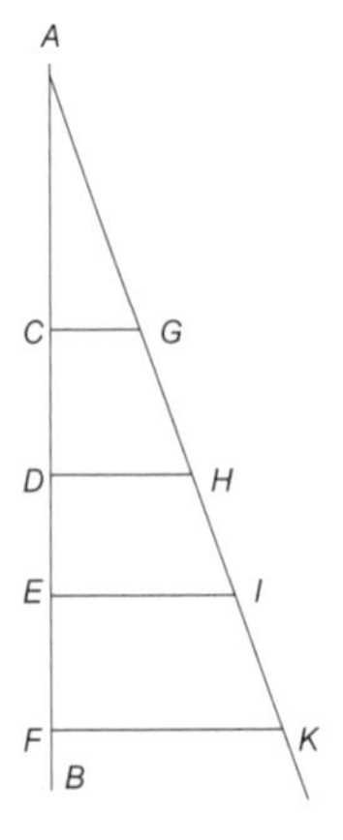

The wording here provides confirmation that Galileo's concept of velocity in 1604 is the Aristotelian one: it is "the time of the motion through a given line," not the instantaneous velocity at the end of the motion. In keeping with this concept he reasons that the times are inversely as the overall velocities, so that, since the velocities are as the square roots of the distances, the times are as the square roots of the distances, and "the distances from the point of departure are as the squares of the times." This is a mistake, since what should follow is that the times are *inversely* as the square roots of the distances. This can be seen more clearly using (modern) proportions: $t_{ad} : t_{ac} = v_{ac} : v_{ad} = \sqrt{ac} : \sqrt{ad}$. $\therefore$ $ac : ad = t_{ad}^2 : t_{ac}^2$. But Galileo does not notice this error, seduced, no doubt, by the obviousness of the fact that as the body falls, its distance and time from the starting point both continually increase.[24]

Subsequently, though, Galileo did come to realize that there is an incompatibility between his t^2 law and the principle on which he had based it, that the velocity of the falling body is proportional to its distance from the beginning of its motion. Guiding him in finding the right principle, as he explains in the *Discorsi*, was above all the Aristotelian notion of "the intimate relationship between time and motion"; this, together with the idea that "we find no addition or increment simpler than that which repeats itself always in the same manner,"[25] he was led to the correct principle that a motion would be uniformly and continuously accelerated "when, during any equal intervals of time whatever, equal increments of velocity are given to it."[26] The degrees of velocity, that is, must be conceived as increasing at successive instants of the time of fall, and not at successive distances from the beginning of the motion.

The subtlety of these considerations and awareness of his previous error account for the circumspection with which Galileo treats the topic of uniformly accelerated motion in the *Discorsi*. He is careful not to talk of the velocities of a body moving with uniform acceleration, but the degrees of velocity (*velocitatis gradus*) it has at each of the instants of its motion, thus staying close to the medieval tradition. His famous so-called Mean Speed Theorem, Theorem 1 of his account of accelerated motion on the Third Day of the *Discorsi*, does not refer to the mean as one between "the highest speed and the speed just before acceleration began," as Crew and de Salvio translate it (1954, 173). The comparison is between a body in uniformly accelerated motion and the same body moving uniformly with a *degree of velocity* equal to one-half of the final degree of velocity of the accelerated one: "The time in which a given space is traversed by a moving body with a motion uniformly accelerated from rest is equal to the time in which the same space would be traversed by the same body moving with an equable motion whose degree of velocity is one half of the last and greatest degree of velocity of the preceding accelerated motion."[27]

This repeats the same argument Galileo had given earlier in the *Dialogo*: there a body moving with an equable motion whose constant degree of velocity is equal to "the greatest degree of velocity acquired by the moving body in the accelerated motion" will cover twice the space in the same time.[28] This should be compared to the analogous application of the Merton rule to uniformly decelerated motion by Nicolas Oresme: "If *a* is moved uniformly for an hour and *b* is uniformly decelerated in the same hour from a degree [of velocity] twice [that of *a*] and terminating at no degree, then they will traverse equal distances, as can easily be proved. Therefore, by the definition of velocity, it ought to be conceded that they were moved equally quickly for the whole hour. Therefore, the whole motion of *b* ought not to be said to be as fast as the maximum degree of velocity."[29] Here Damerow and Freudenthal make some cautions, following Anneliese Maier (Damerow et al. 1992, 18–19). In Oresme, and in the later representations, this is simply a method of graphical representation that shows how the variation of qualities can be depicted, rather than a method of calculation. Something similar seems to be implied in Galileo's usage: if the area represents the velocity, the lines represent the degrees. There is no question of "integrating" infinitesimally thin lines to make a surface.

With these considerations in mind, let us examine a typical argument of one of Galileo's predecessors, before we proceed to his criticisms. In his *De motu tractatus* of 1584, Michel Varron wrote:

> The spaces of this motion conserve this proportion of celerity, so that whatever the ratio of the whole space through which the motion is made to the part (the beginning on both sides is assumed to be where the beginning of the motion is), the ratio of celerity to celerity is the same. For example, if some force will move through the line ABE, and AB is part of this line, then the ratio of AE to AB will be the same as that of the celerity of the motion at the point E to the celerity at the point B.
>
> A proportion of this kind is observed in the parallels cutting a triangle . . . So if the space is divided into aliquot parts, at the end of the second space it will be carried twice as fast as at the end of the first.[30]

There is ambiguity here: Varron talks of the body being carried twice as fast at the end of the second space, which certainly looks like an instantaneous velocity. Nevertheless, he is (like Galileo) applying the theory of proportions: if it is carried twice as fast through the second space of its motion, it will complete that space in half the time. This requires the celerity to be taken in the Aristotelian sense as the swiftness of the motion taken as a whole.

Now let us turn to the way in which Galileo criticizes his predecessors. The passage in question runs as follows:

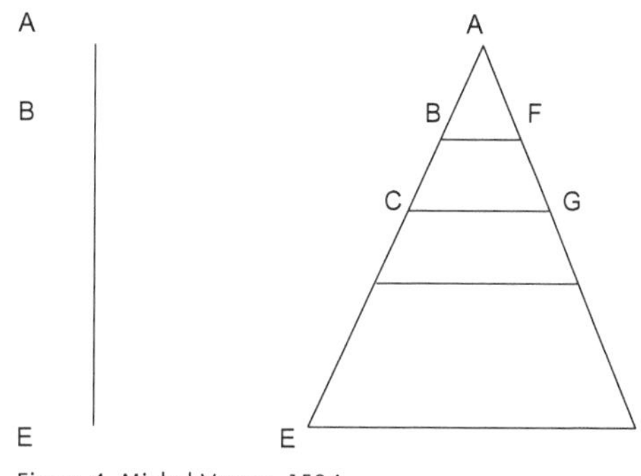

Figure 4. Michel Varron, 1584.

> When the velocities have the same proportion to the spaces traversed, or to be traversed, such spaces come to be traversed in equal times; for if the [velocity][31] with which the falling body traverses the space of four cubits be twice the [velocity] with which it traverses the first two cubits (seeing as the former space is double the latter), then the times of such traversals will be equal; but for the same mobile to traverse the four cubits and the two in the same time could take place only in an instantaneous motion; but we have seen that the heavy body makes its falling motion in time, and traverses the two cubits in a smaller time than the four; therefore it is false that the velocity increases as the space.[32]

Alexandre Koyré (1978) calls this a "specious argument" and a "thoroughly mistaken" refutation of his predecessors' views. Again citing Mach and Tannery, he writes: "The argument contains a similar error to that which we found in the argument discussed earlier: Galileo applies to motion, of which the increase of speed is proportional to the distance covered a calculation which is only applicable to uniformly (in relation to time) accelerated motion."[33]

We may explicate Koyré's criticism as follows. Galileo applies his mean speed theorem (valid only for motion uniformly accelerating with time) to obtain $s = \frac{1}{2}vt$, where v is the final velocity. This yields $t = 2s/v$. So if t_1 and t_2 are the times of fall through 2 and 4 cubits respectively, $t_1 = 4/v_1$, and $t_2 = 8/v_2$. But since $v_2 = 2v_1$, $t_2 = 8/2v_1 = 4/v_1$, we have $t_2 = t_1$, and the time of fall through the second two cubits is 0. So, according to Koyré, Galileo has generated this contradictory conclusion by applying to the case of a velocity increasing with distance a calculation valid only for a velocity increasing with time.

But I think much more sense can be made of Galileo's refutation. Even though it is very tempting to read Varron, as did Mach, to intend by his expression "the celerity of motion at E" the *instantaneous velocity* at that point, there is no independent evidence to suggest that Varron has made this conceptual advance—one that, I stress, Galileo himself did not make. It is not obvious that Varron is using the word *celeritas* as a synonym for intensity of motion or degree of speed, so even if there is something of that in his conception, Galileo seems to be interpreting *celeritas* as a synonym for *velocitas*, the swiftness of the whole motion calculated as terminating at E.

So, I contend, Galileo is not misapplying his mean speed theorem. He is interpreting his predecessors as holding that the overall *velocitas* of one motion to another is proportional to the respective spaces traversed in the same times, as he himself had previously done. The more quickly a given space is covered, or the more space that is completed in a given time, the swifter the motion. According to this concept, a motion twice as swift will cover twice the distance in the same time. So if the first two cubits are traversed with an overall velocity v_1, and the whole four cubits with a velocity $v_2 = 2v_1$, then the times of fall will be $t_1 = 2/v_1$, and $t_2 = 4/v_2 = 2/v_1$, and we will have $t_2 = t_1$, so that the time of fall through the second two cubits is 0, as Galileo argued. If this is the correct interpretation of Varron's argument, then Mach's analysis is in error in imputing to him the modern concept of instantaneous velocity. Moreover, on this interpretation Galileo's *reductio* goes through, and no paralogism is involved. There may well be a rhetorical component to his argument, however, if it is the case that Varron is confusing the intensity of the motion at a given instant (Galileo's "degree of velocity") with this concept of overall *velocitas*. In that case, Galileo, having successfully disentangled them, is attributing only the latter to his predecessors in order to make his point.

DESCARTES' MODEL OF FREE FALL

Now let us turn our attention to Descartes. The French savant had first proposed a solution to the problem of a heavy body falling in a vacuum under the influence of a constant force of attraction in response to a very neat formulation of the problem by his Dutch mentor Isaac Beeckman in 1618. I have analyzed that episode elsewhere (see Arthur 2007, 2011). What concerns me here is Descartes' abiding understanding of what he had thought himself to have proved then, as evidenced in his letters to Mersenne of

November 13, 1629, and August 14, 1634 (Descartes 1991, 9, 44).[34] In the first of these he presents a diagram in which AC represents the distance through which the body falls, and also represents the motion of a body that, having received an impulse at A, travels at a uniform speed from A to C for the whole duration. This is in keeping with what Descartes had learned from Beeckman, that "in a vacuum, what has once begun to move keeps on moving at the same speed," and that the action of gravity can be analyzed in terms of impulses given to the falling body at successive moments of the fall. Thus the vertical lines 2, 3, 4, etc. represent the distances that would be traversed by the same body if it moved solely by virtue of the additional impetuses, each equal to the first, received in the second, third, etc. moments of its fall. (Each is shorter than the last, in that the body, traveling at the same speed, would have a slightly shorter time of fall.) Accordingly, the force of the body's motion in the first moment will be proportional to AC; in the second moment this force is maintained, and added to it is a second force represented by the line 2, and so on. The horizontal lines (such as BE) would then represent the force or intensity of motion at each subsequent moment.

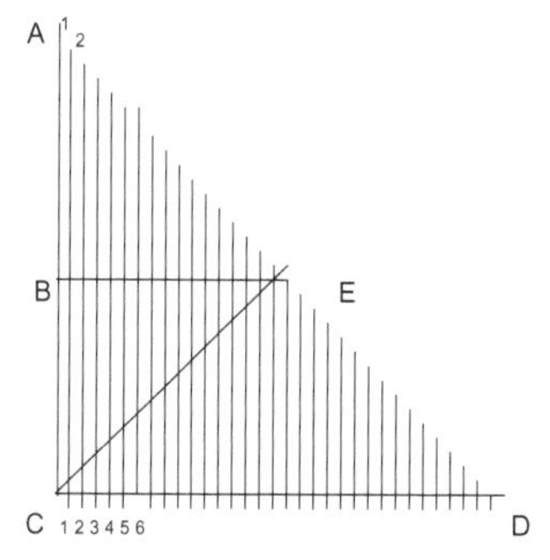

Figure 5. Descartes to Mersenne, 1629.

Says Descartes,

> Thus we get the triangle ACD, which represents the increase in the velocity of the weight as it falls from A to C, the triangle ABE representing the increase in the velocity over the first half of the distance covered, and the trapezium BCDE representing the increase in the velocity over the second half of the distance covered, namely BC. Since the trapezium is three times the size of triangle ABE (as is obvious), it follows that the velocity of the weight as it falls from B to C is three times as great as what it was from A to B. If, for example, it takes 3 seconds to fall from A to B, it will take 1 second to fall from B to C. Again, in 4 seconds it will cover twice the distance it covers in 3, and hence in 12 seconds it will cover twice the distance it does in 9, and in 16 seconds four times the distance it covers in 9, and so on in due order.

Commenting on this passage, the editor of the standard English translation of Descartes' writings (for this section, Dugald Murdoch) says

"Descartes wrongly takes the line ABC to represent the time, instead of the distance travelled; this leads him to take the distance travelled as being proportional, not to the square of the time ($d = \frac{1}{2}gt^2$), but to a power of the time, the exponent of which is log2/log$\frac{1}{3}$" (Descartes 1991, 9n1).

Murdoch does not give the calculation. But here is a reconstruction. Descartes is assuming that the velocity of the falling body is proportional to the distance traversed, and inversely proportional to the time elapsed. Assuming velocity means the instantaneous velocity, and that it is expressible as a function of time elapsed and of distance covered at a given time, we have

$$v = \mathrm{d}s/\mathrm{d}t = \mathrm{k}s/t$$

The solution to this is $s = \mathrm{c}t^{\mathrm{k}}$, where s is the distance fallen, t the time, and c and k are constants. The first two examples Descartes gives indicate that twice the distance is done in 4/3 of the time, and the third example that $4s : s$ is as 16t : 9t. These relations fit the formula $s_1 : s_2 = t_1{}^{\mathrm{k}} : t_2{}^{\mathrm{k}}$, and yield $2 = (4/3)^{\mathrm{k}}$ or k = log2/log$\frac{4}{3}$ = log2/(log4 – log3). (Here the logarithms may be taken to any base; Murdoch's log$\frac{1}{3}$ instead of log$\frac{4}{3}$(= 2log$\frac{1}{3}$) in the denominator appears to be an error.) In a precisely similar calculation, Damerow et al. (1992, 59) arrive at the following formula:[35]

$$s = \mathrm{c}\, t^{\,\log2/(\log4\,-\,\log3)}$$

Neither Damerow et al. nor Murdoch reference Paul Tannery's note in his and Adam's edition of Descartes' works, although this is the probable source, at least of Murdoch's own note. There Tannery remarks that Descartes "therefore comes to a relation essentially different from that of Galileo, since it would amount to considering the space traversed as proportional, not to the square of the time, but to a power of time whose exponent is the ratio of log 2 to log 4/3, that is to say, about 2.4."[36] Jullien and Charrak do reference this note, rightly calling it "a little clumsy, because anachronistic" of Tannery to "construct an exponential function giving the spaces in relation to the variable *time*."[37]

It certainly is anachronistic. Given that Descartes eschewed transcendental curves from his geometry, let alone that he knew nothing of functions or the integral calculus, this attribution to him of an $s = ct^{\mathrm{k}}$ law is unconvincing, to say the least.[38] Moreover, it again attributes to Descartes a conception of instantaneous velocity. But inspection of these passages shows Descartes writing of "the velocity of the weight as it falls from B to C." This is nonsense if one has a conception of velocity as constantly increasing as the body falls, and is only interpretable, I submit, if velocity here, as above, is an affection of the whole motion, the swiftness with which this whole (leg of the) motion is accomplished.

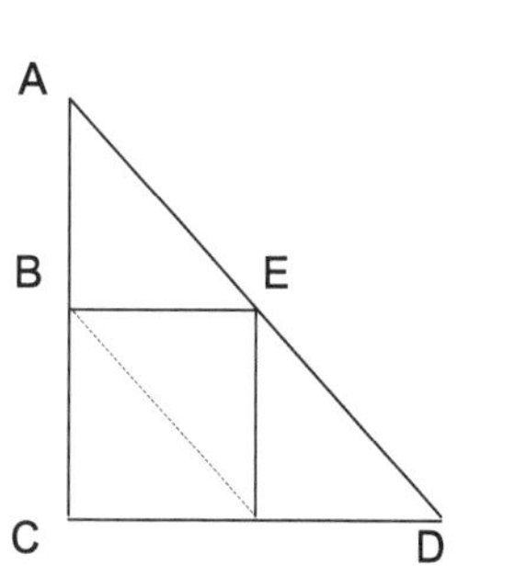

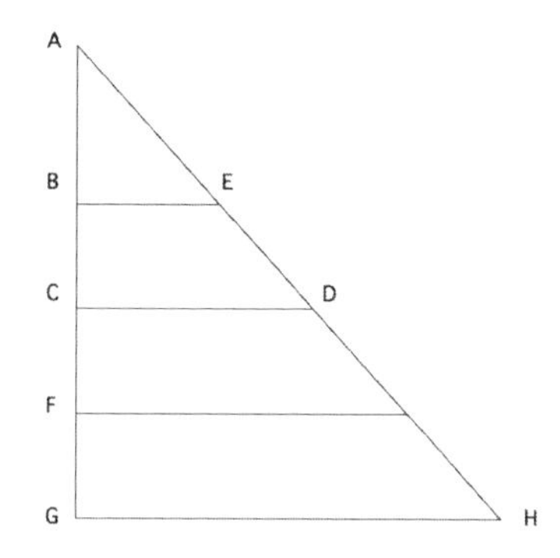

Figure 6. Descartes to Mersenne, 1634.

Accordingly, my interpretation is as follows: a motion that takes three seconds to cover AB is three times less fast (*velox*) than one that takes one second to cover the equal distance BC. Thus if the *velocitas* for AB is 1/3 of that for BC—since ABE = 1/3 BCDE—and it covers AB in 3 seconds, it will cover BC in 1, and AC in 4. If it covers AB in 9 seconds, it will cover BC in 3, and thus AC in 12. And if it covers AB in 9 seconds, and thus AC in 12, since ACD:CDHG is also 1:3, it will cover CG in 4 seconds, and thus ACG in 16.[39]

Thus if the velocity through the space AB, represented by ABE, is 1 unit, that through BC is 3 units, that through CF is 5 units, that through FG is 7 units. This is the odd number rule that Descartes discovered. It is not identical to Galileo's rule, contrary to Descartes' impression, where it is the velocity over each equal part of the uniformly accelerated motion, reckoned timewise, that increases as the odd numbers. The reason Descartes does not see the discrepancy is that he has reasoned as follows: If successive equal spaces are covered by the moving body in times of ratios 1, 1/3, 1/5, 1/7, . . . to 1, then in successive equal times the spaces covered are of ratios 1, 3, 5, 7, . . . to 1.

In other words, he has applied the rule that a body traveling with N times the velocity will traverse an equal space in 1/N of the time; or in an equal time will traverse a space N times as great. But this is just to apply the Aristotelian conception of *velocitas*, the overall velocity discussed above, to successive portions of the motion.

It is, I submit, not surprising that Descartes does not see the discrepancy between his results and Galileo's. For in the case of equal times, they do indeed get identical results. If the individual spaces covered in equal times are in the ratios 1, 3, 5, 7, . . . to 1, then the overall spaces covered are as the squares of the times.

What is wrong with Descartes' conceptual apparatus can be seen by looking at the motion through four equal spaces, say four cubits, in two *spatial* halves. By hypothesis, the moving body covers these in times of ratios 1, 1/3, 1/5, 1/7, to 1. Thus the overall times of fall through 1, 2, 3, and 4 cubits are 1, 4/3, 23/15, 176/105, respectively. Now looking at this in two halves, the first 2 cubits are covered in 4/3 s, the second 2 cubits in $176/105-4/3=36/105=12/35$ s. Thus the times of these two equal halves of the motion are in the ratio $4/3:12/35=35:9$, that is, almost 4 : 1, whereas the velocities are in the ratio 4:12, or 1:3.

In sum, Descartes gets the right result by a compensation of errors, as Koyré pointed out. But this compensation consists in his using the Aristotelian law of *velocitas* to transpose the right result—obtained correctly in 1618 as a result of Beeckman's having set the problem up in terms of times—into a result in terms of equal spaces, and then back again. It is not a misreading of a proportionality between instantaneous velocity and time elapsed as equivalent to a proportion between instantaneous velocity and space covered. Neither Descartes, nor Galileo, nor Varron, made that particular error. For none of them *had* the concept of instantaneous velocity. And none of these historical episodes can be understood using the terms of this modern conceptual apparatus without hopeless distortion.

CLAGETT AND THE MERTON SCHOOL

Now, at this point I anticipate that it might be objected that I have exaggerated the remoteness of seventeenth-century conceptions of motion from ours to such a degree as to obscure the continuity of Galileo's conceptions with the modern understanding. For although Descartes apparently never recognized the error involved in his analysis of the problem of fall, Galileo did correct his analogous error. And his concept of the *degree of velocity* is, even if not technically a *velocity* for the reasons given above, clearly the source for Newton's concept. Implicit in Galileo's concept of degree of velocity, as in the young Newton's idea of a velocity at an instant (what he later called a *fluxion*), is the idea that this is the velocity that a body would move at if it were to continue moving uniformly with the same degree of velocity for longer than an instant. Moreover, as Clagett has observed, this same concept can be found in the earlier Merton School of fourteenth-century Oxford. In a passage from a fragment of *On Motion* attributed to Richard Swineshead, for example, we find: "The reason why the velocity of this motion will be attended by a described line belonging to it is this: to each degree [of velocity] in a local motion there corresponds a certain lineal distance which would be described in some time by a motion with just such a degree [of velocity] throughout."[40] His contemporary William Heytesbury proposed a similar conception in the following passage describing nonuniform motion: "In nonuniform motion, however, the velocity will be attended at each instant by a line belonging to it which would be described by a point moving with the fastest motion if it were moved uniformly for a time with that degree of velocity by which it is moved in the same instant, at any given instant whatever."[41] Clagett translates the beginning of this passage as "In nonuniform motion, however, the velocity at any given instant will be measured by the path which would be described by the most rapidly moving point" (1959, 236).[42] Consequently, he attributes to Heytesbury the concept of velocity at an instant, or *instantaneous velocity*: "For him instantaneous velocity is to be measured or determined by the path which *would* be described by a point if that point were to move during some time interval with a uniform motion of the velocity possessed at the instant" (237). If, however, we bear in mind the concept of velocity I have been urging as the norm, the velocity in question is that of the whole motion: what Heytesbury and Swineshead are both doing in these passages is justifying their representation of the intension of motion at any instant by an extended line representing the degree of velocity at that instant.

Clagett draws attention to Galileo's employment of the terminology and conceptual apparatus of the Merton School (1959, 237), giving an excerpt from the *Discorsi* in Crew and de Salvio's translation, which he has altered in parts to a "very literal" translation so as "to reveal more clearly Galileo's dependence on the Merton vocabulary":

> To put the matter more clearly, if a moving body were to continue its motion with the same degree or moment of velocity [*gradus seu momentum velocitatis*] it acquired in the first time-interval, and continue to move uniformly with that degree of velocity, then its motion would be twice as slow as that which it would have if its velocity [*gradus celeritatis*] had been acquired in two time intervals. And thus, it seems, we shall not be far wrong if we assume that increase in velocity [*intentio velocitatis*] is proportional [*fieri juxta*] to the increase of time [*temporis extension*]. (Clagett 1959, 251)

This is all very revealing. Clagett has correctly substituted "degree or moment of velocity" for Crew and de Salvio's plain "speed," but he has left unchanged their translation of *gradus celeritatis* as "velocity" when a literal translation of the Latin would be "degree of swiftness." He has then changed their "increment of speed . . . proportional to increment of time" to "increase in velocity . . . proportional to increase of time," when a "very literal" translation would be "if we suppose the intension of velocity to occur in proportion to the extension of time." Here Galileo is following the Merton theorists in representing the intension or degree of velocity at each instant by a line proportional to the time of fall—"the extension of time." Clagett, on the other hand, concludes that "Galileo in this passage compared the instantaneous velocities at the end of the first time-period and at the end of the second time-period." As should by now be clear, I believe that this is an anachronistic projection of a modern conceptual understanding back onto Galileo and the Merton School. It echoes, and perhaps has its source in, Koyré's comment on this same passage: "'The intension of the speed' or 'the degree of speed' is the instantaneous speed of the moving body" (Koyré 1978, 124n136). For further discussion of the gulf between the medieval and modern representations of motion, I defer to the excellent analysis given by Damerow et al. (1992).

MERSENNE AND DISCRETIST ACCOUNTS OF FREE FALL

Again, I am not denying that Galileo's ideas are the proximate source of our modern conception. But the transformation of his ideas was a complex historical process involving many aspects that I can at best gesture at here. The most salient consideration is that we have so far been treating motion entirely in isolation from any consideration of its cause. For even if Galileo's mathematical treatment of nonuniform motion is accepted, in the absence of an account of the cause of this motion one is not obliged to agree that this is how free fall actually occurs. Now, if there is one thing that the proponents of the emerging mechanical philosophy did not dispute in Aristotle's physics, it was that all changes of motion are effected by bodies in contact. Given this, whatever the precise cause of gravity, it seemed to most seventeenth-century thinkers that the action on a body resulting in its falling at an accelerated rate should be explained in terms of the impacts on it of other moving particles: hence the heavy emphasis on the rules of collision. Thus one could accept Galileo's principle that increases in velocity are always proportional to the time of fall without having to accept his claim that in actual fact acceleration does "occur continually from moment to moment, and not interruptedly from one quantified part of time to another."[43]

This consideration was particularly germane to the reception of and development of Galileo's account of fall. At the center of the dissemination of Galileo's work in France and beyond was the Minim Father Marin Mersenne, who discussed it in several works of the 1630s and 1640s. As has been demonstrated with great lucidity by Carla Rita Palmerino (1999; 2010), Mersenne's growing skepticism over the truth of Galileo's account of fall during this period was largely conditioned by such worries. In particular, there were two difficulties raised by Descartes that had ever greater influence on Mersenne's skepticism regarding Galileo's account: his conviction that there had to be an initial impact, and that a body could not in fact go through an actual infinity of degrees of speed.[44] These two worries combined to suggest that a more acceptable account of fall might still be consistent with the empirical facts: a very large but finite number of discrete increases in uniform motion caused by the impacts of subtle matter, each lasting for a very small but finite moment.

As I have argued elsewhere (Arthur 2011), precisely such a model had been suggested by Beeckman to Descartes in their first encounter in 1618. Beeckman had conceived the acceleration of the falling body to occur by successive discrete tugs resulting in successive equal increments of the force of motion in equal moments, with the previously acquired increments in the force of motion being conserved. Thus in the second moment it has twice the force of motion as in the first, and so covers twice the space it did in that moment. So as a result of these discrete tugs occurring at the beginning of each successive equal moment or physical indivisible of time, the speeds will be as 1, 2, 3, 4, etc. in the successive moments; and given the equality of moments, the distances will therefore also be as 1, 2, 3, 4, etc. in the successive moments, that is, in arithmetical progression.[45] Thus in two hours each divided into 4 moments, the speeds will be as 1, 2, 3, 4, and 5, 6, 7, 8, with proportional distances covered in each moment. So after one hour the total distance covered would be as $1 + 2 + 3 + 4 = 10$, and after 2 hours, $10 + 5 + 6 + 7 + 8 = 36$, giving the ratio of the space covered in one hour to that covered in two as 10:36. By increasing the number of moments in each hour, the ratio can be made arbitrarily close to the ratio 1:4, as Beeckman noted in his diary in 1618. One can, in fact, derive a general formula. If n is the number of moments into which each hour is divided, the ratio between the distance D_1 covered in the first hour to the distance D_2 covered in two hours will be

$$D_1 : D_2 = 1/2\ n/(n + 1) : n\ (2n + 1) = (1 + 1/\text{n}) : (4 + 2/\text{n})$$

Almost thirty years later, as Palmerino (2010) reports, Fabry made exactly the same argument in his *Tractatus physicus de motu locali*: "*A naturally accelerated motion is not propagated through all degrees of slowness*. For there are as many degrees of this propagation as there are instants through which this motion endures, since in every single instant a new accession of impetus occurs; but there are not infinitely many instants, as we will demonstrate in our *Metaphysics*."[46] Thus each actual moment or instant corresponds to the duration of a uniform motion, with successive increases in this motion caused by the collision or tug of whatever particles are responsible for gravity. Consequently, there was no way to decide empirically between Galileo's continuist law and this discretist rival: the moments could always be supposed small enough to make any discrepancy smaller than what could be empirically discerned. But the discretist law of Beeckman and Fabry had the dual advantage that it did not presuppose an infinity of degrees of speed, and that it conformed to the kind of causes sanctioned by the mechanical philosophy.

CONCLUSION

In conclusion, let me now turn to the historiographical issues I raised in the opening section. What I have tried to show in adequate detail is that the projection back onto Galileo and Descartes of the modern mathematical understanding of the motion of free fall is seriously anachronistic. It supposes that the required concepts for a correct understanding—the modern concept of *instantaneous velocity*, of motion as a *function* relating varying instantaneous velocities to the *independent variable time*, of gravitational acceleration as truly *continuous*—were all available prior to the establishment of classical mechanics. As Damerow, Freudenthal, McLaughlin, and Renn comment, such a "position either contains an implicit denial that conceptual development takes place at all, since the concepts remain the same as before, or else must claim that a conceptual development *preceded* the discovery of the central laws of a new science" (1992, 2). They argue that the classical concepts are "rather the outcomes of the establishment of the law [of free fall] than its prerequisites" (3).

This, however, is still to look at the history in terms of how it leads to the modern understanding of free fall. That is part of what interests us, of course. But once we peeled back the veil of assumed modern concepts, we found revealed a certain way of understanding motion with its own conceptual and mathematical trappings. If velocity is an affection of motion, and motion is a change of place occurring through some stretch of time, then there can be no such thing as an instantaneous velocity. Also, if velocity is the swiftness with which a particular motion is accomplished, then the velocities of two motions through equal spaces can be compared, and so can the velocities of two motions through equal times. For such a conception the theory of proportions seemed to be wholly adequate, and as we have seen, this is why neither Galileo (initially) and Descartes thought they could translate back and forth between the velocity of one motion being three times as fast as another through equal spaces to the first motion completing three times as much space as the second in equal times. Of course, this led to intractable difficulties when applied to nonuniform motion. But where these mathematical difficulties led Galileo to a realistic treatment of degrees of velocity, in which infinitely many of them "add" timewise into an overall velocity, to his opponents these same difficulties, coupled with the difficulties of the composition of the continuum facing the Galilean approach, suggested that the only way successfully to treat free fall was in terms of discrete increments of uniform velocity. This, moreover, was in keeping with the mechanical philosophy, where all forces act by discrete tugs or pushes of microparticles, whereas the continuist account of acceleration lacked a plausible causal account.

Again from a retrospective point of view, one could conceive the episodes discussed here in terms of "obstacles" to progress, like the *obstacles épistémologiques* suggested by Gaston Bachelard. Thus, so long as velocity was conceived as an affection of a motion achieved over time, scholars were prevented from forming the concept of instantaneous velocity; likewise, conceiving it as the swiftness with which an overall motion was accomplished, while enabling a treatment through the theory of proportions, was an impediment to a functional interpretation of motion, and to a successful treatment of nonuniform motion. But such a description still seems to presuppose a view of the history of mathematical physics as a linear progression to modern concepts taken as unassailable truths, the finishing line of the hurdler's race. One can say that the theory of proportions that went along with the Aristotelian conception of velocity was an obstacle to the formation of the modern concept only after that modern concept has been formed, but there is no guarantee that this is the only correct way to conceive motion, that there is a unique linear progression from the early seventeenth century to a perfectly correct modern concept. (Indeed, if we look at "motion with respect to cause" in a modern context, we again find the actions of continuous forces explained in terms of the collisions of microparticles;

and in this quantum context, a realistic picture is much more elusive than it was classically.)

I believe the situation is better captured by a notion I have introduced elsewhere (Arthur 2012), but which I have not yet fully developed. This is the notion of "epistemic vectors." These are aspects of theorizing, usually embodying a mathematical framework, that both impel and constrain thinking in a certain direction, without that implying that they have a known outcome or that they are necessarily an impediment. In that respect they differ from Bachelardian *obstacles*. They have something in common with Kuhn's idea of an exemplar, in that they may implicitly involve certain practices or ways of approaching problems. But they differ from exemplars in that they are implicit drivers of thought rather than exemplary components of an established paradigm that usually have a strong, explicit pedagogical function. Thus the law of free fall does become an exemplar after the further contributions of Huygens, Newton, and Leibniz; but in the period we have been investigating, it has not yet become so, because the ideas constituting that exemplar are still in the process of being developed. The conception of velocity as the swiftness with which an overall motion is accomplished, together with the theory of proportions that accompanies it, is an epistemic vector; so is the conception of overall velocity as being representable by an area, with lines in that area representing degrees of velocity; so is the idea that changes in motion must be effected by the impacts of bodies. These vectors do not necessarily push theorizing in the same direction: the idea is that science progresses unevenly by the joint action of such vectors. Positing them helps to explain commonalities in the thinking of historical actors—for example, the fact that the young Galileo and Descartes both make what appears in hindsight to be the same mistake—as well as why what seem to us as right views were opposed—for example the opposition to Galileo's account of fall by many of his contemporaries, whom Drake regarded as merely benighted—why certain options did not occur to them, or why they persisted in proceeding down what appear to us as blind alleys. But I will leave a thorough explication of this notion of epistemic vectors for another occasion.

NOTES

This chapter grew out of research conducted in the Fisher Rare Books Library at the University of Toronto in 2001–2. Previous versions of it were read in Dubrovnik (31st International Conference on the Philosophy of Science April

2004), Winnipeg (Canadian Society for the History and Philosophy of Science, Annual Meeting, June 2004), Pisa (Workshop on the Contested Expanding Role of Applied Mathematics from the Renaissance to the Enlightenment, September 2010), and London, Ontario (at the Workshop on the Language of Nature at the Rotman Institute, October 2012). I am grateful to my audiences there for their comments, and especially to Geoffrey Gorham and the other participants in the Language of Nature workshop for their helpful comments and advice.

1. Here "law" should not be understood to connote a universal law of nature—see Daniel Garber's chapter in this volume.

2. As we shall see later, much the same point has been made by Damerow et al. (1992) and by Jullien and Charrak (2002). My thanks to Dan Garber for drawing my attention to the latter during the workshop.

3. "Ma perchè l'accelerazione si fa continuamente di momento in momento, e non intercisamente di parte quanta di tempo in parte quanta" (*Dialogo*; Galilei 1897, 162). All translations given in this chapter from the Latin, Italian, and French are my own. English translations of all relevant texts may also be found in Damerow et al. (1992).

4. "cioè gli spazzii passati dal moto naturale esser in proporzione doppia dei tempi, et per conseguenza gli spazii passati in tempi eguali esser come i numeri impari ab unitate" (Galilei, *Opere* 10, 115; Koyré 1978, 111).

5. "Et il principio è questo: che il mobile natural vadia crescendo di velocità con quella proportione che si discosta dal principio del suo moto; come, v.g., cadendo il grave dal termine *a* per la linea abcd suppongo che il grado di velocità che ha in *c* al grado di velocità che hebbe in *b* esser come la distanza ca alla distanza ba et così conseguentemente in *d* haver grado di velocità maggiore che in *c* secondo che la distanza *da* è maggiore della *ca*." (Galilei, *Opere* 10, 115; Koyré 1978, 111).

6. See Ernst Mach, *Die Mechanik* ([1883] 1973, 245–46); English translation in Mach 1902, 247–48.

7. The logarithms can be taken to any base, but here, as throughout this chapter, I will take them to base *e* (Naperian logarithms), which, for visual clarity, I write $\log_e x$ rather than the usual ln x.

8. "in tempi eguali si facciano eguali additamenti di velocità; . . . l'accelerazione loro vadia crescendo secondo che cresce il tempo e la durazione del moto" (Galilei 1898, 202).

9. "Mi pare che con chiarezza forse maggiore si fusse potuto definire, senza variare il concetto: Moto uniformemente accelerato esser quello, nel qual la ve-

locità andasse crescendo secondo che cresce lo spazio che si va passando" (ibid.).

10. "Moto equabilmente, ossia uniformemente accelerato, diciamo quello che, a partire dalla quiete, in tempi eguali acquista eguali momenti di velocità" (ibid., 205).

11. See Mach ([1883] 1902, 248; 1973, 246).

12. This passage from Leonardo is quoted from Koyré (1978, 72).

13. "The process which gave rise to modern physics consisted in an attempt to rationalize, in other words, geometrise, space, and to mathematize the laws of nature" (Koyré 1978, 73).

14. Garber (1992); see also Ariew chapter 4 in this volume.

15. Burtt 1924, 95; see also p. 262 on the "de-spiritualization of nature."

16. See Aristotle, *Physics*, 220 a8–a21 (Aristotle 1996, 107–8).

17. Again, cf. Aristotle: "Now, what is before and after is found primarily in place. In that context it depends on position, but because it is found in magnitude, it must also be found, in an analogous fashion, in change. And since time always follows the nature of change, what is before and after applies also to time" (*Physics*, 219 a14–a18; Aristotle 1996, 105).

18. "gli infiniti instanti che sono nel tempo DA, corrispondenti a gli infiniti punti che sono nella linea DA" (Galilei 1897, 162).

19. This notion of velocity as an affection of a body's motion taken as a whole was still current even significantly later. Giusti (1990, xxx) cites Saccheri's *Neostatica* from 1708, and gives the following quotation from the *De legibus gravitatis* of Paolo Frisi, who was professor of mathematics in Pisa from 1756 to 1764: "Celerity is that affection of a moving body which occurs so that more or less space is covered in a given time." Giusti, however, insists that Galileo employs two types of velocity in accelerated motion, the "velocity with which a moving body traverses a given line," and the degree of velocity (xxxiv). My subtle disagreement with him is that I do not think that "degree of velocity" is a velocity (even though in unpublished manuscripts Galileo does sometimes call it a velocity); here we are dealing with a stage in the transition to the concept of instantaneous velocity, but we are not there yet. See Damerow et al. (1992) for detailed arguments to the same effect.

20. As noted earlier, on discovering Jullien's and Charrak's book in the late stages of composition of this chapter, I discovered that many of my points about the anachronism of reading the modern functional view back into the originators of classical mechanics had already been made with great clarity in Damerow et al. (1992). As Jullien and Charrak remark, summarizing Damerow

et al.: "Elle consiste en un vaste anachronism en very duquel on mobilise les concepts «classiques» de vitesse instantanée, de summation intégrale (ou au moins de limite) et de function" (2002, 36–37).

21. Again, much the same point has been lucidly argued by Damerow, Freudenthal, McLaughlin, and Renn, who write: "From the point of view of Aristotelian natural philosophy—where no equivalent to a functional dependence of motion on a certain parameter exists, and where the velocity of motion always refers to its overall extension in space and time—the alternative: velocity stands either in such a specified relation to space or else to time, cannot sensibly be posed" (1992, 1–2).

22. "Vanno dunque continuamente crescendo i gradi di velocità in tutti i punti della linea af secondo l'incremento delle parallele tirate da tutti i medesimi punti. In oltre, perché la velocità con la quale il mobile è venuto da a in d è composta di tutti i gradi di velocità auti in tutti i punti della linea ad, e la velocità con che ha passata la linea ac è composta di tutti i gradi di velocità che ha auti in tutti i punti della linea ac, adunque la velocità con che ha passata la linea ad, alla velocità con che ha passata la linea ac, ha quella proporzione che hanno tutte le linee parallel tirate da tutti i punti della linea ad sino alla ah, a tutte le parallele tirate da tutti i punti della linea ac sino alla ag" (Galilei 1898, 373).

23. "Adunque la velocità con che si è passata la linea ad, alla la velocità con che si è passata la linea ac, ha doppia proporzione di quell ache ha da a ca. E perché la velocità alla velocità ha contraria proporzione di quella che ha il tempo al tempo (imperò che il medesimo è crescere la velocità che sciemare a tempo), adunque il tempo del moto in ad al tempo del moto in ac ha subduplicata proporzione di quella che ha la distanza ad alla distanza ac. Le distanze dunque dal principio del moto sono come i quadrati de i tempi, e, dividendo, gli spazi passati in tempi eguali sono come i numeri impari ab unitate" (Galilei 1898, 373–74).

24. My analysis here agrees with that of Giusti, who criticizes Galileo for an equivocation in his use of the term "inverse proportion (contraria proporzione)" (Giusti 1990, xxxvi); but it is at odds with that of Jürgen Renn in Damerow et al. (1992). Renn claims that Galileo's reasoning has "the advantage [over Descartes'] of yielding the correct result." For, unlike Descartes, Galileo "inverts . . . the double proportionality of the velocities by a transition to a 'half' or 'mean' proportionality of the times thus obtaining the law of fall in its mean proportional form" (167). I do not see that this re-expressing of the proportion also involves an inversion.

25. "Ora, se consideriamo attentamente la cosa, non troveremo nessun aumento o incremento più semplice di quello che aumenta sempre nel medesimo modo. Il che facilmente intenderemo considerando la stretta connessione tra tempo e moto" (Galilei 1898, 197).

26. "[lo possiamo] in quanto stabiliamo in astratto che risulti uniformemente e, nel medesimo modo, continuamente accelerato, quel moto che in tempi eguali, comunque presi, acquista eguali aumenti di velocità" (Galilei 1898, 197–98).

27. "Il tempo in cui uno spazio dato è percorso da un mobile con moto uniformemente accelerato a partire dalla quiete, è eguale al tempo in cui quel medesimo spazio sarebbe percorso dal medesimo mobile mosso di moto equabile, il cui grado di velocità sia sudduplo [la metà] del grado di velocità ultimo e massimo [raggiunto dal mobile] nel precedente moto uniformemente accelerato" (Galilei 1898, 208).

28. "E sì come la BC era massima delle infinite del triangolo, rappresentanteci il massimo grado di velocità acquistato dal mobile nel moto accelerato . . . passi con moto equabile nel medesimo tempo spazio doppio al passato dal moto accelerato" (Galilei 1897, 163; 1953, 229).

29. Oresme 1968, 559–61; quoted from Damerow et al. 1992, 18.

30. Varron 1584, 12ff.

31. Here I am reading the singular 'la velocità' and 'della velocità' for the text's plurals, 'le velocità' and 'delle velocità,' which do not appear to make sense in the context, although it is perhaps explicable in terms of an implicit comparison of pairs of velocities in a proportion.

32. "Quando le velocità hanno la medesima proporzione che gli spazii passati o da passarsi, tali spazii vengon passati in tempi eguali; se dunque le velocità con le quali il cadente passò lo spazio di quattro braccia, furon doppie delle velocità con le quali passò le due prime braccia (sì come lo spazio è doppio dello spazio), adunque i tempi di tali passaggi sono eguali: ma passare il medesimo mobile le quattro braccia e le due nell'istesso tempo, non può aver luogo fuor che nel moto instantaneo: ma noi veggiamo che il grave cadente fa suo moto in tempo, ed in minore passa le due braccia che le quattro; adunque è falso che la velocità sua cresca come lo spazio" (Galilei 1898, 203–4).

33. Koyré [1939] 1978, 116n51. Koyré cites in this connection Mach's *Mechanik* ([1883] 1973, 245) and also Paul Tannery (1926, 400ff).

34. Since writing this section I have discovered the detailed analyses of Descartes' mathematical treatments of the law of fall in Damerow et al. (1992, chapter 2), and Jullien and Charrak (2002). But my analysis of the "4/3 proportion" differs from theirs, as we shall see.

35. I have no explanation of why, having correctly identified the Aristotelian theory of proportions at work in Descartes' mathematization, these authors go on to give an explanation of Descartes' reasoning that appeals to the functional interpretation they earlier decry.

36. "Il aboutit donc à une relation essentiellement différent de celle de Galilée, puisqu'elle reviendrait à considérer l'espace parcouru comme proportionnel, non pas au carré du temps, mais à une puissance du temps dont l'exposant est la rapport de log2 à log4/3, c'est à dire environ 2,4 (AT I 75)." Quoted from Jullien and Charrak (2002, 121–22).

37. "C'est le sens d'une note maladroit, car anachonique, de Tannery, qui construit une function exponentielle donnant les espaces par rapport à la variable temps" (Jullien and Charrak 2002, 121).

38. In a letter to Mersenne in October 1631 Descartes writes that if a void and constant action of gravity are accepted, "there would be no means of explaining the speed of this movement by numbers other than the ones I have sent you, at least ones that are rational; and I do not even see that it would be easy to find irrational ones, nor any line in geometry which would explicate them better." Tannery apparently took Descartes' remark to license an interpretation in terms of exponential powers, but as Jullien and Charrak argue (1992, 123), it is more natural to see him as rejecting any such an interpretation as unphysical.

39. Here the first diagram is from the 1634 letter, with the lettering of the 1629 letter. The second is my own.

40. "Causa autem quare penes lineam descriptam velocitas illius motus attendit, est hoc: cuicunque gradui in motu locali correspondet certa distantia linealis quae in tanto tempore et in tanto cum partibus tali gradu describeretur" (Clagett 1959, 245).

41. "In motu autem difformi, in quocunque instanti attendetur velocitas penes lineam quam describeret punctus velocissime motus, si per tempus moveretur uniformiter illo gradu velocitatis quo movetur in eodem instanti, quocunque dato" (ibid., 240).

42. The expression "by the most rapidly moving point" occurs because these authors are considering the motion of a rotating radius. See Clagett (1959, 216).

43. "Ma perchè l'accelerazione si fa continuamente di momento in momento, e non intercisamente di parte quanta di tempo in parte quanta" (Galilei 1897, 162).

44. See Palmerino 2010 for details.

45. See Arthur 2011 for details.

46. "*Motus naturaliter accelerato non propagatur per omnes tarditatis gradus;* quia tot sunt huius propagationis gradus, quot sunt instantia, quibus durat

hic motus, cum singulis instantibus nova fiat impetus accessio, sed non sunt infinita instantia, ut demonstrabimus in Metaphysica."

REFERENCES

Aristotle. 1996. *Physics*. Trans. R. Waterfield. Oxford: Oxford University Press.

Arriaga, R. 1632. *Cursus Philosophicus*. Antwerp: Jacob Quesnel.

Arthur, R. T. W. 2007. "Beeckman, Descartes and the Force of Motion." *Journal for the History of Philosophy* 45 (1): 1–28.

———. 2011. "Beeckman's Discrete Moments and Descartes' Disdain." *Intellectual History Review* 22 (1): 69–90.

———. 2012. "Time Atomism and Ash'arite Origins for Cartesian Occasionalism, Revisited." In *Asia, Europe and the Emergence of Modern Science: Knowledge Crossing Boundaries*, ed. A. Bala, 73–92. New York: Palgrave Macmillan.

Burtt, E. A. [1924] 1954. *The Metaphysical Foundations of Modern Science*. Garden City, N.Y.: Doubleday Anchor.

Clagett, M. 1959. *The Science of Mechanics in the Middle Ages*. Madison: University of Wisconsin Press.

Damerow, P., G. Freudenthal, P. McLaughlin, and J. Renn. 1992. *Exploring the Limits of Preclassical Mechanics*. Dordrecht: Springer.

Descartes, R. 1644. *Principles of Philosophy*. Vol. 8 of *Oeuvres de Descartes*, ed. C. Adam and P. Tannery. Paris: Libraire Philosophique J. Vrin, 1973.

———. 1991. *Philosophical Writings*. Vol. 3 of *The Correspondence*, ed. and trans. J. Cottingham, R. Stoothof, D. Murdoch, and A. Kenny. Cambridge: Cambridge University Press.

Drake, S. 1989. *History of Free Fall*. Toronto: Wall & Thompson.

Fabry, H. 1646. *Tractatus physicus De Motu Locali . . . Auctore Petro Mousnerio cuncta excerpta ex praelectionibus R. P. Honorati Fabry*. Lyon: Joannes Champion.

Galilei, G. 1632. *Dialogo sopra i due Massimi Sistemi del Mondo, Tolemaico, e Copernicano*. Florence: Gio-Batista Landini.

———. 1638. *Discorsi e dimostrazioni matematiche: intorno a due nuove scienze attinenti alla mecanica ed i movimenti locali*.

———. 1897. *Le Opere di Galileo Galilei*. Edizione Nazionale. Vol. 7. Firenze: Tipografia di g. Barbèra.

———. 1898. *Le Opere di Galileo Galilei*. Edizione Nazionale. Vol 8. Firenze: Tipografia di g. Barbèra.

———. [1914] 1954. *Dialogues Concerning Two New Sciences.* Trans. by Henry Crew and Alfonso de Salvio. New York: Dover.

———. [1953] 1967. *Dialogue Concerning the Two Chief World Systems.* Trans. by Stillman Drake. Berkeley: University of California Press.

Garber, D. 1992. *Descartes' Metaphysical Physics.* Chicago: Chicago University Press.

Giusti, E., trans. 1990. "Galilei e le leggi del moto." In *Discorsi e dimostrazioni matematiche: intorno a due nuove scienze attinenti alla mecanica ed i movimenti locali,* ed. G. Galilei, ix–xxvii. Torino: G. Einaudi.

Jullien, V., and A. Charrak. 2002. *Ce que dit Descartes touchant la chute des graves.: de 1618 à 1646, étude d'un indicateur de la philosophie naturelle cartésienne.* Paris: Presses Universitaires du Septentrion.

Koyré, A. [1939] 1978. *Galileo Studies.* Trans. J. Mepham. Atlantic Highlands, N.J.: Humanities Press.

Mach, E. [1883] 1973. *Die Mechanik.* Darmstadt: Wissenschaftliche Buchgesellschaft.

———. [1883] 1902. *The Science of Mechanics: A Critical and Historical Account of its Development.* 4th ed. Trans. T. J. McCormack. Chicago: Open Court.

Oresme, N. 1968. *Nicole Oresme and the Medieval Geometry of Motions.* Ed. and trans. M. Clagett. Madison: University of Wisconsin Press.

Palmerino, C. R. 1999. "Infinite Degrees of Speed: Marin Mersenne and the Debate over Galileo's Law of Free Fall." *Early Science and Medicine* 4: 269–328.

———. 2010. "Experiments, Mathematics, Physical Causes: How Mersenne Came to Doubt the Validity of Galileo's Law of Free Fall." *Perspectives on Science* 18 (1): 50–76.

Tannery, P. 1926. *Mémoires Scientifiques.* Vol. 6 of *Sciences Modernes 1883–1904.* Publiés par J.L. Heiberg & H.G. Zeuthen.

Varron, M. 1584. *De motu tractatus.* Geneva: Jacob Stoer.

4

THE MATHEMATIZATION OF NATURE IN DESCARTES AND THE FIRST CARTESIANS

ROGER ARIEW

Some of the motivation for this volume is the reevaluation of a prominent historiographical orientation of twentieth-century research on the scientific revolution, in light of the proliferation of novel methodological orientations and studies in the last generation of scholars. The historiographical orientation at issue is what is called the mathematization of nature; its exemplary proponents are Alexandre Koyré, Eduard Jan Dijksterhuis, and Edwin Arthur Burtt.[1] This should be a welcome reevaluation, especially since the position has held fairly strongly for almost a century. Burtt published the first edition of his *Metaphysical Foundations of Modern Science* in 1924.[2] Dijksterhuis reiterated in large part the historical-philosophical accounts implicit in Burtt's work in 1950.[3] And the views of Burtt and Dijksterhuis found their historiographical champion in Koyré's Husserlian- and Bachelardian-inspired position.[4] In 1950 Dijksterhuis already knew and cited a number of Koyré's theses from his publications available in the 1940s, such as Koyré's thesis on the mathematization of physical space and his devaluation of experimental approaches. As Dijksterhuis states with respect to the first thesis, "The substitution of the world-picture of classical physics for that of Aristotle involved a radical change in the conception of space in which the phenomena of nature occur. Without explicitly saying so, scientists had always thought of the latter as physical space to distinguish it from the geometrical space to which the reasonings of mathematics related . . . In the sixteenth and seventeenth centuries, however, this distinction was becoming blurred . . . Koyré characterized this by the term 'mathematization of physical space'" (Dijksterhuis 1969, 377).[5]

The reference Dijksterhuis gives is to Koyré's 1939 *Études galiléennes.* There Koyré does state that one of the major changes between classical and modern science is "the geometrisation of space," that is to say, "the substitution for the concrete space of pre-Galileo physics of the abstract space of Euclidean geometry. It was this substitution that made the invention of the law of inertia possible" (1978, 3).[6]

The second Koyré thesis referred to by Dijksterhuis is exemplified in his judgment of Francis Bacon's lack of importance for early modern science. Dijksterhuis cites again from Koyré's *Études galiléennes*, in large part approving of Koyré's view that Bacon "did not make a single positive contribution to science, and in some cases he entirely failed to recognize the merits of others who *had* done so. This is why Koyré calls it a *mauvaise plaisanterie* to regard him as one of the founders of modern science" (Dijksterhuis 1969, 396–97).[7] And, indeed, Koyré does say that the role of Bacon "in the history of the scientific revolution was completely negligible," and reinforces this judgment by asserting in a footnote, as Dijksterhuis said, that "'Bacon, the founder of modern science' is a joke, and a bad one at that." Koyré continues, stating: "In fact, Bacon understood nothing about science. His manner of thinking was closer to alchemy and magic . . . to a thinker of the Renaissance than to that of a Galileo or even a Scholastic" (Koyré 1978, 39).

The two theses go hand in hand, as can readily be shown by articles on Galileo published by Koyré in 1943.[8] For Koyré, experience is an *obstacle* in the establishment of modern science. As Koyré says in one article: "We are so well acquainted with, or rather so well accustomed to, the concepts which form the basis of modern science, that it is nearly impossible for us to appreciate rightly either the *obstacles* that had to be overcome for their establishment, or the difficulties that they imply and encompass" (1968, 3; my emphasis).[9] Koyré then states in another article: "One must not forget that observation and experience, in the sense of brute, common-sense experience, did not play a major role—or if it did, it was a negative one, the role of *obstacle*—in the foundation of modern science" (1968, 18; my emphasis). The connection between Koyré's two theses—the devaluation of experience and the mathematization of nature—lies in his distinction between experience and experiment:

> It is not experience but experiment which played—but only later—a great positive role. Experimentation is the methodical interrogation of nature, an

> interrogation which presupposes and implies a language in which to formulate the questions, and a dictionary which enables us to read and interpret the answers. For Galileo it was in curves and circles and triangles, in mathematical or even more precisely, in geometrical language—not in the language of common sense or in that of pure symbols—that we must speak to Nature and receive her answers. Yet obviously the choice of language, the decision to employ it, could not be determined by the experience which its use was to make possible. (Koyré 1968, 18–19)

Koyré's devaluation of experience has the further consequence that Galileo's observations and experiments are likewise devalued; as Koyré states: "It is obvious that the Galileo experiments are completely worthless: the very perfection of their results is a rigorous proof of their incorrection" (1968, 94).[10]

So what is one to do; that is, how is one to reevaluate this prominent historiographical orientation? Well, one could point out that Koyré is simply historically wrong about his devaluation of experience, wrong about Francis Bacon, and wrong about Galileo's experiments.[11] Worse yet, his perspective fails to appreciate the rise of scientific societies, the social nature of the epistemology of observation, and so forth. But such an objection is easily countered by simply casting off what Koyré thinks of as an integral part of his view (the necessity of the mathematization of nature as preliminary to any genuine experimental culture) and accepting just a portion of his historiographical orientation, for example, mathematization of nature by itself, as one of the crucial elements of early modern science as contrasted with past science, along with experimental culture by itself. We can see this rejoinder given numerous times, as in Floris Cohen's review of Joella Yoder's book *Christian Huygens and the Mathematization of Nature*. Cohen states: "It is well known that, on the topic of Galileo's experiments, Koyré has been proved simply wrong—to the extent that he declined to take literally in Galileo's own statements what we now know should indeed be taken literally. Granting so much does not, however, render all Koyré had to say on the mathematization of nature worthless" (Cohen 1991, 83). Still, the historical studies of the "last generation of scholars" about the culture of experiment require the mathematization of nature to become a less global historiographical thesis, but perhaps to remain an important one.

A more promising reevaluation would consist in reconsidering the mathematization of nature from the perspective of "novel methodological orientations," and in particular, of contextual history. Here one could point out that when we talk about the "mathematization of nature" we mean different things with regard to different thinkers. When one focuses on what mathematizing nature would be for Galileo, for Descartes, Huygens, or Newton, we find that these are radically different activities; what Galileo tried to do in this regard is clearly different from what Descartes tried to do, or what Huygens, Newton, and others tried to do. The mathematization of nature as an account of the scientific revolution or early modern science begins to look like our twentieth-century invention, perhaps a construction we are forcing on the past.[12]

Such a thesis, however right it seems to me, would be beyond the scope of a single chapter. Instead I propose to examine the mathematization of nature for Descartes and the Cartesians. I will try to show that, from the perspective of a more contextual history, the thesis that mathematization of nature refers to radically different elements in different thinkers can be recovered in a single thinker; it applies to Descartes by himself. Basically, there is no one thing one can call the mathematization of nature in Descartes, perhaps no mathematization of nature at all, if the concept is considered narrowly. I will then corroborate this historical position by demonstrating that various Cartesians in the seventeenth century understood Descartes differently on these issues. The Cartesians have little to say about the mathematization of nature when viewed as a grand narrative for the scientific revolution, though their remarks on the relations between mathematics and physics advance various aspects of Descartes' understanding about those relations and contrast with the way we conceive of them as part of that grand narrative. But first, I need to discuss the views of Burtt and Dijksterhuis on Descartes and the mathematization of nature.

BURTT AND DIJKSTERHUIS ON DESCARTES' MATHEMATIZATION OF NATURE

In his *Metaphysical Foundations of Modern Science,* Burtt's chapter on Descartes proceeds somewhat chronologically. Burtt refers to Descartes' early interest in mathematics, including what he calls the "remarkable experience" of November 10, 1619, which confirmed for Descartes the "trend of his previous thinking and gave the inspiration and the guiding principle for his whole life-work," namely, according to Burtt, the conviction "that mathematics was the sole key needed to unlock the secrets of nature" (Burtt [1924] 1954, 105). Burtt follows this introduction with a section called

"Mathematics as the Key to Knowledge." There, Burtt details Descartes' *Rules*, which he calls "a series of specific rules for the application of his all-consuming idea," starting with Rule 1, "that all the sciences form an organic unity" and interpolating something from the *Search for Truth* that "all the sciences must be studied together and by a method that applies to all" (106–7). He asserts that "the method must be that of mathematics, for all that we know in a science is the order and measurement revealed in its phenomenon" and, citing Rule 4, that "mathematics is just that universal science that deals with order and measurement generally." Burtt also asserts that "Descartes is at pains carefully to illustrate his thesis that exact knowledge in any science is always mathematical knowledge" (107). He refers to the Rule 3 concepts of intuition and deduction (deduction now considered as mathematical deduction) as two steps of this mathematical method and introduces the simple natures of Rule 14 "as discoveries of intuitions" (108). However this is where Burtt thinks Descartes goes astray: "As he proceeds from this point he is on the verge of the most far-reaching discoveries, but his failure to keep his thought from wandering, and his inability to work out the exceedingly pregnant suggestions that occur to him make them barren for both his later accomplishments and those of science in general . . . [A]t the crucial points his thoughts wander, and as a consequence Cartesian physics had to be supplanted by that of the Galileo-Newton tradition" (109). The rest of Burtt's (clearly whiggish) account consists of a generally negative report of Descartes' views in the *Principles*—Descartes' "soaring speculations"—as failing to live up to his initial fundamental mathematical intuition, producing something of mere historical, not scientific, significance.

It is ironic that Burtt is so firmly convinced of his interpretation of Descartes, based on a defective reading of a juvenile unfinished manuscript, that he cannot make much sense of the *Principles*, Descartes' mature published treatise, on which his reputation rested during the seventeenth century. The situation, however, is not much better with Dijksterhuis, though the latter is somewhat more sober than the former. Again, we assert that in seventeenth-century science "the structure of the external world was essentially mathematical in character and a natural harmony existed between the universe and the mathematical thought of the human mind." Dijksterhuis adds that "the standpoint taken by Descartes cannot be better described than by saying that by carrying this conception to its extreme he virtually identified mathematics and natural science" (Dijksterhuis 1969, 404). He then refers to Descartes' tree of philosophy from the preface to the French edition of the *Principles.* He recognizes that physics there is depicted as rooted in metaphysics, and that "Mathematics is not referred to," but he adds that the foundation on which metaphysics is based is also not referred to, and ends with a rhetorical question: "Cannot the explanation of this be that it is mathematical thought, considered not with regard to its contents but to its form, which constitutes this foundation?" (404). I suppose rhetorical questions should not be answered, though the answer is clearly "no."[13] Still, to his credit, Dijksterhuis knows that one cannot find the thesis of the mathematization of nature in Descartes' first published work, the *Discourse on Method.* He rightly states that "the formulation of the four famous rules which are recommended as guiding principles for scientific thought is immediately preceded by the statement that the author failed to find the method he needed in the Analysis of the Ancients and the Algebra of the Moderns" (404). He opines that Rule 1 of the *Discourse*, about evidence, "was apparently inspired by the style of mathematical thought," and adds that "the other three rules have been kept so vague and general that in the first place they admit of different interpretations and secondly they contain little that is of specifically mathematical character" (405). Dijksterhuis cites with approbation Leibniz's calling them vacuous and mocking them, describing the method as like advising a chemist to "take what you have to take, do with it what you have to do, and you will get what you desire" (404–5).

However, Dijksterhuis is quickly over his disappointment with Descartes' *Discourse*: "in order to become really acquainted with the method of Descartes one should not read in the first place the charming *Discours*, which is a *causerie*, rather than a treatise, but the . . . *Rules for the Direction of the Mind*, which was already composed in 1629" (405).[14] And, naturally, we now get the fact that the *Rules* contains an exposition of *Mathesis Universalis*, which, Dijksterhuis asserts, "Descartes always regarded as one of his greatest methodological discoveries." At this juncture Dijksterhuis claims that Descartes wanted to see *Mathesis Universalis* applied in all the natural sciences, by which he means that Descartes prescribes the application of algebraic methods to all those branches of science that admit of quantitative treatment. He adds that Descartes also admits the possibility of "arranging propositions in deductive chains," so he concludes that "the aim of the Cartesian method is indeed to cause all scientific thinking to take place in the manner of mathematics, namely by deduction from axioms and by algebraic calculation" (405). Dijksterhuis in this way rejoins the thesis of the mathematization of nature and Burtt's account. He shares Burtt's disappointment with Descartes: "Descartes did not get very far in carrying out the concrete program of universal mathematics in science," though he asserts that "his metaphysical as well as his scientific thinking always followed a mathematical pattern" (406). The rest is a litany of Descartes' failures, that is, more analyses of Descartes from the perspective of the present: "if Descartes could have foreseen the future of mathematics . . ."; "Descartes never produced . . ."; "The modern reader, who is accustomed to find more and more trouble expended on this part of the process of forming scientific concepts, may have some difficulty in looking upon the Cartesian way of studying science as a serious contribution to the methodology of scientific thought." For Dijksterhuis, as for Burtt, Cartesian physics is of mere historical importance; for Dijksterhuis, it was "an illusion" that enabled Descartes "to put before his contemporaries the transparent ideal of a rational system for the interpretation of nature that was to rely on none but mathematical and mechanical conceptions" (409).

Of course, Descartes did not put before his contemporaries any such ideal as described by Dijksterhuis and Burtt. He put before his contemporaries the arguments of the *Discourse*, *Meditations*, and *Principles*, but not those of the *Rules*. Simply put, the *Rules* was not generally available in the seventeenth century, though a few Cartesians had access to various small portions of it, as was obvious in the fourth edition of the *Port-Royal Logic* (1674, 42) the work itself was first published in Latin in Descartes' *Opuscula Posthuma* only in 1701, with a Dutch-language version published in 1684. The main Cartesians published their works before the publication of the *Rules*, from circa 1654 to circa 1694, without any knowledge of its views. One might be able to argue that an analysis of the *Rules* in the fashion of Burtt and Dijksterhuis could reveal Descartes' deepest intuitions, but such an analysis cannot provide any understanding of Descartes' significance or influence for Cartesians or for anti-Cartesians or for seventeenth-century science in general. This is the important point to make.

A subsidiary point is that the interpretations of Dijksterhuis and Burtt about the *Rules* are deeply flawed. Take Burtt's assertion that "all that we know in a science is the order and measurement revealed in its phenomenon" or "Descartes is at pains carefully to illustrate his thesis that exact knowledge in any science is always mathematical knowledge." When Descartes gives an example of his method in the *Rules*, he talks about the problem of determining the anaclastic line, in which parallel rays are refracted in such a fashion that they all meet at a point. He does explain that those who limit themselves to mathematics alone cannot investigate the problem, "since it does not belong to mathematics, but to physics" (Descartes 1964–76; AT, 10:394). A person who "looks for the truth in any subject" will not fall into the same difficulty. That person can perceive clearly by intuition both mathematical and physical matters, about the proportion of the angles of incidence and angles of refraction depending on the variation of these angles in virtue of the difference of the media and about the manner in which rays penetrate into a transparent body. The latter presupposes that the nature of illumination is known and what a natural power is in general. As Descartes says: "this is the last and most absolute term in this whole sequence" (AT, 10:395). It is the intuition from which the problem will be solved, from which evident knowledge of the anaclastic line is derived, according to Descartes' method (see Garber 2001, 85–110). I fail to see how the intuition about the nature of illumination or of a natural power would not be considered knowledge or has to be thought as mathematical knowledge, as Burtt would want it. Nor do I see how any of this can license Dijksterhuis's claim that "the aim of the Cartesian method is indeed to cause all scientific thinking to take place in the manner of mathematics, namely by deduction from axioms and by algebraic calculation."[15] Burtt and Dijksterhuis are so sure of their general thesis about the mathematization of nature that they construct their own Descartes from an unfinished manuscript that Descartes himself never refers to; they then mostly neglect what he says in his mature published works. Worse yet, they are so sure of the mathematization of nature as the endpoint for physics that they criticize Descartes for failing to see what they think they perceive in present science. In this process, they cannot provide any understanding of Descartes' views nor of what the Cartesians saw in Descartes. They cannot provide an account of early modern science in relation to what preceded it or in relation to what succeeded it.

DESCARTES ON THE RELATIONS BETWEEN MATHEMATICS AND PHYSICS

It is not as if Descartes does not issue enough statements about what he considers the relation among physics, metaphysics, and mathematics in his published writings, the *Discourse* (1637), *Meditations* (1641–42), and *Principles* (1644–47), as well as in his *Correspondence* (published posthumously in three volumes, 1657–67). There are numerous pronouncements that

Descartes is looking for certainty at least equal to that of mathematics; in the *Discourse*, he intimates that the real use that can be made of mathematics is to extend its method into other realms (AT, 6:7), and to prepare the mind to follow the real philosophical method, which mathematics presupposes (19–22). This last statement can lead one to consider that Descartes does not accept mathematics as the *foundation* for all knowledge. In fact, early on he claims that his metaphysical demonstrations are more certain than geometrical demonstrations (AT, 1:145). But Descartes does say that all his "physics is nothing other than geometry" (AT, 2:68), and he speaks of "having reduced physics to the laws of mathematics" (AT, 3:39). That is how Descartes' anonymous correspondent in the letters used as preface to the *Passions of the Soul* seems to have understood Descartes: "the [Scholastic] Philosophers accept mathematics as part of their physics, because almost all of them are ignorant of it; but on the contrary, the true physics is a part of mathematics" (AT, 11:314–15).[16]

There is also the last article of *Principles* (II, art. 64, AT, 8:78): "The only principles I accept or desire in physics are those of geometry or abstract mathematics, because these explain all natural phenomena and enable us to provide certain demonstrations of them" (AT, 7:78; 11:101).[17] It is an important article indeed, which Descartes published in a prominent work. What the text of the article explains is that Descartes recognizes "no matter in corporeal things apart from what can be divided, shaped, and moved in all sorts of ways, that is, the one the geometers call quantity"—that "he considers in such matter only its divisions, shapes, and motions"—because he does not want to admit anything as true "other than what has been deduced from [these] indubitable common notions so evidently that it can stand for a mathematical demonstration." Descartes ends his article by asserting that "since all natural phenomena can be explained in this way, I do not think that any other principles are either admissible or desirable in physics [than the ones that are here explained]." It is important to note that the properties of matter that Descartes accepts, the divisions, shapes, and motions of corporeal things, are not accepted *because* they are geometrical or mathematical, but because they are the modes of extension that can be distinctly known. In Part I of the *Principles*, that is, the "metaphysical" portion, representing the *Meditations*, Descartes asserts that extended substance can be clearly and distinctly understood as constituting the nature of body (*Principles* I, art. 63, AT, 8:30-1) and that extension as a mode of substance can be no less clearly and distinctly understood as substance itself (*Principles* I, art.

64, AT, 8:31). Descartes then lists the properties or attributes of extension as their shapes, the situation of their parts, and their motions (*Principles* I, art. 65, AT, 8:32). It happens that these properties are what "the geometers call quantity." But that is because mathematicians rely on some of the same clear and distinct perceptions as natural philosophers do. Descartes roots his physics in a metaphysics that produces, at first,[18] a physics that looks the same as mathematics, not because it is rooted in mathematics, but because it is rooted in a metaphysics of clear and distinct ideas.[19] But I do not think many scholars would have been tempted to call this the mathematization of nature or have considered it as an integral part of the scientific revolution.

CARTESIANS AND THE RELATIONS BETWEEN PHYSICS AND MATHEMATICS

It may be one thing to write about Descartes' deepest intuitions as we understand them and another to explicate his influence on his followers; that is, how he was understood by others. When the issue is the scientific revolution, one's account should resonate with the latter. It thus becomes relevant to understand how Descartes was understood by his followers. The Cartesians are a diverse group. Let me limit myself to a few representative thinkers: Du Roure, Rohault, Le Grand, and Régis. Le Grand and Régis are famous for their attempts to publish multivolume Cartesian textbooks that would mirror what was taught in the schools, containing treatises on Cartesian logic, metaphysics, physics, and ethics. Le Grand initially published a popular version of Descartes' philosophy in the form of a scholastic textbook (1671), expanding it in the 1670s and 1680s. The work, *Institution of Philosophy*, as it was called then, was then translated into English together with other texts by Le Grand and printed in two large volumes as part of *An Entire Body of Philosophy according to the Principles of the famous Renate Des Cartes* (1694). On the continent, Régis issued his three-volume *Système Général selon les Principes de Descartes* at about the same time (1691). The difficulties Régis encountered in obtaining permission to publish considerably delayed its publication. The various portions of this work embody Régis's adaptations of diverse works, both Cartesian and non-Cartesian: Antoine Arnauld's *Port-Royal* logic (mostly excerpted); Robert Desgabet's peculiar metaphysics;[20] Rohault's physics; and an amalgam of Gassendist, Hobbesian, and especially Pufendorfian ethics.[21] Ultimately, Régis's unsystematic (and very often un-Cartesian) *Système* set the standard for Cartesian textbooks. Early attempts at setting out a complete Cartesian system before those of Le Grand and Régis included Du Roure's multivolume *La Philosophie divisée en toutes ses parties* (1654) and its (less Cartesian) successor *Abrégé de la vraye philosophie* (1665). Du Roure was one of the first followers of Descartes, belonging to the group that formed around Descartes' literary executor, Claude Clerselier.

Unlike Du Roure, Le Grand, and Régis, who tried to publish complete "systems" of Cartesian philosophy, Rohault limited himself to natural philosophy. He was the foremost proponent of Cartesian physics in the decades immediately following the death of Descartes. In the mid-1650s he began to hold weekly lectures at his house in Paris; these "mercredis de Rohault" brought him to the attention of prominent Cartesians. He became Régis's teacher and won him over to the cause of Cartesianism. Rohault was best known for his 1671 *Traité de physique*, which went through numerous editions and remained a standard textbook in Cartesian natural philosophy well into the eighteenth century, long after Rohault's death in 1672. The *Traité de physique* was initially translated into Latin in 1682 and then again, with annotations by Samuel Clarke, in 1697. Clarke's Latin edition was translated into English in 1723 by his younger brother John and published as *Rohault's System of Natural Philosophy*. As the work went through multiple editions, Samuel Clarke increasingly "illustrated" it with "notes taken mostly out of Sr. Isaac Newton's Philosophy."

First, I need to touch on the question of what the Cartesians take as Descartes' method in general. The answer is as expected: What they consider as method varies somewhat, but does not involve *Mathesis Universalis* or anything from the *Rules*. For example, Du Roure's section of the multivolume *Philosophy* on Cartesian logic consists of a summary of *Discourse* Part II, with a commentary on Descartes' rules of method, in succeeding chapters. Du Roure's view of the usefulness of those precepts is influenced by Descartes' preface to the French edition of the *Principles*. He recommends Descartes' logic in order for us to conduct our reason well: "But because it depends considerably on usage, it is extremely advantageous to practice the rules on simple and easy questions, such as those of mathematics. And when we will have acquired some habit in discovering the truth, we must apply ourselves with care to Philosophy" (Du Roure 1654, 183–84).[22] This is the view of logic and mathematics as tools for sharpening the mind, much like solving crossword puzzles, not as the foundation of physics.

The *Port-Royal Logic*, which dominated in the second half of the seventeenth century, thus also Régis's *Logique* and Le Grand's *Logick*, all end with a section called "Method." By method, however, these writers mean analysis and synthesis—which does not have to be anything particularly Cartesian[23]—though we do find Descartes' rules of method enumerated in the chapters on analysis. The *Port-Royal Logic* lists Descartes' four rules, saying that they are "general to all sorts of methods and not particular to the method of analysis alone" (Arnauld 1674, 375) but then moves on to give five rules of composition, focusing on these and enlarging them by chapter 10 to eight principal ones (428–31). Régis follows suit. He adds a chapter on "the advantages we draw from observing the four precepts of analysis" (Régis 1691, 152–54) and abbreviates the lengthy *Port-Royal* discussion of synthesis into a single small chapter and just three brief rules: leave no term ambiguous; use clear and evident principles; and demonstrate all propositions (56). Part IV of Le Grand's *Logick*, "Concerning Method, or the Orderly Disposition of Thoughts," deals with the analytic and synthetic methods, that is, resolution and composition. As part of the analytic method, Le Grand asserts that this method is the art that guides reason in the search for truth; because we cannot proceed to something unknown except by means of something known and questions are propositions that include something known and something unknown, whenever the nature or cause of anything is proposed, we must

> in the first place accurately examine all the Conditions of the question propounded, without minding things as are Extraneous, and do not belong to the Question. Secondly, We are to separate those things which are certain and manifest from those that include any thing of Confusion or Doubt . . . Thirdly, Every Difficulty we meet with is to be divided into Parts . . . Fourthly, We are orderly to dispose of our Perceptions, and the Judgments we frame thence; so that beginning from the most easie, we may proceed by degrees to those that are more difficult . . . Fifthly, That the Thing in question, be furnished with some Note or other that may determine it, and make us judge it to be the same, whenever we meet with it. (Le Grand 1694, 1:46)

This seems to be Le Grand's version of Descartes' four rules of method, restricted to what is useful to analysis. Le Grand ends his *Logick* with chapters on composition, giving various rules of definition, axiom, and demonstration similar to the ones given by the *Port-Royal Logic*. As could be expected, the seventeenth-century Cartesians construct accounts of Descartes' method based on their understanding of Descartes' assertions in the *Discourse* and *Principles*.

The Cartesians also integrate Descartes' various comments about the relation between mathematics and physics into their accounts. Rohault discusses some of these relations in the preface to his *Treatise on Physics*, beginning with a rebuke of scholastics for not teaching mathematics in their schools: "The Fourth Defect that I observed in the Method of the [School] Philosophers, is the neglecting Mathematicks to that Degree, that the very first Elements therof are not so much as taught in their Schools. And yet, which I very much wonder at, in the Division which they make of a Body of Philosophy, they never fail to make Mathematicks one Part of it" (Rohault 1729, n.p.; roughly 13–16). He then formulates the argument we have already seen in Du Roure about the use of mathematics in general:

> Now this Part of Philosophy is perhaps the most useful of all others, at least it is capable of being apply'd more Ways than all others: For besides that Mathematicks teach us a very great number of truths which may be of great Use to those who know how to apply them: They have this further very considerable advantage, that by exercising the Mind in a Multitude of Demonstrations, they form it by Degrees and accustom it to discern Truth from Falsehood infinitely better, than all the Precepts of Logick without Use can do. And thus those who study Mathematicks find themselves perpetually convinced by such Arguments as it is impossible to resist, and learn insensibly to know Truth and to yield to Reason.

In large part, this is Rohault's take on Descartes' justification for mathematics outside the tree of philosophy: exercising the mind—crossword puzzles and all that. But Rohault goes a bit further, justifying the use of mathematics in natural philosophy—indeed, in all arts—with two additional arguments:

> First, that as there is a natural Logick in all Men, so is there also natural Mathematicks, which according as their Genius's are disposed, make them more or less capable of Invention. Secondly, That if their Genius alone, conducted only by natural Light, will carry them so far, we cannot but hope Greater Things from the same Genius if the study of Mathematicks be added to its natural Light, than if that study be neglected. And indeed all the prop-

> ositions in Mathematicks are only so many truths, which those, who apply themselves to them, come to the Knowledge of by good Sense.

This offers a positive role for mathematics that does not refer expressly to Cartesian metaphysics. It demonstrates Rohault's recognition (shared by Descartes) that mathematics and physics rely on the same intellectual faculties (Du Roure [1654, 188] expresses a similar sentiment). But it is not an argument to the effect that the method of physics is the same as the method of mathematics or that mathematical truth or mathematical properties are the basis for physical truth or physical properties.

Rohault's generally positive view is not reflected in the work of his follower. Régis (1691) demarcates between mathematics and physics, specifically asserting that he has avoided all mathematical questions in his philosophy:

> Those who read this book will more easily experience its flaws if they do not stop at equivocal words, ambiguous definitions, or any idea that is foreign to Philosophy, given that we have even purposely avoided Mathematical questions, both because they are little understood by the majority of those who want to apply themselves to Philosophy, and because we all too often confuse them with purely Physical questions, though they are of an entirely different nature. For one is not satisfied in Mathematics by knowing that some things have greater magnitude than some other things; we claim also to know with evidence the exact ratios holding between them, or precisely by how much they are greater, which does not at all concern Physics. (Régis 1691, preface)

Régis continues his demarcation between physics and mathematics, accepting the usefulness but denying the importance of mathematics to physics, stressing the experiential basis of physics, in contrast with how geometry is usually practiced: "one can be a good Physicist without being a great Geometer, but one cannot be a great Geometer without being a good Physicist, at least if we have Geometry consist (as we must) in demonstrations based upon facts, or on constant truths; for if we have it consist (as is usually done) in demonstrations based on arbitrary assumptions, nothing prevents a bad physicist from being a good geometer" (n.p.).

Unlike Rohault and Régis, who emphasize the empirical aspects of natural philosophy (see Ariew 2013; Dobre 2013), Le Grand is interested in the standard question of the certainty of natural philosophy (what he also calls physiology). He proceeds very much in the spirit of a scholastic, substituting Cartesian terminology and doctrines. He has considered the nature of God and inquired into his attributes: "Physiology comes next to be considered by us, which contemplates *Natural Things*, and deduceth their *Causes* from the *first Original* . . . Now that *Physiology* is a Species of *Science*, and is conversant with things that are True and Necessary, appears from the Demonstrations that are made of *Natural Things*; the Certainty whereof depends on the Stability of Things that are defined, and supposeth their determinate Essence" (Le Grand 1694, 1:91).

Le Grand then attempts to answer the objection: since bodies are only perceived by the senses and the senses may represent false things to the understanding, how can the certainty required for science be had in natural things? His answer is that "It is False that *Material Things* are known by the *Senses* . . . to speak properly, nothing is conveigh'd from things without us, by the *Organs of Sense*, to our *Minds*, save only some *Bodily motions*, by which the *Ideas* of *Objects* are offer'd to them . . . Wherefore, *Bodily things* are not known by the *Senses*, but by the *Understanding* alone: So that to be sensible of a *Material Substance*, is nothing else, but to have an *Idea* of it, which is not the work of the outward *Senses*, but of *Cogitation*" (Le Grand 1694, 1:92).

The further objection is that natural philosophy treats material things as changeable, which seems inconsistent with the notion of science as certain and perpetual knowledge. Le Grand's answer is that "Nevertheless we must say, that *Natural Philosophy* is indeed a *Science*, because the *Nature* of a *Science* is not consider'd with respect to the things it treats of, but according to its *Axioms* of an undoubted *Eternal Truth*. For tho' the things which *Physiology* handles, be changeable; yet the *Judgments* we make of them are stable and firm; and consequently the Truth we have of them is *Eternal* and unchangeable" (ibid.). Le Grand (1694) gives as examples of these indubitable and constant truths propositions such as "all that is bodily is changeable" and "every mixed body is dissoluble." In this way, he rejoins here Descartes' view from in the end of *Principles*, Part II:

> Forasmuch as every *Science* hath a *Subject*, about which it is conversant, and to which, whatsoever is handled in the same may be attributed either as *Principles*, *Parts* or *Affections*, we say that the *Material Subjects of Physiology*, are natural *things*, and that *Magnitudes*, *Figures*, *Situation*, *Motion*, and *Rest*

> are the *Formal Subject* of it; . . . Wherefore, if a *Natural Philosopher* considers nothing in matter besides these *Divisions, Figures* and *Motions*, and admit nothing for *Truth* concerning them, which is not evidently deducible from common Notions, whose *Truth* is unquestionable, it is altogether manifest, that no other *Principles* are to be looked for in *Natural Philosophy*, than in *Geometry* or *abstract Mathematicks*; and consequently that we may have as well Demonstrations of *Natural Things*, as of *Mathematical*. (1:92)

Let us repeat the last thought: as long as we limit ourselves to what is deducible from common notions, we may have demonstrations of natural things as well as those of mathematical things. Régis (1691) has an exemplary exposition of the same Cartesian view, delineating carefully among metaphysics, mathematics, and physics:

> Metaphysics not only serves the soul to make itself known to itself, it is also necessary for it in order to know things outside it, all natural sciences depending on metaphysics: mathematics, Physics, and Morals are founded on its principles. In fact, if Geometers are certain that the three angles of a triangle are equal to two right angles, they received this certainty from Metaphysics, which has taught them that everything they conceive clearly is true and that it is so because all their ideas must have an exemplary cause that contains formally all the properties these ideas represent. If Physicists are certain that extended substance exists and that it is divided into several bodies, they know this through Metaphysics, which teaches them not only that the idea they have of extension must have an exemplary cause, which can only be extension itself, but also that the different sensations they have must have diverse efficient causes that correspond to them and can only be the particular bodies that have resulted from the division of matter. (64)[24]

The Cartesians found Descartes' philosophy enormously important for the seventeenth century. The verso of Du Roure's title page from his *Philosophy* tells the story very well; while he appreciates Gassendi and Hobbes and quotes them at times, his admiration for Descartes knows no bounds: "One can oppose Hobbes, Gassendi, and Descartes against all those whom are glorified by Rome and Greece . . . Those who would take the trouble to read this philosophy will find numerous opinions of these three wise philosophers, but principally those of Descartes. This is why I want to show the extent he is esteemed by the following testimony." The six subsequent paragraphs are superlative praise for Descartes, including: "Descartes is the premier philosopher of all time." When we tell the story of the seventeenth century, we need to capture what these thinkers found so appealing about Descartes (and what the anti-Cartesians found so dangerous). And when we do so, we find also many different views about the relations between mathematics and natural philosophy: that natural philosophy can develop a method similar to that of mathematics; that propositions in natural philosophy can be as certain as those of mathematics; that mathematics can be of use in sharpening one's mind for the practice of philosophy; that mathematics has a mode of exposition that is particularly persuasive; that philosophy can be based on the same clear and distinct ideas as those on which mathematics are based. But we do not find the view that the method of philosophy is reducible to the method of mathematics or that philosophy is founded in mathematics. The generally positive views of mathematics in Descartes and the Cartesians do not legitimate a historical or historiographical thesis of the mathematization of nature in the fashion of Burtt, Dijksterhuis, and Koyré.

ABBREVIATIONS

AT Descartes, R. 1964–76. *Oeuvres de Descartes*

NOTES

1. I take the language of the motivation for this volume from the original prospectus of the workshop on which the volume is based.

2. The work had a second edition in 1932. Burtt indicates that the second edition contains no changes in his narrative before Newton: "No historical researches during the last six years with which I have become acquainted seem to require any essential changes in the survey here embodied, so far as it reaches" ([1924] 1954, preface).

3. *Mechanisering van het wereldbeeld*, 1950; I will be citing the English translation by Dikshoorn (Dijksterhuis 1969).

4. Koyré studied with Husserl at Gottingen. As Sophie Roux (2010) states: "Husserl claimed that Galileo was the first to mathematize nature, i.e., according to Husserl, to surreptitiously substitute mathematical idealities for the concrete things of the intuitively given surrounding world" (1n). For Koyré's Bachelardian inspiration, see, for example, Iliffe (1995).

5. See also Koyré's (1962, 11–12) reiteration of this view. The 1957 English version of that work and Koyré's *A Documentary History of the Problem of Fall from Kepler to Newton* (1955) find their way into the bibliography of Dijksterhuis's 1961 English translation, but, obviously, not in the text itself.

6. For Koyré, the second major change was the "dissolution of the cosmos," with all that entails. An aside: To the extent that I think that the law of inertia was first formulated by Descartes in his 1632 *Le Monde*, I do not think correct Koyré's view that the mathematization of nature made the law of inertia possible. But I will not pursue this train of thought here.

7. Dijksterhuis also cites with approbation another opinion referring to Bacon as one of the "great creative spirits of the seventeenth century." He then asserts that both "parties are right up to a point. Perhaps more the first than the second: if Bacon with all his writings were to be removed from history, not a single scientific result would be lost" (1969, 397).

8. "Galileo and the scientific revolution of the seventeenth century" and "Galileo and Plato" reprinted as chapters 1 and 2 of Koyré (1968), translated into French and reprinted in Koyré (1966). Dijksterhuis also knows Koyré's "Galileo and Plato."

9. Iliffe refers to "obstacle" and "mutation" as Bachelardian concepts.

10. From "An experiment in measurement" of 1953. In this way Koyré positions himself against both Pierre Duhem's internal continuous and Marxist external social accounts.

11. The literature showing Koyré wrong about Galileo's experiments is large and now fairly old; see the works of Stillman Drake et al. A more historical appreciation of Francis Bacon's scientific method and views on experiments is perhaps as extensive, but more recent; see the essays of Dana Jalobeanu et al. and chapter 2 in this book.

12. See also the excellent other suggestions about contextualizing the mathematization of nature in Roux (2010).

13. I will not go into the details of this answer. It should suffice to refer to Descartes on the creation of the eternal truths and the fact that, for Descartes, metaphysical truths are more certain than mathematical truths.

14. Dijksterhuis actually knows that the *Rules* was not published until 1701 in the *Opera Posthuma*.

15. A word about *Mathesis Universalis*: It has been pointed out (Weber 1964) that Rule 4 has two dissonant parts, the second of which contains Descartes' views on *Mathesis Universalis*. While some able commentators (Marion 1975, for example) have argued that one can provide a reading of Rule 4 that

takes both parts into account, others have argued that *Mathesis Universalis* is either a later or an earlier version of Rule 4 or even that it does not belong at all in the manuscript. I think that these issues can be settled in favor of *Mathesis Universalis* being a later interpolation and I am confirmed in this by the fact that the recently discovered Cambridge manuscript of the *Rules* is missing part 2 of Rule 4, containing *Mathesis Universalis*. See the edition of the *Rules* by Serjeantson and Edwards (forthcoming). Thus, for Descartes, *Mathesis Universalis* is not the "guiding principle for his whole life-work," and mathematics was not "the sole key needed to unlock the secrets of nature."

16. The last few assertions differ from the first few in that they reveal something like a metaphysical thesis about the relations between mathematics and physics, as opposed to an epistemological or methodological one. There is, of course, also the notion of geometric order (*more geometrico*) in Descartes' appendix to *Second Replies*. But as Descartes makes clear, this is not a method, but a mode of exposition not applying solely to narratives proceeding by axioms, postulates, and theorems. For a development of this view in a Cartesian, see Lodewijk Meyer's preface to Spinoza's *Descartes' Principles of Philosophy*. The issue is complex, exemplars of it span such diverse thinkers as Jean-Baptiste Morin's *Quod Deus sit* and Nicolaus Steno's *Elementorum myologiae specimen seu Musculi description geometrica*.

17. The French version is almost the same: "I do not accept any principles in physics that are not also accepted in mathematics, so that I may prove by demonstration everything I would deduce from them; these principles are sufficient, inasmuch as all natural phenomena can be explained by means of them" (AT, 9:101).

18. By Book 3 of the *Principles*, Descartes will be invoking hypotheses or supposition that he knows cannot be reduced to the principles of Part I or their deductions in Part II; as he asserts: "I dare say that you would find at least some logical connection and coherence in it, such that everything contained in the last two parts [that is, *Principles* 3 and 4] would have to be rejected and taken only as a pure hypothesis or even as a fable, or else it all has to be accepted. And even if it were taken only as a hypothesis, as I have proposed, nevertheless it seems to me that, until another is found more capable of explaining all the phenomena of nature, it should not be rejected" (AT, 4:216–17). See Ariew (2010, 31–46).

19. Let me put the same point somewhat differently. Descartes is no atomist, but supposing he was, he would refer all natural phenomena to his two fundamental principles, atoms and the void. The properties of bodies then would be "what the geometers call quantity," namely, size, shape, and motion.

20. For an account of the peculiarities of the Cartesian metaphysics of Desgabets and Régis, see Schmaltz (2002).

21. A contemporary description of the work, from a letter by Simon Foucher to Leibniz, confirms this impression of eclecticism: "You know that I think Regis has given the public a great system of philosophy in 3 quarto volumes with several figures. This work contains many very important treatises, such as the one on percussion by Mariotte, chemistry by l'Emeri, medicine by Vieuxsang and by d'Uvernai. He even speaks of my treatise on Hygrometers, although he does not name it. There is in it a good portion of the physics of Rohault and he refutes there Malbranche, Perraut, Varignon—the first concerning ideas, the second concerning weight, and the third, who has recently been received by the Académie Royale des Sciences, also concerning weight. The *Metheores* of l'Ami also in part adorn this work, and the remainder is from Descartes. Regis conducted himself rather skillfully in his system, especially in his ethics" (Leibniz 1890, 1:398–400).

22. In the prelude to his tree of Philosophy, Descartes asserts: "a man who as yet has merely the common and imperfect knowledge . . . should above all try to form for himself a code of morals sufficient to regulate the actions of his life. . . . After that he should likewise study . . . the logic that teaches us how best to direct our reason in order to discover those truths of which we are ignorant. And since this is very dependent on custom, it is good for him to practice the rules for a long time on easy and simple questions such as those of mathematics. Then, when he has acquired a certain skill in discovering the truth in these questions, he should begin seriously to apply himself to the true philosophy" (AT, 10b:13–14). This is, of course, related to the statement in the *Discourse* cited earlier that "the real use that can be made of mathematics is . . . to prepare the mind to follow the real philosophical method."

23. There are numerous methods called analysis and synthesis in early modern philosophy, most of which have nothing to do with the various things Descartes called analysis and synthesis: resolution and composition within the method of the *Rules*, the two modes of demonstration of the *Second Replies*, or the analysis (and synthesis) of the ancients. The notion originates at least from Aristotle's *Posterior Analytics* and is found in seventeenth-century scholastic textbooks in the portion of their Logic texts dealing with scientific method. Du Roure's analysis and synthesis follow the same lines as scholastic authors such as Scipion Dupleix: "Method is the order of the sciences and of their discourse: where one makes several things out of one, which is called the analytic method, or from several one, which is called the synthetic or compositional method"

(Du Roure 1665, section 2). For more on Scholastic and Cartesian Logic, see Ariew (2014).

24. Régis continues: "Metaphysics not only serves as foundation for all natural Sciences, it is yet simpler and easier to acquire than all of them; the mind's access to this science is common to all kinds of native intelligences, because there is nothing in life or in the society of men which does not dispose or lead itself to it. Every occasion all needs contribute incessantly to the material of Metaphysics which concerns the knowledge of the soul and we experience in ourselves all the proofs of the things that are the object of this knowledge. In contrast, with the other sciences we are required to go out from ourselves in order to consider the objects we examine. For example, we go out from ourselves in Geometry in order to contemplate shapes, we go out from ourselves in Physics to consider motions, and we go out from ourselves in Morals in order to observe the conduct of other men."

REFERENCES

Ariew, R. 2010. "The New Matter Theory and Its Epistemology: Descartes (and Late Scholastics) on Hypotheses and Moral Certainty." In *Vanishing Matter and the Laws of Nature: Descartes and Beyond*, ed. D. Jalobeanu and P. Anstey, 31–46. London: Routledge.

———. 2013. "Censorship, Condemnations, and the Spread of Cartesianism." In *Cartesian Empiricisms*, ed. M. Dobre and T. Nyden, 25–46. New York: Springer.

———. 2014. *Descartes and the First Cartesians*. Oxford: Oxford University Press.

Arnauld, A. 1674. *La logique*. Paris: G. Desprez.

Burtt, E. A. [1924] 1954. *Metaphysical Foundations of Modern Science*. Garden City, N.Y.: Doubleday Anchor Books.

Cohen, H. F. 1991. "How Christiaan Huygens Mathematized Nature. Unrolling Time: *Christiaan Huygens and the Mathematization of Nature* by Joella G. Yoder." *British Journal for the History of Science* 24: 79–84.

Descartes, R. 1964–76. *Oeuvres de Descartes*. Ed. C. Adam and P. Tannery. 12 vols. 2nd ed. Paris: Vrin.

Descartes, R. Forthcoming. *Regulae*. Ed. and trans. by R. Serjeanston and M. Edwards. Oxford: Oxford University Press.

Dijksterhuis, E. J. 1969. *The Mechanization of the World Picture*. Translated by C. Dikshoorn. Oxford: Oxford University Press.

Dobre, M. 2013. "Rohault's Cartesian Physics." In *Cartesian Empiricisms*, ed. M. Dobre and T. Nyden, 203–26. New York: Springer.

Du Roure, J. 1654. *La Philosophie divisée en toutes ses parties.* Paris: Chez François Clouzier.

———. 1665. *Abrégé de la vraye philosophie.* Paris: Chez l'Auteur.

Garber, D. 2001. *Descartes Embodied.* Oxford: Oxford University Press.

Iliffe, R. 1995. "Theory, Experiment and Society in French and Anglo-Saxon History of Science." *European Review of History* 2: 65–77.

Koyré, A. 1955. *A Documentary History of the Problem of Fall from Kepler to Newton.* Philadelphia: American Philosophical Society.

———. 1962. *Du monde clos à l'univers infini.* Paris: PUF.

———. 1966. *Etudes d'histoire de pensée scientifique.* Paris: PUF.

———. 1968. *Metaphysics and Measurement: Essays in Scientific Revolution.* Cambridge, Mass.: Harvard University Press.

———. 1978. *Galileo Studies.* Atlantic Highlands, N.J.: Humanities Press.

Le Grand, A. 1671. *Philosophia veterum e mente Renati Descartes, more scholastica breviter digesta.* F. Martyn.

———. 1694. *Entire Body of Philosophy according to the Principles of the famous Renate Des Cartes.* London: Richard Blome.

Leibniz, G.W. 1890. *Philosophische Scriften.* Ed. C. Gerhardt. Berlin: Weidmannsche Buchhandlung.

Marion, J.-L. 1975. *Sur l'ontologie grise de Descartes.* Paris: Vrin.

Régis, P.-S. 1691. *Système Général selon les Principes de Descartes.* Amsterdam: Huguetan.

Rohault, J. 1729. *Rohault's System of Natural Philosophy.* Trans. by J. Clarke. London: James and John Knapton.

Roux, S. 2010. "Forms of Mathematization." *Early Science and Medicine* 15: 1–11.

Schmaltz, T. 2002. *Radical Cartesianism.* Cambridge: Cambridge University Press.

Weber, J. P. 1964. *La constitution du texte des Regulae.* Paris: Société d'édition d'enseignement supérieur.

5

LAWS OF NATURE AND THE MATHEMATICS OF MOTION

DANIEL GARBER

Nature came to be understood through mathematics in the seventeenth century, when Galileo (1890) famously wrote: "Philosophy is written in this grand book, the universe, which stands continually open to our gaze. But the book cannot be understood unless one first learns to comprehend the language and read the letters in which it is composed. It is written in the language of mathematics, and its characters are triangles, circles, and other geometric figures without which it is humanly impossible to understand a single word of it; without these, one wanders about in a dark labyrinth" (6:232; Galilei 1957, 237–38). This, in a way, can be understood as a motto for the century as a whole, or, at least, for those figures in the century who are now recognized as the ancestors of modern science. But it is also the century in which the laws of motion as we know them are first articulated. People before the seventeenth century had certainly talked about nature as being governed by overarching laws. But even so, it is only in this century that specific laws were proposed, and their consequences explored.[1]

It is quite natural to see these two trends as being closely linked. In contemporary physics, after all, mathematically expressed laws of nature such as quantum theory or general relativity are at the heart of mathematical physics, which would be unimaginable without such structures. One might imagine that this close link between mathematics and laws goes back to the origin of both in early modern natural philosophy. But, I shall argue, however plausible such an assumption might be, the story is more complicated than that. Though both the mathematization of nature and the idea of a law of nature are important to the early modern vision of the physical world, they are to a large extent distinct. In making my case, I would like to explore three key figures from the period: Descartes, Galileo, and Hobbes. In Descartes we certainly have laws of nature: these are central to the project of his physics, both in the early *Le monde* and in the later *Principia philosophiae*. But, I shall argue, despite what Descartes sometimes says, mathematics is only marginal to his program. In Galileo, on the other hand, mathematics is central: his application of mathematics to motion was one of the great accomplishments of early modern science. But even so, I would argue, Galileo made no substantive use of anything that could properly be called a law of nature. And finally, with Hobbes we have something of an ambiguous situation. For Hobbes, the subject matter of geometry is body, including its motion, and so there is a sense in which the kind of general statements about motion that Descartes identifies as laws are a part of mathematics, but a mathematics that is very different from what anyone else in the period recognized as such. And while Hobbes offered a number of important general statements about bodies in motion, he was very careful *not* to call them laws.

DESCARTES AND LAWS OF NATURE

Descartes may not have invented the idea of a law of nature, the idea that nature is structured in accordance with some general laws that order things in the world. But he may well be the first who actually tried to articulate the laws of nature in such a way that their consequences for how nature works can be set out and evaluated.[2]

For Descartes, of course, body consists only of extension. As a consequence, everything in nature must be explicable in terms of the size, shape, and motion of bodies and the smaller parts that make them up. What he calls the rules or laws of nature govern the motion of bodies, one of its modes. These laws of nature are first given in chapter 7 of his early *Le monde*, which Descartes suppressed after finding out about the condemnation of Galileo in 1633. But they appear later in his *Principia philosophiae*, suitably rethought and reorganized. It is in this form that they were best known by his contemporaries.

The laws as given in the *Principia* are as follows:

> [Law 1] Each and every thing, in so far as it can, always continues in the same state; and thus what is once in motion always continues to move. (PP 2.37)[3]

> [Law 2] All motion is in itself rectilinear; and hence any body moving in a circle always tends to move away from the center of the circle which it describes (PP 2.39).
>
> [Law 3] If a body collides with another body that is stronger than itself, it loses none of its motion; but if it collides with a weaker body, it loses a quantity of motion equal to that which it imparts to the other body (PP 2.40).

Preceding the statement of these laws in the *Principia*, Descartes proposes a principle in accordance with which the total quantity of motion in the world must remain constant:

> In the beginning [God] created matter, along with its motion and rest; and now, merely by his regular concurrence, he preserves the same amount of motion and rest in the material universe as he put there in the beginning . . . For we understand that God's perfection involves not only his being immutable in himself, but also his operating in a manner that is always utterly constant and immutable. Now there are some changes whose occurrence is guaranteed either by our own plain experience or by divine revelation, and either our perception or our faith shows us that these take place without any change in the creator; but apart from these we should not suppose that any other changes occur in God's works, in case this suggests some inconstancy in God. Thus, God imparted various motions to the parts of matter when he first created them, and he now preserves all this matter in the same way, and by the same process by which he originally created it; and it follows from what we have said that this fact alone makes it most reasonable to think that God likewise always preserves the same quantity of motion in matter (PP 2.36).

Though this principle is not called a law, it is an important constraint on the behavior of bodies in motion. Following the statement of the third law, Descartes works out a series of examples in which he shows the outcome of direct collisions for two bodies with different sizes and speeds (PP 2.45ff).[4] The conservation principle is a key tool that Descartes uses in deriving those supposed consequences of the third law: Descartes treats collisions in such a way that the total quantity of motion in the system of colliding bodies remains the same before and after the collision. From those examples, it is clear that the quantity of motion is measured jointly by the size and the speed of the bodies. Size rather than mass, since Descartes does not have a conception of quantity of matter distinct from size, and speed rather than velocity, since directionality is treated distinctly from the magnitude of the speed, and the conservation principle does not govern it.[5]

The argument for the conservation of quantity of motion is grounded in a theological doctrine that dates from long before Descartes' time, the view that God must sustain the world from moment to moment for it to continue to exist. Descartes' claim is that because God is immutable, and acts in a constant way, the total quantity of motion must remain constant in the world. This same divine immutability is also what he claims grounds the three explicitly named laws of nature. After the second law he notes that "the reason for this second rule is the same as the reason for the first rule, namely the immutability and simplicity of the operation by which God preserves motion in matter. For he always preserves the motion in the precise form in which it is occurring at the very moment when he preserves it, without taking any account of the motion which was occurring a little while earlier" (PP 2.39). And similarly for the third law, which follows from the fact that "since God preserves the world by the selfsame action and in accordance with the selfsame laws as when he created it, the motion which he preserves is not something permanently fixed in given pieces of matter, but something which is mutually transferred when collisions occur" (PP 2.42). The laws of nature are thus grounded in divine immutability and the fact that the created world depends from moment to moment on the power by which he keeps the world in existence.

In what sense are Descartes' laws of nature intended to be *laws* of *nature*? First of all, one can point to their generality: they are true not merely of this or that group of bodies, but of *all* bodies, of bodies *as such*. In this way their scope is over nature as a whole. In both *Le monde* and in the later *Principia philosophiae*, the general laws are taken to apply to the cosmos as a whole, and are appealed to in a broad and hand-waving way to explain the general features of the cosmos as a whole, such as the fact that there are infinite suns, each of which is a source of light.[6] Now, in *Le monde* Descartes refers to them as "the laws God imposed on it [i.e., nature]" (AT 11 36). This suggests that the laws are chosen by God and then "imposed" on bodies. But this is clearly not exactly what is going on, even in *Le monde*, which, as in the *Principia philosophiae*, grounds the laws in the constant and continuous activity of God on bodies in sustaining them in existence. God does not impose them on bodies in the way in which a monarch might formulate laws and then impose them on the citizenry. But there is another sense in which the laws might be said to be "imposed" on bodies. For Descartes bodies are essentially extended. As I understand that, it means bodies contain only their geometrical properties: they are the objects of geometry, made real, made concrete. But as such, considered as the objects of geometry, there is nothing in them that determines that they must obey the rules that Descartes calls the laws of nature. Taken by themselves, they are completely indifferent to motion of any kind. However, when they are realized in nature, as real, existent things, they must behave in a certain way because of the way in which an immutable God sustains them. What Descartes calls the laws of nature are really just a way of formulating how it is that God keeps them in existence. But insofar as these laws do not follow from the nature of body as such, and insofar as they require God's intervention, we might say that they are, indeed "imposed" on body by God. This can be so even though Descartes claims that "if God had created many worlds, they [i.e., the laws] would be as true in those worlds as they are in this one" (AT 11 47). To that extent, one might want to say that the laws of nature are, in a sense, necessary for Descartes. But if they are necessary it is not because of any necessity in body, or any necessity in the laws themselves, but only because God exists necessarily, and is necessarily immutable, so necessarily will act immutably in any possible world he creates. In this way the laws of nature pertain not to the essence of body but to its continued existence, insofar as the continued existence derives directly from God.

It is interesting to note here that as Descartes understands them, the laws of nature by themselves do not entail that bodies are heavy, that is, that they tend to fall toward the center of the earth. Cartesian bodies, as I earlier noted, are completely indifferent to motion, including gravitational motion. For Descartes, bodies tend to fall to the center of the earth only because of the particular configuration of the vortex of subtle matter that surrounds the earth and the make-up of the gross bodies of our experience.[7]

Descartes sometimes talks as if his entire physics were just a kind of mathematics. For example, writing to Mersenne on March 11, 1640, Descartes remarked: "I would think I knew nothing in physics if I could say only how things could be, without demonstrating that they could not be otherwise. This is perfectly possible once one has reduced physics to the laws of mathematics. I think I can do it for the small area to which my knowledge extends" (AT 3 39). And at the end of Part II of the *Principia philosophiae*, he wrote: "The only principles which I accept, or require, in physics are those of geometry and pure mathematics; these principles explain all natural phenomena, and enable us to provide quite certain demonstrations regarding them" (PP 2.64).[8] And, indeed, much of Descartes' writings about nature do involve serious attempts at applying mathematics to the physical world. Famously, the young Descartes and the young Isaac Beeckman attempted to make physics mathematical, in some sense or another. In a famous passage in his journals, which bears the marginal note "Physico-mathematici paucissimi," Beeckman wrote, with pride, that the young Descartes had told him that "he had never found a man, beside me [i.e., Beeckman] who . . . had accurately joined physics with mathematics in this way" (AT 10 52). So inspired, Descartes' early work bristles with various attempts to combine physics and mathematics, including attempts to treat the problem of free fall that Galileo would solve in mathematical terms.[9] Later, of course, Descartes will use mathematics essentially in his derivation of the law of refraction in optics, in a famous argument he will give in discourse 2 of the *Dioptrique*. Sophisticated mathematical arguments will also play a central role in Descartes' account of the rainbow in discourse 8 of the *Météores*. In his correspondence in the late 1630s, there are attempts to apply very serious mathematics to a number of problems in the motion of bodies (on this, see Garber 2000).

But for all of that, and despite the statements in which he claims that all his natural philosophy is mathematics, there is no serious mathematics at all either in *Le monde*, or in the later *Principia philosophiae*, the canonical presentation of his natural philosophy. As he notes in discussing his thought with Frans Burman in 1648, "You do not . . . need mathematics in order to understand the author's philosophical writings [e.g., his physics], with the possible exception of a few mathematical points in the *Dioptrique*" (AT 5 177). This is largely true of his laws of nature. The laws themselves are given in purely qualitative terms in the text. It is true that a bit of arithmetic enters into the example he works out of the application of the third law, the law of collision to the case of direct collision, as mentioned earlier. There he works out some solutions to the problem of direct collision by applying the principle of the conservation of quantity of motion to the various combinations of size and speed of two colliding bodies. But the mathematics is trivial in comparison with other attempts to join physics and mathematics in Descartes' corpus and hardly counts as serious "physico-mathematics."

And their application later in the *Principia philosophiae* is qualitative as well. For example, in PP 3.59, Descartes addresses the question as to force (*vis*) associated with the striving (*conatus*) that a body, rotating, has to escape along the tangent of a circle in which it is rotating, a striving derived from

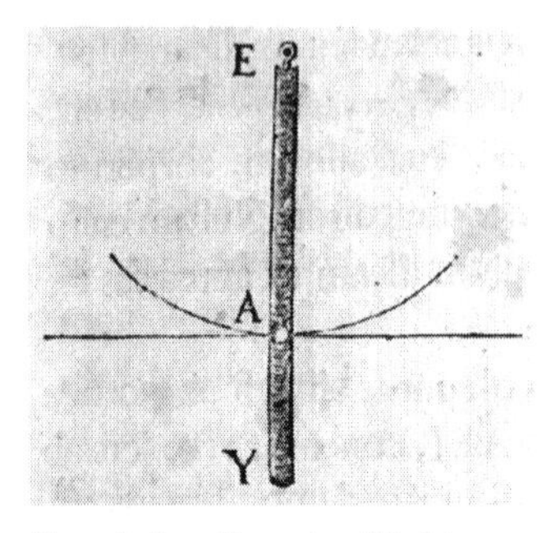

Figure 1. From Descartes, *Principia philosophiae* (Amsterdam: Elzevir, 1644), 101.

his second law. To treat this striving, he imagines a hollow tube, EY, fixed at E and rotating, with a ball A in the tube (see Figure 1). He writes: "When we first begin to rotate this tube around the center E, the globe will advance only slowly toward Y. But in the next instant it will advance a bit faster, because in addition to retaining its original force, it will acquire new force from its new striving to recede from E: because this striving continues as long as the circular motion lasts and is, as it were, renewed constantly" (PP 3.59).[10] It is interesting here—and in radical contrast with a similar analysis in Galileo, as we shall see—that Descartes nowhere ever attempts to represent the motion of A or its acceleration in mathematical terms.

GALILEO AND THE *SINTONI* OF MOTION

While mathematics enters into Descartes' natural philosophy from time to time, his treatment of the laws of motion seems to be quite independent of any attempts that he may have made to understand nature mathematically. With Galileo, on the other hand, the application of mathematics to nature is quite central to his project. However, it is not clear that the laws of nature play any substantive role in his account of the physical world. Now, there is no doubt that in some sense Galileo did recognize the idea of the laws of nature. In his important "Letter to the Grand Duchess Christina," in which he discusses Copernicanism and the Bible, he wrote: "Nature . . . is inexorable and immutable; she never transgresses the laws [*leggi*] imposed upon her, or cares a whit whether her abstruse reasons and methods of operation or understandable to men" (EN 5 316).[11] But even though in a very general sense he may have recognized the importance of laws of nature, I will argue that his own mathematical science of motion would seem to involve nothing that one can call a general law of nature.

Galileo worried about the behavior of bodies in motion for virtually his entire career, from the time that he was a young professor until his last years under house arrest. It would not be appropriate in this modest essay to try to survey Galileo's thought throughout his long career. Instead, I would like to begin by looking at the treatment of bodies in motion in his last work, the *Discorsi e dimostrazioni matematiche, intorno à due nuove scienze* (*Two New Sciences*).[12]

The second of the two new sciences treated in the book is the science of motion. This is introduced at the beginning of the Third Day of the dialogue, in which the interlocutors gather to read and discuss a Latin treatise by "the Academician," Galileo, of course. The treatise is most likely the result of work on motion done during the first decade of the seventeenth century, before Galileo got sidetracked by the astronomical project of the *Starry Messenger*. It begins as follows: "We bring forward a brand new science concerning a very old subject. There is perhaps nothing in nature older than MOTION, about which volumes neither few nor small have been written by philosophers; yet I find many essentials of it that are worth knowing which have not even been remarked, let alone demonstrated" (EN 8 190). This, then, is the subject matter of the new science: motion, treated in the Third and Fourth days of the dialogue.

In the Third Day, Galileo begins with a treatment of uniform motion. But the centerpiece is the treatment of naturally accelerated motion. For most readers the featured result is often referred to as the "law of free fall":

> Proposition II. Theorem II
>
> If a moveable descends from rest in uniformly accelerated motion, the spaces run through in any times whatever are to each other as the duplicate ratio of their times; that is, are as the squares of those times. (EN 8 209)

According to this theorem, the distance fallen by a body in free fall is proportional to the square of the time. Also important is the so-called odd-number rule, a corollary to this theorem, in accordance with which the distances fallen in equal successive times are proportional to the sequence of odd numbers (210).

Interestingly, though, Galileo's own presentation is somewhat different and not focused on the times-square rule. He begins with a question as to what the proper definition of accelerated motion is. The definition that he proposes is the following: "I say that motion is equably or uniformly accelerated which, abandoning rest, adds on to itself equal momenta of swiftness in equal times" (198, 205). But, Galileo wonders, is this definition the correct definition for falling bodies as they actually accelerate in nature? He writes: "And first, it is appropriate to seek out and clarify the definition that best agrees with that which nature employs. Not that there is anything wrong with inventing at pleasure some kind of motion and theorizing about its consequent properties . . . But since nature does employ a certain kind of acceleration for descending heavy things, we decided to look into their properties so that we might be sure that the definition of accelerated motion which we are about to adduce agrees with the essence of naturally accelerated motion" (197). This, then, is the question that he investigates.

The theorem generally called the law of free fall is presented as a direct mathematical consequence of that definition.[13] Galileo then goes to nature, and sees if falling bodies actually satisfy that consequence, that is, he goes to nature to see if the distance fallen is proportional to the square of the time. (Actually, it is somewhat more complicated than that. Galileo does not have the means to measure that directly, so he has to slow free fall down by rolling balls down inclined planes, so that he can actually measure the time and compare it with the distance fallen. But he needs to establish that the relation between time and distance fallen in free fall will be the same for a ball rolling down an inclined plane.) Galileo describes in some detail the experiments that he performed: "In a wooden beam or rafter about twelve braccia long, half a braccio wide, and three inches thick, a channel was rabbeted in along the narrowest dimension" (212). Balls were rolled down in the channel, and the relation between distance and time noted. Galileo even felt the need to inform the reader about how exactly the time was measured by way of a water clock. His conclusion is that the balls do, indeed, observe that ratio between distance and time that Galileo's definition of acceleration requires (212–13). His conclusion is that the definition of acceleration that begins the discussion—"this first and chief foundation upon which rests an immense framework of infinitely many conclusions"—is, indeed, the kind of acceleration that is at issue in naturally falling bodies.

The way in which Galileo derives the mathematical account of free fall from the definition of acceleration shows an interesting contrast with Descartes. Descartes' account of the acceleration of the ball in the rotating tube, discussed earlier, is entirely qualitative: there is no mathematical reasoning and no mathematical treatment of accelerated motion. Galileo, on the other hand, starts with the idea that in accelerated motion, equal momenta of speed are added in equal times, similar to Descartes' starting place. But he then represents the relation between time and distance fallen in geometrical terms, and then uses geometrical reasoning to derive an exact geometrical expression of the relation between time and distance.

In the remaining pages of the Third Day of the dialogue, Galileo goes on to draw further consequences from the definition of natural acceleration that he establishes early in the dialogue.[14] Among the consequences that Galileo noted was one that we may consider of special interest. In the middle of a scholium to Proposition XXIII Problem IX, Galileo makes the following observation: "It may also be noted that whatever degree of speed is found in the moveable, this is by its nature indelibly impressed on it when external causes of acceleration or retardation are removed, which occurs only on the horizontal plane: for on declining planes there is cause of more acceleration, and on rising planes, of retardation. From this it likewise follows that motion in the horizontal is also eternal, since if it is indeed equable it is not weakened or remitted, much less removed" (243). This looks very much like Descartes' first and second laws of nature from the *Principia philosophiae*, the so-called (but incorrectly named) principle of inertia.[15] But there are differences. It is important to note here that what Galileo is talking about is not rectilinear motion but *horizontal* motion: motion on a plane all of whose points remain equidistant from some point toward which the heavy body is attracted. That is to say, what persists is circular motion around the point to which a heavy body tends to fall. And it is important to note that we are dealing with motion *on a plane*: if the plane were eliminated, the body would simply continue to fall toward the center of attraction. And finally, it should be noted that as salient as this observation is to us, in the context of the Third Day of the *Discorsi*, Galileo presents this simply as an observation in passing in the course of a scholium. In the context of the Third Day, it does not even get separate designation as a theorem or proposition. Though we may consider it of special interest, in the context of the Third Day, Galileo felt otherwise.

But this observation has a special role to play in the central proposition of the Fourth Day. The very first proposition of the Fourth Day is Galileo's account of projectile motion: "When a projectile is carried in motion compounded from equable horizontal and from naturally accelerated downward [motion], it describes a semiparabolic line in its movement" (EN 8 269). This proposition is proved quite simply by putting together the uniform motion of a body on a horizontal plane with the accelerated motion of a body in free fall: when you combine the two, it follows straightforwardly that the projectile describes a semiparabola.

In this way, Galileo's new theory of motion is able to give very sophisticated mathematical descriptions of the motion of bodies in various important circumstances, in free fall, on a horizontal plane, and in projectile motion. But are these *laws of nature*? I think not. But why not? Fundamentally it is a question of scope.

The facts about motion that we have been examining depend strongly on the assumption that we are dealing with heavy bodies, bodies that have a tendency to fall toward a particular point. In the *Discorsi*, that point is, of course, the center of the earth. In his earlier book, the *Dialogo sopra i due massimi sistemi del mondo* (*Dialogue Concerning the Two Chief World Systems*), Galileo attempts to generalize this. In the First Day of that dialogue he attempts to articulate an alternative to the Aristotelian cosmology that takes the center of the earth as the center of the universe. In that context he writes: "Every body constituted in a state of rest but naturally capable of motion will move when set at liberty only if it has a natural tendency toward some particular place; for if it were indifferent to all places it would remain at rest, having no more cause to move one way than another. Having such a tendency, it naturally follows that in its motion it will be continually accelerating" (EN 7 44).[16] It is evident that the center of the earth is such a particular place for bodies on the earth: "the parts of the earth do not move so as to go toward the center of the universe, but so as to unite with the whole earth (and that consequently they have a natural tendency toward the center of the terrestrial globe, by which tendency they cooperate to form and preserve it)" (57–58; TCWS 33). But it is not at all clear how to generalize this to other bodies outside of the earth. In the TCWS, Galileo does offer his famous Platonic hypothesis about the formation of the planetary system:

> Let us suppose that among the decrees of the divine Architect was the thought of creating in the universe those globes which we behold continually revolving, and of establishing a center of their rotations in which the sun was located immovably. Next, suppose all the said globes to have been created in the same place, and there assigned tendencies of motion, descending toward the center until they had acquired those degrees of velocity which originally seemed good to the Divine mind. These velocities being acquired, we lastly suppose that the globes were set in rotation, each retaining in its orbit its predetermined velocity. (EN 7 53; TCWS 29)

Galileo claims that if we assume such a hypothesis, we can find a single place from which, should each of the planets be imagined to fall toward the sun and their final speed converted into rotational motion, we would get a system that agrees very closely with the observed speeds of the actual planets. In this way Galileo hypothesizes the sun as a center of tendency for planets, and in this way, in principle, extends his account of motion to the planets with respect to the sun. But this is far from a systematic generalization of the accounts of free fall, the persistence of horizontal motion, and the behavior of projectiles, which remain particular claims about heavy bodies near the surface of the earth. What would happen if a piece of Earth were released ten feet over the surface of Mars, or a Martian rock ten feet over the surface of Earth? What happens to bodies in interplanetary space, or beyond the orbit of the last planet?

Lacking an obvious general applicability outside of a fairly narrow context, it is difficult to see how Galileo's accounts of motion could play anything like the role in organizing the Galilean world that Descartes' laws of nature play in his, where they are used to explain the structure of the whole universe. In writing to Mersenne about his impressions of Galileo's *Discorsi*, Descartes (1638) noted the following:

> Generally speaking, I find that he philosophizes much more ably than is usual, in that, so far as he can, he abandons the errors of the Schools and tries to use mathematical methods in the investigation of physical questions. On that score, I am completely at one with him, for I hold that there is no other way to discover the truth. But he continually digresses, and he does not take time to explain matters fully. This, in my view, is a mistake: it shows that he has not investigated matters in an orderly way, and has merely sought explanations for some particular effects, without going into the primary causes in nature; hence his building lacks a foundation. (Descartes to Mersenne, 11 Oct. 1638, AT 2 380)

It seems true to say that "his building lacks a foundation": the generalizations that Galileo presents explain things on Earth, but they fail to treat nature as a whole. For that he would need to articulate general laws that unite the terrestrial and cosmological domains.

Now, this may not be entirely fair to Galileo: it is not clear to me that he wanted to do the kind of thing that Descartes was doing, and may have been quite happy to work piecemeal, one problem at a time. In a letter Galileo wrote to Belisario Vinta in 1610, describing the project that would become Days Three and Four of the *Discorsi*, he projected "three books on local motion—an entirely new science in which no one else, ancient or modern, has discovered any of the most remarkable characteristics [*sintomi*] which I demonstrate to exist in both natural and violent movement (EN 10 351–52).[17] Similarly, in a letter from January 1639, shortly after the publication of the *Discorsi* in 1638, Galileo describes the project in similar terms: "I'm interested in examining what might be the characteristics [*sintomi*] which accompany the motion of a moving body, which, starting from a state of rest, it goes on moving with a speed that constantly increases in the same way" (EN 18 12).[18] As we saw earlier in the "Letter to the Grand Duchess Christina," Galileo has the concept of an overarching law of nature, something that governs reality as a whole. But his mathematical account of the motion of bodies is not conceived in those terms: his aim is just to give some of the most interesting "sintomi" of accelerated motion. What we have in Galileo, in essence, is a thoroughly mathematical account of at least some aspects of the motion of bodies, but without laws of nature.

HOBBES AND THE GEOMETRY OF MOTION

Motion plays a central role in Hobbes's natural philosophy: it is the sole determinant of change in his materialistic world of body. And mathematics is central as well. Hobbes, like Galileo before him, was interested in a mathematical account of motion. In fact, for Hobbes, motion is an integral element of his geometry. However, while Hobbes put forward a number of general statements about bodies in motion, his relation to the tradition of laws of motion is not entirely clear.

Hobbes was a great admirer of Galileo, particularly in regard to his treatment of motion. In the 1660 dialogue, *Examinatio et Emendatio Mathematicae Hodiernae*, Hobbes wrote: "the doctrine of motion is known to very few, notwithstanding the fact that the whole of nature, not merely that which is studied in physics, but also in mathematics, proceeds by motion. Galileo was the first who wrote anything on motion that was worth reading" (Hobbes 1660 quoted in Jesseph 2004). In *De corpore* (Hobbes 1655) his praise was even stronger, advancing Galileo as the founder of natural philosophy: "After him [i.e., Copernicus] the Doctrine of the Motion of the Earth being now received, and a difficult Question thereupon arising concerning the Descent of Heavy Bodies, *Galileus* in our time striving with that difficulty, was the first that opened to us the gates of Natural Philosophy Universal, which is the knowledge of the Nature of Motion. So that neither can the Age of Natural Philosophy be reckoned higher than to him"

(DC, Epistle Dedicatory, n.p.).[19] But despite his high praise for Galileo, his own treatment of motion was radically different from that of his hero, and much closer to that of Descartes, whose philosophy he generally rejected.

While there are a number of treatments of motion in Hobbes's writings, I will concentrate on what is arguably the canonical treatment in his *De corpore* of 1655, the treatise on body that begins his *Elementa philosophiae*, the three-part philosophical project that begins with a physics, is followed by an account of the human being (*De homine* 1658), and is completed by a politics (*De cive* 1642). *De corpore* is divided into four parts. Part I, "Computation or Logique," is a preface to the *Elementa* project as a whole, and contains a treatise on logic. Parts II, III, and IV constitute a natural philosophy. Part II is called "The First Grounds of Philosophy." This First Philosophy consists in "universal definitions . . . the most common notions [distinguished] by accurate definition, for the avoiding of confusion and obscurity" (DC 6.17, "The Author's Epistle to the Reader," n.p.). In one place, Hobbes characterizes Part III as concerning "the expansion of space, that is, geometry" (DC, "The Author's Epistle to the Reader," n.p.). But elsewhere he is more expansive. He writes: "Next [i.e., after the First Philosophy], those things which may be demonstrated by simple motion (in which Geometry consists). After Geometry, such things as may be taught or shewed by manifest action, that is, but thrusting from, or pulling towards" (DC 6.17). The final part contains the investigation of "the motion or mutation of the invisible parts of things, and the doctrine of sense and imagination" (ibid.). Unlike Parts II and III, which involve, in principle, only definitions and that which follows directly from definitions, in Part IV, "Physiques," Hobbes argues from physical effects observed by the senses to conjectured underlying causes, "the finding out by the appearances or effects of nature which we know by senses, some ways and means by which they may be (I do not say, they are) generated" (DC 25.1).

Part II does, indeed, contain a number of important definitions, including definitions of space (7.2), time (7.3), body (8.1), accident (8.2), place (8.5), motion (8.10), and rest (8.11), among other things. And it is in this context that Hobbes introduces certain general truths about bodies in motion. After offering his basic definitions, Hobbes advances a statement very much like Descartes' first law of nature: "Whatsoever is at rest, will always be at rest, unless there be some other body besides it, which, by endeavouring to get into its place by motion, suffers it no longer to remain at rest" (DC 8.19). This statement is defended as follows:

> For suppose that some finite body exist and be at rest, and that all space besides be empty; if now this body begin to be moved, it will certainly be moved some way; seeing therefore there was nothing in that body which did not dispose it to rest, the reason why it is moved this way is in something out of it; and in like manner, if it had been moved any other way, the reason of motion that way had also been in something out of it; but seeing it was supposed that nothing is out of it, the reason of its motion one way would be the same with the reason of its motion every other way, wherefore it would be moved alike all ways at once; which is impossible. (DC 8.19)

That is, if a body at rest were to begin to move, it would have to move in some direction or another, and there is no reason why it should move one way rather than another. And for a similar reason, Hobbes holds that a body in motion will remain in motion: "In like manner, *whatsoever is moved, will always be moved, except there be some other body besides it, which causeth it to rest*. For if we suppose nothing to be without it, there will be no reason why it should rest now, rather than at another time; wherefore its motion would cease in every particle of time alike; which is not intelligible" (DC 8.19). Here the argument is very similar: if a body in motion were to come to rest there is no reason why it should come to rest in any one moment in preference to any other moment. From this Hobbes infers a more general principle, that the only thing that can cause motion is another motion: "There can be no cause of motion, except in a body contiguous and moved" (DC 9.7).[20] Unlike Descartes' arguments, which appeal to a God who sustains the world from moment to moment, Hobbes appeals to something like a principle of sufficient reason.

These accounts of motion occur in Part II of *De corpore*, ostensibly about definitions. But in Part III, his "Geometry," Hobbes offers a general statement that looks a great deal like Descartes' second law. To understand that law we need to understand Hobbes's notion of endeavor (*conatus* in the Latin version): "I define ENDEAVOUR *to be motion made in less space and time than can be given; that is, less than can be determined or assigned by exposition or number*; that is, *motion made through the length of a point, and in an instant or point of time*" (DC 15.2). Despite appearances, this is not an infinitesimal motion:

> For the explaining of which definition it must be remembered, that by a point is not to be understood that which has no quantity, or which cannot by any

> means be divided; for there is no such thing in nature; but that, whose quantity is not at all considered, that is, whereof neither quantity nor any part is computed in demonstration; so that a point is not to be taken for an indivisible, but for an undivided thing; as also an instant is to be taken for an undivided, and not for an indivisible time. (DC 15.2)

Endeavor, then, is a genuine motion, the motion of a body through a finite (though "inconsiderable") distance in a finite time.

It is in terms of this notion of endeavor that Hobbes characterizes (without supporting argument) the motion of a body that is moved simultaneously by two different causes (motions):

> And whatsoever the line be, in which a body has its motion from the concourse of two movents, as soon as in any point thereof the force of one of the movents ceases, there immediately the former endeavour of that body will be changed into an endeavour in the line of the other movent. Wherefore, when any body is carried on by the concourse of two winds, one of those winds ceasing, the endeavour and motion of that body will be in that line, in which it would have been carried by that wind alone which blows still. (DC 15.5–6)

And from this he draws the following consequence:

> And in the describing of a circle, where that which is moved has its motion determined by a movent in a tangent, and by the radius which keeps it in a certain distance from the centre, if the retention of the radius cease, that endeavour, which was in the circumference of the circle, will now be in the tangent, that is, in a straight line. For, seeing endeavour is computed in a less part of the circumference than can be given, that is, in a point, the way by which a body is moved in the circumference is compounded of innumerable strait lines, of which every one is less than can be given; which are therefore called points. Wherefore when any body, which is moved in the circumference of a circle, is freed from the retention of the radius, it will proceed in one of those strait lines, that is, in a tangent. (DC 15.6; cf. DC 21.9)

This closely resembles what Descartes puts forward in his second law. As in the case of the earlier general statements about bodies in motion, this one does not involve God and his continual conservation. In this case it is taken to follow from an apparently self-evident principle about the combination of motions.

Hobbes only touches on the problem of collision, and unlike Descartes, does not really offer a developed account of impact, or any arguments for his account.[21] But even more significantly, Hobbes does not seem to present any kind of conservation principle at all that corresponds to Descartes' principle of the conservation of quantity of motion. This, presumably, cannot be done without God. Or, at least, Hobbes, I suspect, was unable to figure out how to do it without God. Or simply chose not to.[22]

There are a number of important ways in which this account of motion is like that of Descartes—and unlike Galileo—despite Hobbes's extravagant praise of the latter. Like Descartes, Hobbes is treating body as such, and not just heavy bodies: his statements are intended to follow in some way or another from the notions of body and motion, and to hold for all bodies. For Hobbes, as for Descartes, heaviness is not essential to body, but is only introduced later, after the general truths about body and motion are given. In the *De corpore*, heaviness appears as a physical phenomenon, to be explained in terms of a conjectured underlying physical mechanism, given in terms of bodies in motion that satisfy the constraints Hobbes had set out earlier in that work.[23]

But, in the context of our questions, are these general constraints on motion *laws*? And are they mathematical? In both cases it is not entirely clear what to think.

The general statements about motion are not called "laws" by Hobbes, unlike Descartes did in the *Principia philosophiae*. Descartes' work was published in 1644, and there is no doubt that Hobbes knew that publication, and knew it well. There are references to Hobbes's reaction to it in the correspondence in his circle, and direct references to it in the *De corpore*, though not explicitly by name. One can suppose that the avoidance of the term "law" in this connection was an explicit decision on Hobbes's part, one that was intended to express a difference between his view and Descartes'. And, as I noted earlier, unlike Descartes, Hobbes very self-consciously does not appeal to God in his account of natural philosophy in general, and these general statements about motion in particular. This, for him, was a matter of principle. In the *De corpore*, Hobbes argues explicitly that God can play no role in natural philosophy. He wrote:

> The *subject* of [natural] Philosophy, or the matter it treats of, is every body of which we can conceive any generation, and which we may, by any consider-

> ation thereof, compare with other bodies, or which is capable of composition and resolution; that is to say, every body of whose generation or properties we can have any knowledge . . . Therefore, where there is no generation or property, there is no philosophy. Therefore it excludes *Theology*, I mean the doctrine of God, eternal, ingenerable, incomprehensible, and in whom there is nothing neither to divide nor compound, nor any generation to be conceived. (DC 1.8)[24]

The question of God comes up in chapter 26 of *De corpore*, where Hobbes takes up the question of creation and the infinity of the world. Such questions, he argues, are beyond reason to resolve. He wrote:

> The questions therefore about the magnitude and beginning of the world, are not to be determined by philosophers, but by those that are lawfully authorized to order the worship of God. For as Almighty God, when he had brought his people into Judæa, allowed the priests the first fruits reserved to himself; so when he had delivered up the world to the disputations of men, it was his pleasure that all opinions concerning the nature of infinite and eternal, known only to himself, should, as the first fruits of wisdom, be judged by those whose ministry he meant to use in the ordering of religion. (DC 26.1)

The questions, in short, are theological and not philosophical. And so, he concludes: "Wherefore I purposely pass over the questions of infinite and eternal; contenting myself with that doctrine concerning the beginning and magnitude of the world, which I have been persuaded to by the holy Scriptures and fame of the miracles which confirm them; and by the custom of my country, and reverence due to the laws" (DC 26.1). Though he does not say so in the *De corpore*, I suspect that Hobbes's attitude to Descartes' grounding of the laws of nature would be the same, insofar as it requires us to know features of God, such as his immutability, that go beyond what we can know through reason.[25]

Another reason to wonder whether they are laws comes from their role in Hobbes's natural philosophy. For Descartes the laws are central constraints on the behavior of bodies as such: they are isolated as special principles, and designated as propositions of special importance. But while Hobbes presents these general statements about bodies in motion, he does not have the same ambitions for them. The general statements are presented almost in passing, in chapters entitled "Of Body and Accident," "Of Cause and

Effect," "Of the Nature, Properties, and divers considerations, of Motion and Endeavour." There is no sense of these as principles that are intended to structure nature, in any real sense.

Furthermore, their status is very close, if not identical, to that of geometrical truths. Here there is another contrast with Descartes. For Descartes, geometrical truths hold for extension as such, whether it is the extension of purely geometrical bodies that do not exist in the real world of created things, or for the objects of pure geometry, independent of real existence. But the laws of nature hold only for bodies that are created—and sustained—by God: they are truths that depend on God in a way in which geometrical truths do not.[26] But their status in Hobbes is rather different. As noted earlier, for Hobbes, natural philosophy begins in first philosophy, and first philosophy begins in definitions. After the definitions, though, "we should first demonstrate those things which are proximate to the most universal definitions (in which consists that part of philosophy which is called "First Philosophy"), and then those things which can be demonstrated through motion *simpliciter*, in which consists geometry" (DC 6.17).[27] Which is to say, these facts about motion are taken to be general truths about motion on a par with geometrical theorems, eternal truths of a sort, either things that follow directly from definitions, or what he calls geometry. It should be noted here that Hobbes's conception of geometry is somewhat idiosyncratic. Motion, for Hobbes, is part of the subject matter of geometry. Furthermore, for Hobbes geometry is just the science of extended body: unlike Descartes, he recognizes no radical distinction between geometrical bodies and physical bodies. Writing in the *Six Lessons to the Professors of the Mathematiques*, published in 1656 with the English translation of the *De corpore*, Hobbes writes: "there is no Subject of Quantity, or of Equality, or of any other accident but Body" (15). That is to say, all mathematical notions pertain to body. And this, for Hobbes, is especially true of motion. This, for Hobbes, determines the proper way of interpreting Euclidean geometry: "And by all these a man may easily perceive that Euclide in the definitions of a Point, a Line, and a Superficies, did not intend that a Point should be Nothing, or a Line be without Latitude, or a Superficies without Thickness . . . For Lines are not drawn but by Motion; and Motion is of Body only" (Hobbes 1656, 9). For Hobbes, in short, the objects of geometry are bodies, strictly speaking, and there is no real distinction between natural philosophy and geometry, at least at the level of the general and foundational part of natural philosophy.[28]

That said, there remains some uncertainty about how to understand what Hobbes is doing. Hobbes certainly did not advance laws in the sense that Descartes did. But can we say that these general assertions about the behavior of bodies in motion are not laws of nature, strictly speaking? One might make such a judgment on the basis of a philosophical conception of what constitutes a law of nature, perhaps. But I now feel somewhat reluctant to do so. Any such philosophical conception would seem to be a priori and perhaps a bit arbitrary, and certainly historically suspect. To look at these Hobbesian texts at the moment when the notion of a law of nature in this sense is just being articulated suggests to me that there may not be a clear answer to this question. And is Hobbes's account of motion mathematical in any sense? Well, it is certainly mathematical in the *Hobbesian* sense: these general statements about bodies in motion are part of mathematics as *Hobbes* understood it. But it is hard to ignore the fact that Hobbes's conception of mathematics is highly idiosyncratic, hardly a conception that we would recognize as mathematical. In this respect, the situation with respect to Hobbes is quite different from the situation with respect to Galileo, whose geometrical treatment of bodies in motion is mathematical in a very classical sense. Which is to say, it is unclear whether Hobbes's account of motion involves laws, and whether we should say that it is mathematical.

CONCLUSIONS

And so, in the end, the relation between the mathematization of nature and the discovery of the laws that govern the natural world would seem to be more complicated than expected. While there are, no doubt, ways in which they are connected, they are also in many ways independent of one another, as the cases of Descartes, Galileo, and Hobbes show. More generally, I think that there is a temptation to suppress the complexity of the so-called scientific revolution of the early modern period in favor of a simpler narrative. In broad brush, there are a number of important developments in the period, including the mathematization of nature and the development of laws, which we have examined here, but also the development of mechanical models, the growth of experimentalism, the invention of new instruments, including the microscope and telescope, the foundation of new institutions, such as the Royal Society and the Académie royale des sciences, the development of the learned journal as a means of communication, among other innovations. (There are also many "innovations" that arose in the period that did not survive.) There is a strong temptation to think that these different elements march together to produce something that we can call The New Science, which replaced the Aristotelian natural philosophy that dominated the intellectual world in the period before. When we look at these more carefully, I think that we will realize that the transition to modern science was much more complicated than that.[29] But that is an argument for another occasion.

ABBREVIATIONS

AT Descartes, R. 1964–76. *Oeuvres de Descartes*

DC Hobbes, T. 1655. *De corpore* (Quotations will be taken from the English translation in Hobbes 1656. Citations reference chapter and section where possible)

EN Galileo, G. 1890. *Le opere di Galileo Galilei*

PP Descartes, R. *Principia philosophiae*, in AT 8A (Latin) and AT 9B (French), with a partial English translation in Descartes 1985–91, vol. 1 and a full English translation in Descartes 1983

TCWS Galileo, G. 1967. *Dialogue Concerning the Two Chief World Systems*

NOTES

I would like to thank the participants in the discussion of my paper at the conference, "The Language of Nature," for a lively discussion. I would especially like to thank the organizers, Geoff Gorham, Ed Slowik, Ben Hill, and Ken Waters, both for organizing the conference and for the very detailed comments on my essay. I owe a special debt to Ursula Goldenbaum, my commentator during the conference session, both for her comments and for the very helpful exchanges we had on the paper after the conference. Because of all of these interventions, this paper is much changed, and, I hope, much improved from the version that I had originally submitted.

1. The pioneering study of the history of the laws of nature is Zilsel (1942). Another important earlier study is Milton (1981). Milton develops his views further in Milton (1998). For more recent studies see, for example, Steinle (1995), Roux (2001), Henry (2004), and the essays collected in Daston and Stolleis (2008). Dana Jalobeanu has emphasized to me for many years that the view of nature as governed by overarching law is a central feature of Stoic thought. See her essay (Jalobeanu n.p.). Henry (2004, 79), in defending the priority of Des-

cartes on the laws of nature, makes an important contrast between laws of nature as "merely references to the regularity of nature," as opposed to "the concept of a law of nature as a specific and precise statement which codifies observed regularities in nature but which is also assumed to denote an underlying causal connection, and therefore can be said to carry explanatory force."

2. See references cited above in note 1.

3. All translations from Descartes are taken from Descartes 1985–91, except where otherwise noted.

4. For a discussion of Descartes' account of collision, see Garber (1992, chapter 8).

5. On directionality (what Descartes calls determination) see Garber (1992, 188ff). On the conservation principle, see Garber (1992, 204ff).

6. The argument here is that light is pressure in the ether, which derives from the rotation of the vortices around each sun by way of the second law.

7. On this see Descartes, *Le monde*, chap. 11 (AT 11:72ff) and PP 4.23ff.

8. Cf. Descartes to Plempius, October 3, 1637 (AT I 410–11 and 420–21).

9. For a discussion of Descartes' attempts to deal with the problem of free fall, see Koyré (1978) and Jullien and Charrak (2002).

10. Translated in Descartes (1983).

11. Translated in Galilei (1957, 182). Thanks to Ursula Goldenbaum for pointing out this passage to me.

12. References to the *Discorsi* will be given in EN 8. All translations are from Galilei (1974). Since this translation is keyed to the pagination in EN 8, I won't cite it separately.

13. The argument goes roughly as follows: In uniformly accelerated motion as Galileo defines it, the speed is proportional to the time. Consequently the terminal speed is proportional to the time. But by the so-called mean-speed theorem, proven in prop. I theorem I (EN 8:208ff), a body uniformly accelerated (on Galileo's definition) will move a distance in a given time equal to the distance that it would go in the same time were it moving at half the terminal speed. So the distance fallen is proportional to one half of the terminal speed times the time. But by the definition of uniform acceleration, the terminal speed is proportional to the time. And so the distance fallen is proportional to the square of the time.

14. At least one of Galileo's readers—René Descartes—was not impressed and could not find the patience to read them: "I shall say nothing of the geometrical demonstrations of which the book is full, for I could not summon the patience to read them, and I am prepared to believe they are all correct. When

looking at his propositions, it simply struck me that you do not need to be a great geometrician to discover them; and he does not always take the shortest possible route, which leaves something to be desired (Descartes to Mersenne, October 11, 1638, AT 2 388).

15. On the notion of "inertia" in the early seventeenth century, see Garber (1992, 253ff).

16. Translated in Galilei (TCWS 20). Cf. EN 7 56, TCWS 31–32.

17. Translated in Galileo (1957, 63), slightly altered, as discussed in the following note.

18. A key question, of course, is the proper translation of the term "*sintomi*" in Galileo. In the previously cited passage, Drake translates it as "laws." This seems clearly wrong. Galileo does not use the term often, but in two passages he pairs it with "*accidenti*," suggesting that they are virtual synonyms. See *Intorno alle cose che stanno in su l'acqua . . .* (EN 4 115) and *Discorso del flusso e reflusso del mare . . .* (EN 5 377). It also appears in the *Letters on Sunspots*, where Reeves and Van Helden translate it as "characteristics." (EN 5 117, Galilei and Scheiner [2010]. See also EN 4 698 and EN 7 189 for similar uses of the term.) I am grateful for help on this tricky issue from Eileen Reeves.

19. Throughout, the English is quoted from the anonymous English translation (Hobbes 1656).

20. Though Hobbes goes on at some length in 9.7 to establish this, so far as I can see it follows pretty directly from the considerations in 8.19.

21. See DC 15.8, for example. Hobbes's account of collision is made particularly complicated by the fact that only motion resists motion, so that a body at rest does not resist the acquisition of new motion. For a very helpful discussion of what amounts to the elements of Hobbes's account of impact in *De corpore*, see Morris (2007).

22. In his essay, Gorham (2013) presents a somewhat different view of Hobbes's physics and its relation to God. Gorham takes seriously Hobbes's various statements that God is body, arguing that for Hobbes, God is a fluid body that infuses the universe, "most pure, most simple corporeal spirit" (Gorham 2013, 254). He then argues that God, understood in this way, "is the perpetual source of motion, and hence diversity, in a material world governed by mechanical principles" (252). This, in part, would seem to address the problem in Hobbes's natural philosophy that "any motion in the world must dissipate in no time, like shock waves" (251). And so, Gorham suggests, Hobbes's physics, like Descartes', would seem to involve divine sustenance and a kind of conservation principle. This is a fascinating suggestion. But it is worth pointing out a few

things. First of all, even if the material God supports motion in this way, Gorham does not suggest that the general statements about motion that correspond to Descartes' laws are in any way derived from God as cause of motion. Second, even though we might see the appeal to a material God in the passages that Gorham cites as addressing the continual diminution of motion in the world, there is no place where Hobbes articulates a conservation principle. And finally, the support for Gorham's position is almost exclusively in texts significantly later than the 1655 publication of the *De corpore*, particularly the "Answer to Bramhall," written probably in 1668 but not published until 1682, after Hobbes's death. My focus in this text is on the doctrine in the much more widely read and influential *De corpore*.

23. See the account of heaviness in DC 30.

24. Gorham (2013) would disagree with this, of course. See note 22.

25. Here, again, we may be dealing with a view that Hobbes gave up in his later writings, if Gorham (2013) is right in its interpretation.

26. This is actually a bit subtle since even geometrical truths for Descartes depend on God by way of his (in)famous doctrine of the creation of the eternal truths. But geometrical truths depend on God for their creation, as do all eternal truths, while the laws of nature depend on God in his moment-by-moment sustenance of bodies in the material world.

27. My translation of the 1655 Latin.

28. On this, see Jesseph (1999), chapter 3. There is, however, a distinction between the project of Parts II and III of the *De corpore*, his first philosophy and his geometry, which are grounded in definitions and what can be drawn from definitions, and the project of Part IV, which he calls physics proper, which involves conjectured mechanisms.

29. I have tried to sketch out an alternative model of intellectual change in the period in Garber (2016).

REFERENCES

Daston, L., and M. Stolleis, eds. 2008. *Natural Law and Laws of Nature in Early Modern Europe: Jurisprudence, Theology, Moral and Natural Philosophy*. Burlington, Vt.: Ashgate.

Descartes, R. 1964–76. *Oeuvres de Descartes*. Ed. C. Adam and P. Tannery. 11 vols. 2nd. ed. Paris: Vrin.

———. 1983. *Principles of Philosophy*. Trans. V. Miller and R. Miller. Dordrecht: D. Reidel.

———. 1985–91. *The Philosophical Writings of Descartes.* Ed. and trans. J. Cottingham, R. Stoothoff, D. Murdoch, and A. Kenny (vol. 3). 3 vols. Cambridge: Cambridge University Press.

Galilei, G. 1890. *Le opere di Galileo Galilei.* Ed. A. Favaro. 20 vols. Florence: G. Barbèra.

———. 1957. *Discoveries and Opinions of Galileo.* Trans. S. Drake. Garden City, N.Y.: Doubleday.

———. 1967. *Dialogue Concerning the Two Chief World Systems.* Trans. S. Drake. Berkeley: University of California Press.

———. 1974. *Two New Sciences.* Trans. S. Drake. Madison: University of Wisconsin Press.

Galilei, G., and C. Scheiner. 2010. *On Sunspots.* Ed. and trans. E. Reeves and A. Van Helden. Chicago: University of Chicago Press.

Garber, D. 1992. *Descartes' Metaphysical Physics.* Chicago: University of Chicago Press.

———. 2000. "A Different Descartes: Descartes and the Program for a Mathematical Physics in the *Correspondence.*" In *Descartes' Natural Philosophy,* ed. S. Gaukroger, J. Schuster, and J. Sutton, 113–30. London: Routledge.

———. Forthcoming. "Why the Scientific Revolution Wasn't a Scientific Revolution." In *Kuhn's Structure of Scientific Revolutions at Fifty,* ed. R. Richards and L. Daston. Chicago: University of Chicago Press.

Gorham, G. 2013. "The Theological Foundation of Hobbesian Physics: A Defence of Corporeal God." *British Journal for the History of Philosophy* 21: 240–61.

Henry, J. 2004. "Metaphysics and the Origins of Modern Science: Descartes and the Importance of Laws of Nature." *Early Science and Medicine* 9: 73–114.

Hobbes, T. 1655. *Elementorum philosophiae section primia de corpore.* London: Andrew Crook.

———. 1656. *Elements of Philosophy, the First Section Concerning Body.* London: Andrew Crook. [This is an anonymous English translation of Hobbes 1655.]

———. 1656. *Six Lessons to the Professors of the Mathematiques.* London: Andrew Crook.

———. 1660. *Examinatio et Emendatio Mathematicae Hodiennae, Qualis Explicatur in Libris Johannis Wallisii,* 53. London: Andrew Crooke.

Jalobeanu, D. Unpublished. "Order and Laws of Nature in the Stoic Tradition."

Jesseph, D. 1999. *Squaring the Circle: The War between Hobbes and Wallis.* Chicago: University of Chicago Press.

———. 2004. "Galileo, Hobbes and the Book of Nature." *Perspectives on Science* 12: 191–211, esp. 203.

Jullien, V., and A. Charrak. 2002. *Ce que dit Descartes sur la chute des graves.* Villeneuve-d'Ascq: Presses universitaires du septentrion.

Koyré, A. 1978. *Galileo Studies.* Atlantic Highlands, N.J.: Humanities Press.

Milton, J. R. 1981. "The Origin and Development of the Concept of the Laws of Nature." *Archives Européennes de Sociologie* 22: 173–95.

———. 1998. "Laws of Nature." In *Cambridge History of Seventeenth Century Philosophy,* ed. D. Garber and M. Ayers, 680–701. Cambridge: Cambridge University Press.

Morris, K. 2007. "Descartes and Hobbes on the Physics and Metaphysics of Resistance." In *Descartes and the Modern,* ed. N. Robertson, G. McOuat, and T. Vinci, 103–26. Cambridge: Cambridge Scholars Publishing.

Roux, S. 2001. "Les lois de la nature aux XXVIIe siècle: le problème terminologique." *Revue de synthèse* 4e Series, 2–4: 531–76.

Steinle, F. 1995. "The Amalgamation of a Concept—Laws of Nature in the New Sciences." In *Laws of Nature: Essays on the Philosophical, Scientific and Historical Dimensions,* ed. F. Weinert, 316–68. Berlin: Walter de Gruyter.

Zilsel, E. 1942. "The Genesis of the Concept of Physical Law." *Philosophical Review* 51: 245–79.

6

RATIOS, QUOTIENTS, AND THE LANGUAGE OF NATURE

DOUGLAS JESSEPH

In his 1623 essay "The Assayer," Galileo notoriously claimed that the "book of nature" was written in the language of mathematics.[1] Yet when we consider his actual formulation of the laws of nature (most notably the law of free fall in the *Two New Sciences*) it becomes apparent that he took the language of mathematics to be something rather different than the mathematical formulations we typically use today. As is well known, Galileo used the Euclidean-Eudoxian language of proportions to express the law of free fall, formulating it in terms of ratios between distances and the squares of elapsed times, rather than as a second-degree equation linking distance covered to elapsed time.[2] Because the 1637 publication of Descartes' *Géométrie* marked the first appearance of analytic geometry, it is no surprise that Galileo did not employ its techniques to state his results. Nevertheless, the Galilean preference for the traditional language of ratios and proportions reminds us that the mathematics employed by seventeenth-century natural philosophers is, in many cases at least, firmly rooted in classical Greek doctrines. Even Newton, who developed his calculus of fluxions some two decades before the publication of the *Principia*, chose to develop his celestial mechanics in the classical language of ratios and proportions drawn from book 5 of the Euclidean *Elements*.[3]

My purpose here is to investigate a seventeenth-century dispute over how best to interpret the classical doctrine of ratios and to link these differences to alternative programs for mechanics. In particular, I wish to focus on the doctrines of Isaac Barrow (first Lucasian Professor of Mathematics at Cambridge) and John Wallis (Savilian Professor of Geometry at Oxford from 1649 until his death in 1703). These two professors took quite different approaches to the account of ratios developed in the Euclidean *Elements*. Wallis identified ratios with quotients arising from division, seeking thereby to place the theory of ratios within a very general algebraic theory that he identified as the *mathesis universalis*, or universal mathematics.[4] Barrow, in contrast, insisted that the doctrine of ratios could only be properly understood when it was taken as grounded in essentially geometric concepts with no algebraic content. A significant part of this dispute dealt with the prospects of applying the theory of ratios to the study of the behavior of bodies in motion. In Part I of his 1670 *Mechanica*, Wallis insisted that the only way to develop a truly general mechanics was to follow the algebraic approach. In contrast, Barrow insisted that such physico-geometrical concepts as space, body, and motion were the only appropriate foundation for a mathematics that could be applied to nature. I begin with a general overview of the classical account of ratio and proportion, then turn to a consideration of Wallis's and Barrow's interpretations of the theory. I close by examining the connection between the theory of ratios and the foundations of mechanics, focusing primarily on Wallis's *Mechanica*.

THE CLASSICAL THEORY OF RATIOS

Seventeenth-century treatments of the theory of ratio and proportion all arise from the interpretation of a series of definitions in Euclid's *Elements*. The relevant definitions from book 5, which introduced the concepts of ratio and proportion, are these:

> 3. A *ratio* is a sort of relation in respect of size between two magnitudes of the same kind.
> 4. Magnitudes are said to *have a ratio* to one another which are capable, when multiplied, of exceeding one another.
> 5. Magnitudes are said to *be in the same ratio*, the first to the second and the third to the fourth, when, if any equimultiples whatever be taken of the first and third, and any equimultiples whatever of the second and fourth, the former equimultiples alike exceed, are alike equal to, or alike fall short of, the latter equimultiples respectively taken in corresponding order.
> 6. Let magnitudes which have the same ratio be called *proportional*. (*Elements*, bk. 5, defs. 3–6)[5]

Thus understood, a ratio is not a quotient formed by the division of one number by another, but rather a relation that holds between geometric magnitudes. Magnitudes are grouped into species or kinds, and the third and fourth definitions guarantee that it is only within species that a ratio can be constructed or magnitudes compared.

To take an example: lines, angles, surfaces, and solids are fundamentally distinct kinds of magnitudes, and there is no way to compare directly the magnitude of one kind (a line, say) with the magnitude belonging to another (such as an angle). This is because no number of lines could ever exceed an angle, as required by the fourth definition. To inquire into how many lines might amount to an angle is a nonsense question, on par with seeking to determine how many potatoes could equal a symphony. Thus, there is no "relation in respect of size" holding between heterogeneous magnitudes. Nevertheless, definitions 5 and 6 *do* permit the comparison of ratios across species of magnitude in a proportion, so it makes sense to say that the ratio $L_1 : L_2$ between the length of two lines is the same as the ratio $V_1 : V_2$ between the volumes of two spheres. In other words, the proportion $L_1 : L_2 :: V_1 : V_2$ is legitimate, even though the magnitudes V_1 or V_2 cannot be directly compared with L_1 or L_2. Likewise, the definition of equality of ratios (definition 5) does not assert that $\alpha : \beta :: \gamma : \delta$ whenever $\alpha \times \delta = \gamma \times \beta$ because the relevant magnitudes may be heterogeneous and incapable of being multiplied together. Instead, sameness of ratio is defined in definition 5 by the preservation of order relations under arbitrary equimultiples.

Although the classical theory of ratios had an impeccable Euclidean pedigree and was often put forward as a paradigm of rigorous mathematics, a great many seventeenth-century authors sought to introduce an alternative understanding of ratios. Much of the motivation for moving beyond the Euclidean scheme arose from concerns about the status of definition 5: it seemed too prolix and intricate to be a true first principle of geometry, and although its truth was never challenged it was thought that there must be simpler and more elegant principles from which the theory of ratios could be developed.[6] This alternative approach to ratio and proportion can be usefully termed the "numerical" treatment of ratios, in contrast with the classical "relational" theory.

The fundamental difference between these two approaches can be brought to light by asking whether ratios themselves are quantities; that is, things that can be greater or less. According to the relational theory, the answer is no: ratios are not quantities, but rather relations that hold between two quantities. Just as it would be nonsense to assert that such numerical relations as "greater than" or "divisible by" are themselves some sort of number or magnitude, the relational theory of ratios holds that a ratio is radically distinct from the quantities that stand in a ratio. From the standpoint of the numerical theory, however, it makes perfect sense to say that one ratio could be greater than another. On the numerical understanding, each ratio is taken to have a size (or "denomination" or "exponent"), and the sameness of two ratios amounts to their having the same size, denomination, or exponent. This approach assimilates ratios into a general domain of magnitudes, and it avoids the complex Euclidean definition of the sameness of ratio in terms of the preservation of order relations under arbitrary equimultiples.

In point of fact, the Euclidean doctrine admits the comparison of ratios as to greater and less, which makes it seem plausible that ratios themselves should count as quantities. In the seventh definition of *Elements* book 5, Euclid states, "When, of the equimultiples, the multiple of the first magnitude exceeds the multiple of the second magnitude, but the multiple of the third magnitude does not exceed the multiple of the fourth, then the first is said to have a greater ratio to the second than the third has to the fourth." So, the ratio 5:3 exceeds the ratio 7:8, because by multiplying the first and third terms by 2 and the second and fourth terms by 3, we discover that $(5 \times 2) > (3 \times 3)$ while $(7 \times 2) < (8 \times 3)$. But if anything capable of a "greater than" comparison is a quantity in its own right, there seems to be a solid case for attributing quantities to ratios.

Notwithstanding its appealing simplicity, the numerical theory nevertheless faces its own difficulties. It is natural to assume that the criterion for sameness of ratio in the numerical theory should be expressed in the principle that the ratio $\alpha : \beta$ is the same as the ratio $\gamma : \delta$ just in case $\alpha \times \delta = \gamma \times \beta$. However, if the quantities α and δ are different species of magnitudes, there is no clear sense to be made of the notion that they could be multiplied together. Indeed, this is precisely why the Euclidean definition requires that two magnitudes have a ratio to one another only if each can exceed the other by being multiplied.

The principal consequence here is that the numerical theory requires all ratios to be homogeneous, or capable of direct comparison with one another. One natural way to do this would be to characterize the denomination of a ratio as a quotient formed by dividing the antecedent of the ratio by its consequent. But doing this raises the difficulty of understanding how the quotient of two incommensurable magnitudes can be understood. Classically conceived, the quotient is a fraction that arises from the division of integers—a fact reflected in the etymological observation that the root of the term is the Latin *quoties*, or "how many." In effect, quotients are simply rational numbers that express how many common units of the denominator are contained in the numerator. Incommensurable magnitudes cannot, of course, be understood as quotients in this sense, so the numerical theory of ratios seems committed to expanding the traditional concept of a quotient to include quotients of irrational magnitudes, that is, to making sense of expressions such as $\sqrt{7}\pi / \sqrt[3]{11}$. As a result, the development of the numerical theory of ratios required a fundamental reconsideration of the traditional concept of number, namely one that expands the traditional Greek notion of number (ἁριτημόζ), conceived as a collection of units, to include all magnitudes in an abstract general theory of magnitudes that is fundamentally algebraic.[7]

The differences between these two approaches to the Euclidean theory of ratios came into sharper focus in the thorny issue of compounding ratios—a much-disputed point that traces back to a pseudo-Euclidean definition that appeared in some editions of book 6. The (now generally regarded as spurious) fifth definition of book 6 reads, "A ratio is said to be *compounded of ratios* when the sizes [πηλικότητεζ] of the ratios multiplied together make some (ratio, or size)." The definition is unusual for its apparent reference to the 'sizes' of the compounded ratios, as well as its employment of the arithmetical operation of multiplication to construct a ratio from the sizes of two given ratios. None of this makes any sense on the relational theory, because a ratio is not a quantity and there is no sense in which it might have a size. For these reasons the definition is now regarded as a late emendation to the Euclidean text, although it was still regarded as canonical in the seventeenth century.[8] This fact led to some creative interpretations of the text. Barrow noted that "ratios, as they lack all quantity, can neither be added nor multiplied,"[9] and took this to indicate that the Greek term πηλικότητεζ should be understood as indicating the quantities contained in the compounded ratios rather than any quantities pertaining to the ratios themselves. A further oddity with this definition is the fact that it is never used in the *Elements*, even in the one place where Euclid speaks of compounded ratios (*Elements* 6 23).

Whatever conceptual problems the definition may present for the relational theory of ratios, it makes perfect sense on the numerical theory of ratios. If ratios have sizes, or if they are identified with quotients, then there is no obstacle to accepting the notion that multiplication of two ratios can form a third. For instance, if the ratios 3 : 8 and 9 : 11 are compounded, the new ratio will arise from the multiplication of the quotients 3/8 and 9/11, yielding 27 : 88 as the compounded ratio.

The two competing accounts of the nature of ratios did not originate in the seventeenth century; in fact, the differences between the relational and numerical theory were discussed among medieval authors.[10] For my purposes, however, the most important exchanges over this topic occurred in the seventeenth century in the works of John Wallis and Isaac Barrow. It is to an analysis of their doctrines that I now turn.

WALLIS AND THE NUMERICAL THEORY OF RATIOS

Wallis was one of the most prominent advocates of the numerical theory of ratios. He argued for it at length in his 1657 treatise *Mathesis Universalis*, which originated as Savilian lectures and is largely devoted to making the case that the principles of geometry are subordinate to those of arithmetic.[11] Wallis was, in fact, a proponent of a view I term "algebraic foundationalism," according to which all of geometry can and should be developed from arithmetical principles, which in turn can be shown to be special cases of more fundamental principles of algebra, or the "arithmetic of species."[12] In other words, Wallis held that algebraic theory is the proper foundation for all of mathematics.

In the *Mathesis Universalis* Wallis argued that geometrical results can be achieved more perspicuously and naturally by the use of arithmetical arguments. In service of this goal he devoted the twenty-third chapter to a series of "arithmetical" demonstrations of results from the second book of Euclid's *Elements*, an enterprise he took to illustrate his contention that the important results in geometry are ultimately based on arithmetical principles. He argued:

> Because some take the geometric elements for the basis of all mathematics, they even think that all of arithmetic is to be reduced to geometry, and that there is no better way to show the truth of arithmetical theorems than by proving them from geometry. But in fact arithemtical truths are of a higher and more abstract nature than those of geometry. For example, it is not because *a two foot line added to a two foot line makes a four foot line* that *two*

> *and two are four*, but rather because the latter is true, the former follows. (MU 11; OM 1 53)

This led Wallis to conclude:

> The close affinity of arithmetic and geometry comes about, rather, because geometry is as it were subordinate to arithmetic, and applies universal principles of arithmetic to its special objects. For, if someone asserts that a line of three feet added to a line of two feet makes a line five feet long, he asserts this because the numbers two and three added together make five; yet this calculation is not therefore geometrical, but clearly arithmetical, although it is used in geometric measurement. For the assertion of the equality of the number five with the numbers two and three taken together is a general assertion, applicable to any other kinds of things whatever, no less than to geometrical objects. For also two angels and three angels make five angels. And the very same reasoning holds of all arithmetical and especially algebraic operations, which proceed from principles more general than those in geometry, which are restricted to measure. (MU 11; OM 1 156)

The remark that "especially algebraic operations" are abstract and apply to "any kinds of things whatever" indicates Wallis's notion of algebra as a highly general science of quantity with no specific connection to any specific kind of number, magnitude, or measure.

This doctrine leads quite naturally to the numerical theory of ratios. At the very least, the project of interpreting all of mathematics as essentially algebraic is helped along by reducing the entire theory of ratios to a special case of arithmetic, which in turn happens to be a special case of algebra. As Wallis saw the matter, the comparison of magnitudes in ratios renders all ratios homogeneous. In his words: "Where a comparison of quantities according to ratio is made, it frequently happens that the ratio of the compared quantities leaves the genus of magnitude of the compared quantities and is transferred into the genus of number, whatever that genus of the compared quantities may be . . . And this is the principal reason I affirm that the doctrine of ratios belongs rather to the speculations of arithmetic than geometry" (MU 25; OM 1 136). Wallis's reasoning can be summarized as follows: when we construct a proportion between two pairs of magnitudes, we have established that the two ratios are of the same size. But the only way to compare things together in regard to their size is to have a common measure of their sizes. Therefore, there must be some common measure for all ratios, which requires that they be instances of a very general concept of number, more abstract than the traditional Euclidean definition of a number (άριτημόζ) as simply a collection of units.

In fact, Wallis argued that Euclid's treatment of ratios in the fifth book of the *Elements* should be demonstrated "arithmetically," and in chapter 35 of the *Mathesis Universalis* he undertook precisely this task. Not surprisingly, Wallis took Euclid's definition of the sameness of ratios to be defective, and declared:

> We have thought it fit to omit this definition from our demonstrations, although it is indeed true and well enough suited to Euclid's purpose, nor do we examine proportionals according to this criterion. And indeed it seems somewhat complex, and perhaps not perspicuous enough—especially to learners—nor indeed does it immediately respect the nature of proportionals, but rather some remote affection of them. But for us, who earlier judged ratios by how much, it seems sufficient to prove the identity or equality of ratios if there is an equality or identity of quotients. So, for instance, if $a/\alpha = b/\beta$, then $a : \alpha :: b : \beta$, and *vice versa*. (MU 35; OM 1 184)

Wallis's thoroughgoing identification of ratios with quotients is equally apparent in his approach to the question of compounding ratios. He held that the obscurities surrounding the fifth definition of book 6 could be set aside by showing that Euclid himself accepted the numerical theory of ratios. In a Savilian lecture from 1663 (later published in the second volume of his *Opera Mathematica* of 1693) Wallis undertook to make this case. He first argued that the Euclidean definition of the term 'ratio' must be given a slightly different interpretation than the tradition had accorded it. In particular, he held that Euclid's "sort of relation in respect of size between two magnitudes of the same kind" (*Elements*, book 5, def. 3) must be rephrased as "a ratio is that relation or habitude of homogeneous magnitudes to one another in which it is shown how the one is to the other, considered according to quantuplicity" (OM 2 665).

The neologism 'quantuplicity' is Wallis's term for how much one magnitude is in comparison with another, or how many times the one is contained in the other. Specifically, he intended to allow relations of quantuplicity that cannot be expressed as ratios of integers, so that the diameter of a circle would be the $1/\pi$th quantuple part of the periphery. Given this understanding of ratios, the composition of ratios is a simple matter: each ratio has an "exponent" that indicates its quantity, which is the quotient arising from the division of the antecedent by the consequent. Thus, according to Wallis, definition 5 of book 6 should be understood to say "a ratio is said to be compounded of ratios when the exponents of the ratios multiplied together make the exponent of that ratio" (OM 2 666). Having considered Wallis's exposition of the numerical theory of ratios, we can now turn to Barrow's response.

BARROW IN DEFENSE OF THE RELATIONAL THEORY

Barrow termed the theory of ratios "the very soul of mathematics," because he saw in it a doctrine "on which nearly everything remarkable and abstruse demonstrated in mathematics ultimately depends" (LM 16; MW 1 252). His *Lectiones Mathematicae* originated as Lucasian lectures in the 1660s and were principally concerned with defending Euclid's account of ratios. In point of fact Barrow understood that his defense of the classical doctrine is an essential part of a broader program to see geometry established as the one true foundation for all of mathematics. Barrow's announced purpose in vindicating Euclid was to show that "there is nothing in the whole work of the *Elements* more subtly found out, more solidly established, or more accurately treated than this whole doctrine of proportions" (LM 23; MW 1 378). In the end, I think that this defense of the classical approach to ratios is due in large part to Barrow's conception of geometric demonstration as founded in the consideration of true causes that are best understood by attending to the motions by which geometric magnitudes are produced.

Barrow viewed any departure from Greek tradition with suspicion, and he spent many pages of the *Lectiones Mathematicae* defending the relational doctrine of ratios against its modern rivals.[13] Indeed, he asked his audience to "pardon my contentiousness, and not hold it against me that I have been led by a certain piety to undertake to vindicate the father and prince of geometry from the undeserved reproaches that are heaped upon him from every side" (LM 18; MW 1 283).

Barrow's response to Wallis was to turn the tables on his opponent and mount a case for the primacy of geometry over arithmetic. One of the main themes of the *Lectiones Mathematicae* is Barrow's argument for what I call "geometric foundationalism," or the view that all of mathematics is ultimately based on geometric concepts and principles. Barrow argued that arithmetic lacked the kind of determinate content necessary to found a true science. "Any number at all," he declared, "may with equal right denote and denominate any quantity" (LM 3; MW 1 51). Barrow's point here can be illustrated as follows: a given line may be deemed one, one hundred, or one thousand, depending on whether we divide it into meters, centimeters, or millimeters.

Barrow responded to Wallis's argument that the arithmetical fact that $2+2=4$ is too general to be based solely on geometry, from which he had concluded that geometry must be founded on arithmetical or algebraic concepts. To this reasoning, Barrow retorted:

> I respond by asking, How does it happen that a line of two feet added to a line of two palms does not make a line of four feet, four palms, or four of any denomination, if it is abstractly, i.e. universally and absolutely true that two plus two makes four? You will say, This is because the numbers are not applied to the same matter or measure. And I would say the same thing, from which I conclude that it is not from the abstract ratio of numbers that two and two make four, but from the condition of the matter to which they are applied. This is because any magnitude denominated by the name *two* added to a magnitude denominated *two* of the same kind will make a magnitude whose denomination will be *four*. Nor indeed can anything more absurd be imagined than to affirm that the proportions of magnitudes to one another depend upon the relations of the numbers by which they are expressed. (LM 3; MW 1 53)

Consequently, in Barrow's view, there is no arithmetical fact without the specification of a unit, but such a specification is too arbitrary to be the basis of a proper science. This led him to conclude that "mathematical number is not some thing having existence proper to itself, and really distinct from the magnitude which it denominates, but is only a kind of note or sign of magnitude considered in a certain manner" (LM 3, MW 1 56). The result is that "number (at least that which the mathematician contemplates) does not differ in the least from that quantity which is called continuous, but is formed wholly to express and declare it. And neither are arithmetic and geometry conversant about diverse matters, but equally demonstrate properties common to one and the same subject, and from this it will follow that many and great advantages derive to the republic of mathematics" (LM 3; MW 1 47). Barrow took this argument for geometrical foundationalism to the extreme of denying that algebra is a mathematical science at all. In his estimation algebra fails to qualify as an independent science because it is at best a fragment of logic, and at worst a collection of purely formal rules for the manipulating symbols. So, where Wallis and others took algebra to be a highly abstract science that took for its object quantity in general, Barrow dismissed it as unscientific "because it has no object distinct and proper to itself, but only presents a kind of artifice, founded on geometry (or arithmetic), in which magnitudes and numbers are designated by certain notes or symbols, and in which their sums and differences are collected and compared" (LM 2; MW 1 46).

These considerations support Barrow's central objection to the numerical theory of ratios, namely that Wallis is guilty of a kind of category mistake in thinking that ratios are quantities that can be studied by an abstract, algebraic science of quantity. Because a ratio is a "pure, perfect relation," it cannot "pass into another category and become a genus of quantity" (LM 20; MW 1 318). In other words, to treat a ratio itself as a quantity is to confuse a relation with one or another of its relata—the ratio is a way for two quantities to be compared, but it cannot itself be a quantity. Barrow admitted that the classical theory of ratios permits such locutions as "the ratio $\alpha : \beta$ exceeds the ratio $\gamma : \delta$" (in accordance with definition seven of book 5 of the *Elements*). However, he held that this can be understood without requiring ratios to be quantities. Rather, such expressions arise whenever the antecedent of one ratio exceeds the antecedent of another, *provided that the ratios have common consequents*. He explained: "Whatever is commonly attributed to ratios, only truly and properly agrees with the denominators of ratios, that is, to their antecedents reduced to a common consequent. The quantity that others assign to ratios is nothing other than the quantity and ratio of the denominators, and when they think they add or subtract ratios themselves, they only add or subtract these denominators, and this is the same thing when they multiply or compound, divide or resolve them" (LM 20; MW 1 315). The methodological picture that emerges from these considerations is fairly straightforward. Geometry is the foundational science for all of mathematics, in the sense that every mathematical truth is ultimately analyzable as a statement about the properties and relations of continuous magnitudes. The correct method for investigating these properties and relations is by constructions carried out in accordance with the definitions, axioms, and postulates of Euclidean geometry. Such constructions will typically aim to establish ratios and proportions, which constitute the "very soul" of mathematics. Further, these constructions will constitute demonstrations that proceed from true causes, and Barrow devoted the sixth of his *Lectiones Mathematicae* to establishing the "causality" of geometric demonstrations. The causality he has in mind "can be called *formal* causality, from the fact that from one property first taken as given, the remaining affections [of a geometric object] arise as from a form" (LM 6; MW 1 93).[14] There is no significant role for algebra in this scheme, since all of the demonstrative work is accomplished by constructions that trace back to first principles that articulate the essential form or nature of geometric objects.

FROM RATIOS TO MECHANICS

The differences of opinion separating Wallis and Barrow were based in divergent conceptions of how best to interpret Euclidean geometry, and particularly the theory of ratios. Yet these differences were not confined to the realm of pure mathematics. Wallis's algebraic foundationalism led him to take the science of mechanics as an application of the very general algebra or "arithmetic of species" that he regarded as the true *mathesis universalis*. Barrow, in contrast, held that mechanics was essentially a branch of geometry proper, where such concepts as space, time, and motion turn out to be both the foundations of pure geometry and the basis for a science of material bodies.

Wallis developed his account of the connection between the theory of ratios and the science of motion in the *Mechanica*, which begins by declaring mechanics to be "that part of geometry that treats of motion, and investigates through geometric reasonings and demonstrations, by what force any motion is effected" (*Mechanica* 1; OM 1 575). The "geometric reasonings" in the *Mechanica* are taken from the theory of ratios, so that after the definitions of key terms, the first propositions deal with ratios and their composition. Thus, Proposition II reads "When a ratio is composed of two or more [ratios], given the components, the composite is given. That is to say, having multiplied the exponents of the compounds into one another, the exponent of the composite may be determined" (*Mechanica* 1, prop. 2; OM 1 580). In the scholium to this proposition, Wallis explains that "Euclid calls the indices or exponents of compounded ratios πηλικότητεζ, which his interpreters call 'quantities,' but I prefer 'quotients', for it means that which arises from the division of the antecedent term by the consequent" (*Mechanica* 1, prop. 2; OM 1 580).

Although the initial propositions in *Mechanica* are "taken from the doctrine of ratios," Wallis explained that their demonstrations are so contrived that they apply both to the general algebra of species and to the specific case of geometric lines and figures. Stating and demonstrating such principles algebraically renders them "more general," so as to apply to any magnitudes whatever, thereby making ratios of geometric figures "only a single case among many, that are contained within a universal proposition" (*Mechanica* 1, prop. 6, Scholium; OM 1 583). The consequence of Wallis's taking the theory of ratios as part of a universal algebra is that it can deliver results that extend beyond geometry and enable the doctrine of motion to be studied mathematically (i.e., algebraically).

The cornerstone of Wallis's approach is Proposition VII of the *Mechanica*: "Effects are proportional to their adequate causes." This proposition, which says nothing directly about lines, figures, or other geometric objects, permits us to reason about the relation between causes and effects, and specifically to investigate the causes of motions by considering motion as an effect of some motive cause. As Wallis announced, "I have reckoned that this universal proposition should be set out at the beginning, since it opens the way by which, from purely mathematical speculation, one may move on to physical [speculation], or rather that the one is connected to the other" (*Mechanica* 1, prop. 7, Scholium; OM 1 584).

Barrow left no systematic treatise on mechanics, but his views on the subject are easily enough reconstructed from his approach to foundational questions in geometry and his remarks on the nature of "mixed mathematics." He proposed a highly kinematic conception of the origin of geometry in which magnitudes such as lines, angles, and surfaces are generated by motions. Thus, he conceived a line or curve as the path traced by a point in motion through space, while a circle is characterized as something produced by the revolution of a line about one of its endpoints. In his *Lectiones Geometricae* (which were assembled some years after the *Lectiones Mathematicae*) he explained that "among the ways of generating magnitudes, the primary and chief is that performed by local motion, which all [others] must in some sort suppose, because without motion nothing can be generated or produced" (LG 2; MW 2 159). Thus, in Barrow's view, the basic concepts of geometry include space, time, and motion. Further, when a geometrical object such as a curve or surface is defined in terms of the motions that produce it, the definition expresses the true formal cause of the object and allows the deduction of necessary properties of the object.

One notable consequence of this view is that "because local motion in general can scarcely be judged as regards its duration, impetus, intension, direction, or any other of its properties, either in itself or compared with another motion, except by the spaces (that is straight or circular lines) that it can describe or pass through, it follows that most parts of physics . . . are to be judged part of mathematics" (LM 2; MW 1 44). Indeed, Barrow concluded that "mathematics, as it is commonly taken, is so to speak coextended and made equal with physics itself" (LM 2; MW 1 44). But, given Barrow's identification of geometry as the mathematical science *par excellence*, it follows that the whole of physics is to be understood geometrically. Thus, as Barrow conceived of the issue, the science of mechanics is concerned with nothing distinct from the continuous quantities of geometry, and these are to be investigated by attending to the properties of such magnitudes as expressed in the motions that generate them.

As a result, the space of geometry is identical with the space of physics, and we understand the properties of such magnitudes by attending to the motions with which they are produced. Significantly, Barrow used the comparison of compound motions in the *Lectiones Geometricae* to effect the construction of tangents and determination of areas, both of which are essential to any treatment of mechanics. The key to his method was to treat a curve as traced by "composite motions" of a point, and then to decompose the composite motion into two instantaneous rectilinear motions, from which the determination of various properties followed fairly naturally. One particular problem is of interest here, namely the determination of properties of the parabola, and specifically the parabolic arcs of bodies in free fall. Speaking of the success of his method in relation to this problem (which he had succeeded in generalizing), Barrow remarked "I believe that not only this but many other propositions of Galileo connected to this one and related to the matter, howsoever they are demonstrated, can also be rendered more general or extended to all sorts of other curves" (LG 4; MW 2 199). In other words, the study of bodies subject to motive forces (i.e., mechanics) is a branch of geometry that proceeds by determining the ratios of the component motions arising from such forces.

If Barrow himself did not leave a systematic treatise on mechanics, it is arguable that his most successful student did. That student is, of course, Newton, and his *Principia* is developed in precisely the style that Barrow held to be necessary for any proper study of the physics of moving bodies. Rather than expressing his results in terms of algebraic equations (in the style of Wallis), Newton constructed ratios and proportions derived from a consideration of the motions of point masses and their trajectories that arise from the application of forces, all the while avoiding anything that might seem overly algebraic or disconnected from the consideration of continuous magnitudes in physical space.[15]

CONCLUSION

If the account I have been developing is anywhere near the truth, we should read the history of seventeenth-century physics against the background of disputes over the nature of ratios. The two traditions I have identified, namely the relational and numerical accounts of ratios and proportions, are associated with two different ways of constructing the mathematical language to be used in investigating the properties of bodies, and more specifically their mechanical properties. The numerical theory lent itself to a highly algebraic treatment of mechanics in Wallis's *Mechanica*, and was further developed by Leibniz and physicists in the Leibnizian tradition. In contrast, the relational doctrine of ratios was tied to a much more geometric treatment of mechanics that avoided the apparatus of algebra in favor of the more traditional language of ratios and proportions. It is an odd irony of history that what is today taught as Newtonian mechanics uses the conceptual and mathematical apparatus of algebra and the associated notion of real-valued functions, which belong to a tradition that neither Barrow nor Newton would recognize as appropriate for expressing the fundamental principles of mechanics. How that came to pass is, however, a matter for another day.

ABBREVIATIONS

LG Lectiones Geometricae
LM *Lectiones Mathematicae*
MU Mathesis Universalis
MW Barrow, I. 1860. *The Mathematical Works*
OM Wallis, J. 1693–99. *Opera Mathematica*

NOTES

1. Galilei 1890–1909, 6:232.

2. Machamer (1998, 65) notes that "Galileo used a comparative, relativized geometry of ratios as the language of proof and mechanics, which was the lan-

guage in which the book of nature was written. This is very different from what will follow in the eighteenth century and from the way we think of science today."

3. See Guicciardini (1999) on the mathematics behind Newton's formulation of his mechanics and the debates it engendered among his eighteenth-century interpreters.

4. On the concept of a *mathesis universalis*, its origins, and its role in seventeenth-century thought, see Crappuli (1969) and Rabouin (2009).

5. My references to the Euclidean *Elements* are to Euclid ([1925] 1956), in the translation of Heath. References are given in the text to book number and definition or proposition number.

6. See Palmieri (2001) on various attempts to rework the classical theory of ratios, notably those by Galileo and Christopher Clavius.

7. See Whiteside (1960, section 2) on the development of a broader algebraic conception of algebraic number in the seventeenth century, as well as Klein (1968).

8. See Heath's introductory note to book 6 of the *Elements* (Euclid [1925] 1956, 2:189–90) for a summary of the evidence for regarding the definition as an interpolation.

9. This remark appears in the twentieth of Barrow's *Lectiones Mathematicae*. Henceforth, I will give citations to Barrow parenthetically in the text, using the abbreviations LM and LG for his *Lectiones Mathematicae* and *Lectiones Geometricae*, with a reference to the relevant lecture number. I also supply page and volume references to Barrow's *Mathematical Works* (Barrow 1860).

10. The medieval history of the definition and its role in seventeenth-century mathematics is studied in Sylla (1984).

11. My references to the *Mathesis Universalis* are given parenthetically in the text, using the abbreviation MU and the relevant chapter number. I add a parallel citation to Wallis's *Opera Mathematica* (Wallis 1693–99). References to Wallis's *Mechanica* are to chapter and definition or proposition number, with a parallel citation to OM.

12. I contrast algebraic and geometric foundationalism in seventeenth-century philosophy of mathematics in Jesseph (2010).

13. See Mahoney (1990) on Barrow's mathematics and its odd mixture of innovation and methodological conservatism.

14. Barrow denied that the causality characteristic of mathematics could be construed as efficient or final causality. This is because "the connection (at least such as can be understood by us) of an external cause (for instance, of an effi-

cient cause) with its effect cannot be such that, the cause having been posited, the effect must necessarily be granted; nor from a posited effect may some determinate cause, strictly speaking, be shown" (LM 6; MW 1 91–92). This view is a consequence of Barrow's theological voluntarism, which requires that in the case of efficient causality the connection between a cause and its effect "depend upon the most free will and omnipotence of Almighty God, who at his pleasure can prevent the influx and efficacy of any [efficient] cause" (LM 6; MW 1 92). For more on the connection between Barrow's voluntarism, his nominalism, and his approach to mathematics, see Malet (1997) and Sepkowski (2005).

15. See Guicciardini (2009, chapter 13, "Geometry and Mechanics") for details on Newton's account of the relationship between geometry and mechanics.

REFERENCES

Barrow, I. 1860. *The Mathematical Works*. Ed. W. Whewell. 2 vols. Cambridge: Cambridge University Press.

Crappuli, G. 1969. *Mathesis Universalis: Genesi di un'Idea nel XVI Secolo*. Rome: Edizioni dell'Ateneo.

Euclid. [1925] 1956. *The Thirteen Books of Euclid's "Elements" Translated from the Text of Heiberg*. Ed. and trans. T. L. Heath. 3 vols. New York: Dover.

Galilei, G. 1890–1909. *Le Opere de Galileo*. Ed. A. Favaro. 20 vols. Florence: Barbéra.

Guicciardini, N. 1999. *Reading the "Principia": The Debate on Newton's Mathematical Methods for Natural Philosophy from 1687 to 1736*. Cambridge: Cambridge University Press.

———. 2009. *Isaac Newton on Mathematical Certainty and Method*. Cambridge, Mass.: MIT Press.

Jesseph, D. M. 2010. "The 'Merely Mechanical' vs. the 'Scab of Symbols': Seventeenth-century Disputes over the Criteria for Mathematical Rigor." In *Philosophical Aspects of Symbolic Reasoning in Early Modern Mathematics*, ed. A. Heeffer and M. Van Dyck, 273–88. Studies in Logic, vol. 26. London: College Publications.

Klein, J. 1968. *Greek Mathematical Thought and the Origin of Algebra*. Trans. E. Brann. Cambridge, Mass.: MIT Press.

Machamer, P. 1998. "Galileo's Machines, His Mathematics, and His Experiments." In *The Cambridge Companion to Galileo*, ed. P. Machamer, 53–79. Cambridge: Cambridge University Press.

Mahoney, M. S. 1990. "Barrow's Mathematics: Between Ancients and Moderns." In *Before Newton: The Life and Times of Isaac Barrow*, ed. M. Feingold, 179–249. Cambridge: Cambridge University Press.

Malet, A. 1997. "Isaac Barrow on the Mathematization of Nature: Theological Voluntarism and the Rise of Geometrical Optics." *Journal of the History of Ideas* 58: 265–87.

Palmieri, P. 2001. "The Obscurity of the Equimultiples: Clavius' and Galileo's Foundational Studies of Euclid's Theory of Proportion." *Archive for History of the Exact Sciences* 55: 555–97.

Rabouin, D. 2009. *"Mathesis Universalis": L'idée de mathématique universelle d'Aristote à Descartes.* Paris: Presses Universitaires de France.

Sepkowski, D. 2005. "Nominalism and Constructivism in Seventeenth-Century Mathematical Philosophy." *Historia Mathematica* 32: 33–59.

Sylla, E. 1984. "Compounding Ratios: Bradwardine, Oresme, and the First Edition of Newton's *Principia*." In *Transformation and Tradition in the Sciences*, ed. E. Mendelsohn, 11–43. Cambridge: Cambridge University Press.

Wallis, J. 1657. *Mathesis Universalis.* Oxford: Litchfield.

———. 1693–99. *Opera Mathematica.* 3 vols. Oxford: Oxford University Press.

Whiteside, D. T. 1960–62. "Patterns of Mathematical Thought in the Later Seventeenth Century." *Archive for History of the Exact Sciences* 1: 179–338.

7

COLOR BY NUMBERS

The Harmonious Palette in Early Modern Painting

EILEEN REEVES

Hidden amid the standard tales of rollicking adulterers and vigorous cheats of Celio Malaspina's *Two Hundred Novellas*, published in 1609, is the story of a boorish Venetian pigment grinder and his tireless tormentors, a petty dealer in brass and a die cutter connected with the mint. There is neither philandering nor fleecing here: the pigment grinder has nothing but a modest shop of "different sorts of colors, chalks and minerals," an aging mother, an excess of superstition, and a clear deficit of common sense. Much of this story has to do with the pigment grinder's efforts to avoid the die cutter, as he is convinced that the latter, a gifted sketch artist, is not only interested in collecting an unpaid debt, but has also been ordered to depict the twelve most insane men in the city. The brass dealer counsels the pigment grinder that in the interest of avoiding such portraiture, he should have himself shaved and even mutilated by the local barber, with the result that the dupe is initially unrecognizable even to his own mother. Startled, finally, by the die cutter's unexpected appearance in his shop and panicked by the emergent drawing, the pigment grinder plasters his whole head with printer's ink, grimaces to disguise himself further, and bellows, "Now just try to sketch me!" (Malaspina 1609, 1:143–45v).[1]

At stake in this story, clearly, is the social and professional identity of the pigment grinder, a figure so misguided in his affections that he asks his antagonist the brass merchant to be "like a father" to him, so abject in his quotidian activities, and so prone to "rushing barefoot in the rain from home to the shop, filthy, his hands, face, and smock smeared with colors," that he might easily be taken as a madman. The most disconcerting episode of the entire story—the brass merchant's suggestion that the barber "engrave" the pigment grinder as he likes—is the prelude to the first of several disfigurements (Malaspina 1609, 2:144r–v). That the die cutter is a man whose business is to make money and whose special talent is the ability to sketch vivid portraits in chiaroscuro further suggests an asymmetrical division of artistic labor, skilled *disegno* and crafty design being the province of the coiner, and color, for what little it is worth, the concern of the impoverished pigment grinder. Given Malaspina's friendship with the prominent sculptor Leone Leoni, this aspect of the story might also be read as a narrative variant on the more ritualized contestations of the aesthetic merits of canvas painting and the low relief carving characteristic of coins and medals; rather than being judged by polished end products and presented by eloquent defenders, each art is reduced to the materials needed for its initial stages, and defined by the inarticulate remarks and comic gestures in which the protagonists specialize (on Malaspina, see Ghirlanda 1960). And competing notions of naturalness are clearly a focus here as well, for while the primitive lifestyle, earthy products, and gullibility of the pigment grinder make him the embodiment of a *naturale* or simpleton, the crafty die cutter is renowned for his *ritratti naturali* or "life-like portraits" of the alleged madmen of Venice.

But it is the opening incident in this series of ruses that is most revelatory, as it captures something of my concern in this chapter, the vexed relationship of color and number in the early modern period, surely among the most problematic efforts to mathematize nature. Having encouraged the pigment grinder to close up shop and to hide at home in order to avoid the prowling coiner, the brass merchant decides to complicate his victim's life by altering the chalked numbers on the various wooden shutters covering the windows of the *bottega*. Perplexed and then maddened by the mismatch between shutter and window, the pigment grinder proves incapable of fitting the appropriate cover to each aperture, and he struggles with the task from the moment nearby church bells ring ten o'clock at night until they sound the Angelus at dawn (Malaspina 1609, 1:143v–144). Evidently unable to distinguish the openings on the basis of size, position, and shape, he relies on the arbitrary index provided by numbers.

It is not that the ruse provides the occasion for theft—there is little enough in the pigment grinder's shop, and nothing that interests the amused onlookers—but rather that the episode itself exposes a crucial concern of early modern natural philosophers and artists, the shifting and often unintuitive ways in which numbers were connected with colors. In the spectacle of the pigment grinder's rage, the numbers with which he has structured his environment appear meaningless and arbitrary to all observers, and, whatever their original logic, are of unrecoverable significance to the victim himself. The numbers, in short, are talismanic, and useful only insofar as they serve to match what are for the pigment grinder otherwise unrecognizable architectural elements. They embody the twin tendencies of a man defined by both credulity and superstition.

Malaspina's anecdote can be read, as I will argue here, as a vernacular response to the celebrated classical story through which number came to be linked first to sound and subsequently to color. Like Malaspina's novella, this tale, otherwise radically different in tenor and in import, emerges in the workplace. In the version told by Boethius in late Antiquity in his *Fundamentals of Music* (1989) and repeated by countless followers, the study of harmony emerged when the ancient philosopher Pythagoras was inspired by the single consonance emitted by five hammers pounding molten metal in a forge. After initial investigations of the matter, Pythagoras judged one of those implements inharmonious and set it aside; weighing the other four, he found that they differed in a ratio of 12:9:8:6. The various relationships between any pair of these weights, he noted, could thus be transcribed by the first four natural numbers.

Such intervals, Pythagoras further argued, could likewise be translated to those between tones on a monochord. When a string is divided in half and plucked, the diapason that sounds is one octave higher in pitch than that emitted by the open string. The diapente, produced when two of three equally divided sections are played, is one-fifth higher than the open string; the diatessaron, emerging when three of four sections of the string are struck, is one-fourth higher. In the Pythagorean view, 9:8 or the interval between the fourth and the fifth, a whole tone, was itself dissonant, though the basis of harmony.

These same relations, Pythagoras added, held true for weights suspended on cords. Thus the difference between a cord bearing twelve pounds and one bearing six pounds would be an octave or diapason; between twelve and eight pounds a fifth or diapente; between eight and six pounds a fourth or diatessaron. The experimentation extended, later writers added, to containers filled with 12, 9, 8, and 6 units of water, to pipes of 12, 9, 8, and 6 units of length, and to bells of 12, 9, 8, and 6 units of volume, and the same consonances always emerged (Boethius 1989).[2]

These proportions—2:1, 3:2, 4:3, and 9:8—are those that matter in this chapter: they were associated, though in somewhat unstable fashion, with the range of hues running from white to black, especially in early formulations of color theory.[3] Once removed from their original musical context, they functioned as terms designating proportions of light to dark, or white to black; entirely remote from painterly practice, they underwent further modifications when fitted to spatial presentations of the spectrum. Until the emergence of the clear distinction between primary and secondary colors, and the simple combinations they offered, the proportions Pythagoras discovered in the forge formed the basis for many philosophical accounts of color.

"MORE COLORS THAN JUST BLACK AND WHITE"

Color is clearly a disastrous business for the pigment grinder of *Two Hundred Novellas,* whose workplace, literally structured by meaningless numbers, yields him very little in the way of profit. Apart from the conventional Venetian *carta azzurra* on which the die caster sketches his incriminating chiaroscuro portraits, other than the deep black of the printer's ink with which the pigment grinder covers himself, and the disappearing whites of his eyes as he takes on this disguise, no hues are mentioned in the story (Malaspina 1609, 1:144, 145). There is the strong suggestion, moreover, in the equivalence of "filth" with the various colors smeared on his hands, face, and smock, of an impoverished, grimy, and monochromatic world. This environment is the parodic legacy of the early modern efforts to connect color with the Pythagorean ratios.

To summarize the problem with which natural philosophers of the sixteenth and seventeenth centuries were confronted, in *On Sense and Sensible Objects* Aristotle had sought to explain the origin of the "intermediate" or mixed colors yellow, red, purple, green, and blue by arguing that they arose through various admixtures of white and black, or of light and darkness.[4] This notion, almost wholly incomprehensible to modern readers, would find its most persuasive instance in the reddish glow of dark clouds struck by sunlight, and allusions to this effect occur regularly in sixteenth-century discussions of color. Aristotle had further argued that only "exactly numerical" ratios of white and black would yield attractive hues. While he preferred the hypothesis that all combinations involved an intermingling of white and black so thorough as to transform those hues, he acknowledged that other thinkers had imagined either a mixing of fine but essentially unaltered black and white particles, or a layering of the two substances. Significantly, these less probable alternatives were distinguished by their kinship with painting, while that favored by Aristotle could only be explained by analogy with musical ratios.[5]

> It is thus possible to believe that there are more colors than just white and black, and that their number is due to the proportion of their components; for these may be grouped in the ratio of three to two, or three to four, or in other numerical ratios, or they may be in no expressible ratio, but in an incommensurable relation of excess and defect, so that these colors are determined like musical intervals. For on this view the colors that depend on simple ratios, like the concords in music, are regarded as the most attractive, e.g., purple and red and a few others like them—few for the same reason that the concords are few—while the other colors are those that have no numerical ratios. (Aristotle 1957, 233)

The Bolognese physician and philosopher Mainetto Mainetti offered one of the most influential commentaries on this Aristotelian text in 1555, tacitly discarding Aristotle's own order—white, yellow, red, violet, green, blue, black—so as to privilege the Pythagorean ratios. He began with the observation that in the Aristotelian *Problemata* the color green was singled out for its restorative qualities, precisely because it was between the extreme points of white and black, the excesses of which disturbed the viewer's eyes. "Two colors are between white and green," Mainetti wrote in accounting for green's attractive nature, "yellow is beyond white and blue before green. They arise in rational proportions, as if a diapason and a diapente. Yellow is indeed in diapason, that is, two to one, since two units of light or brightness and one of earthy darkness generate yellow. Blue is rather in diapente, which is two to three, since blue is born of three units of brightness, and two of opacity" (Mainetti 1555, 80). Mainetti added that matters were similar for purple and red, already identified by Aristotle in *On Sense and Sensible Objects* as especially pleasing to viewers. As an intermediate hue between green and black, red could be compared to the diapason, because it was composed of two units of opacity and one of brightness; purple, having three parts opacity to two of light, was like the diapente.

Returning to the argument later in his commentary, Mainetti further compared yellow to the diapason and brown to the diapente. Here, however, the crucial ratios involved white and black, rather than light and dark: yellow contained two measures of white to one of black; brown, three of black to two of white (152). In both accounts, though, the diapason and diapente were associated with specific ratios; the diatessaron, judged insufficiently pleasing, had no place in this system.

Mainetti's formulation appears to have been adopted in a somewhat condensed version by the Florentine physician Guido Guidi, who likewise identified green as equally composed of clarity and opacity, blue as embodying the ratio 3:2, red as 2:1, and purple as 2:3. Avoiding the terms "diapente" and "diapason," Guidi (1626) merely noted that "where certain proportions are maintained in admixtures, the colors will be pleasing, as in the harmony of voices, but where such ratios are not preserved, they will be unappealing" (162). Likely written in the 1560s but unpublished for decades, this sort of discussion would prove durable. Its most striking feature is its distance from artistic practice: though he was the maternal grandson of Domenico Ghirlandaio, and associated with the Mannerist painter Francesco de' Rossi ("il Salviati"), in this instance Guidi deferred to the traditional and exclusively theoretical explanation of color.[6]

Others who followed Mainetti's lead sometimes sought to mute his overt reliance on musical intervals, and chose simply to refer to the mathematical ratios of the Pythagorean traditions. Thus, for instance, in a discussion of 1581, the Fribourg humanist Sébastien Werro presented black and white as the sole simple colors, and red, rather than Mainetti's green, as the product of equal proportions of these two hues. Pink had a 3:2 ratio of white to black; blue, conversely, a 3:2 ratio of black to white. Saffron had a 2:1 ratio of white to black, while scarlet had the same ratio of black to white. The ratio of black to white in green was 5:4—effectively, the ditone or major third—but Werro did not draw on terms borrowed from the discourse of harmony. Yellow and brown, finally, involved ratios of 2:1, white to red and red to white, respectively, which meant that their proportions of white to black were 6:5 and 5:6, that of the semi-ditone or minor third. This musical interval likewise went unnamed in Werro's account (1581, 124–26).

AS PAINTERS DO

Such discussions, whatever proportions they involved, differed entirely from the actual practices of early modern painters. While artists relied on a variety of substances to obtain green and purple pigments, techniques for blending a saffron-based lake and azurite, or mashed iris petals and Naples yellow, or orpiment and indigo, or saffron and indigo, or for layering a red lake over azurite had been known for well more than a century (Ball 2001; Salazaro

1877, 23–24, 25–26; Merrifield 1849, 2:420–25, 584–87, 610–11; Hall 1992, 15–16, 32).[7] Dyers in Mainetti's day could combine indigo and *giallo santo*, a yellow lake, to obtain green (Ruscelli 1557, 105v–106). To judge from early modern colored woodcuts, by the mid-sixteenth century several different shades of orange were produced through combinations of red lead or vermillion with ochre or lead-tin yellow (Dackerman 2003, 57, 169, 206, 236, 274, 276, 277). This is not to say that such recipes were always reliable: the ambiguity of color terms, the imprecision of measurements and techniques, the variability caused by locale and season, and the tendency of numerous substances to deteriorate over time, or in contact with other substances, guaranteed unpredictable results. But the increasing incidence of concoctions favoring mixtures of blue and yellow, or of red and blue, or of red and yellow, even or rather especially in the case of false combinations, suggests a growing familiarity with knowledge that would soon be codified as a system of primary and secondary colors.[8]

We might regard the monochrome world of Malaspina's pigment grinder as a symptom of the confused account of color offered by natural philosophers in this period. All the chalk, minerals, and pigments on his face, hands, and smock seem reduced to a single "filthy" hue; the protagonist and his enemy the die cutter both produce, in rather different ways, faces rendered only in black and white; a quick chiaroscuro sketch on *carta azzurra* is identified as an extraordinarily lifelike portrait, as if its absent and aberrant colors were of no great importance. These narrative details signal that the elegantly calibrated admixtures of black and white, or light and dark, would result solely in various shades of gray; but it is the relationship of such discussions to the brass vendor's initial trick that merits special consideration.

If we compare, for example, Mainetti's explanation of color with Werro's subsequent elaboration, we can see that both systems can be read as linear spectrums running from light to dark, but that the transition from the original ratios to whole numbers produces peculiar features. Thus while Mainetti's spectrum progresses from white through yellow, blue, green, purple, red, and black, were we to transcribe the harmonic ratios of light to dark as integers on a 100-point scale, the arrangement would suggest something other than a uniform passage from one hue to the next. In such a configuration, white, of course would be rendered as 100, yellow as 66, blue as 60, green as 50, purple as 40, red as 33, and black as 0. Werro's slightly more elaborate version yields a different chromatic ordering and more pronounced clustering; reduced to the same scale and restricted to integers, it runs from white (100) to saffron (66) to pink (60) to yellow (54) to red (50) to brown (45) to green (44) to blue (40) to scarlet (33) and finally to black (0).

It is difficult to find a more apposite image of these reductions of color to number than that evoked by the brass vendor's first trick, where the pigment grinder's difficulties in closing his shop involved his inability to recognize, without the aid offered by numbers, the shutter designed for each aperture. Just as Werro's spectrum associates individual colors with specific numerical values distributed in nonuniform fashion and in a manner that correlates only weakly with hue, so the various windows of the shop, identified by a unique number and sometimes poorly differentiated in size, shape, and position, frame the pigments and display them as disjunct elements in a seemingly arbitrary sequence.

The revelation of the strange role numbers play for the pigment grinder comes from the brass merchant, who in addition to altering "a two to a six, and a six to a four, and so forth," discreetly marks each shutter "at the foot with a sign known to himself," a signature of sorts, in order eventually to close the place up (Malaspina 1609, 1:143v). While the episode implies more the trickster's skepticism concerning the connection of number with color than a systematic effort to explain the phenomena in other terms, his professional identity is telling. In Malaspina's coy phrase, this ruffian "plied his trade by selling various brass objects in the balance-makers' street" (1609, 1:143). His obvious propensity for deception, Malaspina's own excellent credentials as an inveterate forger, and the widespread practice of tampering with mercantile measures suggest that these wares were fraudulent weights, rather than the legitimate metal components of balances and steelyards; as Francis Bacon had observed in 1601, "this fault of using false weights and measures is grown so intolerable and common that if you would build churches, you shall not need for battlements and bells other things than false weights of lead and brass."[9] The more immediate point may be, however, the brass merchant's implicit familiarity with metrology. Two details from his final trick—his perforation of the pail within which the pigment grinder twice pours wine purchased by volume, only to have it twice trickle away, unnoticed, as he carries it through the streets, and the description of the dupe's rage as "unmeasured"—confirm the expected opposition between one who perceives weight, and one who does not (1609, 1:145r–v).

It was within the context of experiments with measures and weights that the increasingly elaborate substructure of Pythagorean proportions became most vulnerable to criticism. Skeptics included the Venetian mathematician and natural philosopher Giovanni Battista Benedetti, who explained consonance and dissonance in 1585 not by the ratios and the string lengths of the monochord, but by the rates of the strings' vibration, the more pleasing sounds being the result of notes concurring with frequency, and the less agreeable ones the product of interrupted or infrequent concurrences. Writing in 1589, Vincenzo Galilei turned to the story of the suspended weights, showing that the ratio needed to be 4:1, not 2:1, to produce an octave or diapason; 9:4, not 3:2, for a fifth or diapente; 16:9, not 4:3, for a fourth or diatessaron. This adjustment of the proportions between weights was not merely the observation that the numbers needed to be squared, but more important, part of a sustained polemic against the prominent composer and music theorist Gioseffo Zarlino's overreliance on Pythagorean ratios to explain all natural phenomena (see Drake 1999; Palisca 2006, 150–51; Peterson, 2011, 170–71; Heller-Roazen 2011, 67–68; Mancosu 2006, 598–604).[10]

Both Zarlino's enthusiastic elaboration of Pythagorean ratios and the sort of empirical knowledge advocated by Benedetti and Galilei had renewed interest in the association of color with number. Given the cultural prominence of Venetian painting, the presence of a well-established textile industry in that city, and the strong interest in color perception among early modern natural philosophers and physicians nearby at the University of Padua, it is not surprising that the setting for these discussions was Venice.[11] Educated in philosophy at Padua before entering a career as a diplomat and a cleric, Filippo Mocenigo addressed the question of color in his *Universal Institutions for the Perfection of Man* of 1581 (on Mocenigo, see Bonora 2011). Such discussion occurred not in his examination of vision, however, where color is hardly mentioned, but rather as an appendage to his presentation of sound and voice.

Mocenigo was strongly influenced by the work of Zarlino. Given that the two men were both associated with the Venetian Academy, and that Zarlino had dedicated another work to Mocenigo's cousin, the Doge of Venice, this engagement is not surprising; the fact that *Universal Institutions for the Perfection of Man* emerged from the press at the moment of Galilei's quickening conflict with Zarlino can only have increased interest in the matter.[12] Of particular relevance here is Zarlino's revision of the Pythagorean system, which made the first six, rather than four, integers the basis of harmony. In addition to the diapason, diapente, and diatessaron, therefore, musicians might draw upon the more modern consonances of the ditone—5:4 or the major third—and the semi-ditone—6:5 or the minor third.[13] As for the tone or 9:8, Zarlino had interpreted the story of Pythagoras and the hammers to mean that this basic unit enjoyed an intermediate status of something neither concordant nor discordant (Zarlino 1558, 61).[14]

Mocenigo (1581) drew on these innovations to describe the spectrum in systematic fashion. "The outermost colors, which in their mutual relationship recall the diapason, are white and black," he began. "The three intermediate ones, which are in fact simple, but bordered by white and black, are red, which is closer to black than to white, yellow or gold, which is nearer to white, and hyacinth." This last color, a bright violet blue, he stated, was "therefore the midpoint, such that with respect to black, it can be compared to the diapente, and with respect to white, the diatessaron. With respect to red, it is like the semi-ditone, and with respect to yellow, the ditone. In the same fashion, yellow with respect to red resembles the diapente, while red with respect to black is like the ditone" (305).

Unlike Mainetti and successors such as Werro, Mocenigo used the harmonic proportions to articulate the spatial relationships of these colors to each other, rather than to indicate notional measures of dark and light, or black and white, in the presumed compositions. The layout of Mocenigo's system of primary colors can readily be mapped onto Zarlino's discussion of the version of the diatonic scale he favored (Zarlino 1558, 120–22).[15]

Figure 1. Gioseffo Zarlino, *Istitutioni Harmoniche* (Venice: 1558), 122.

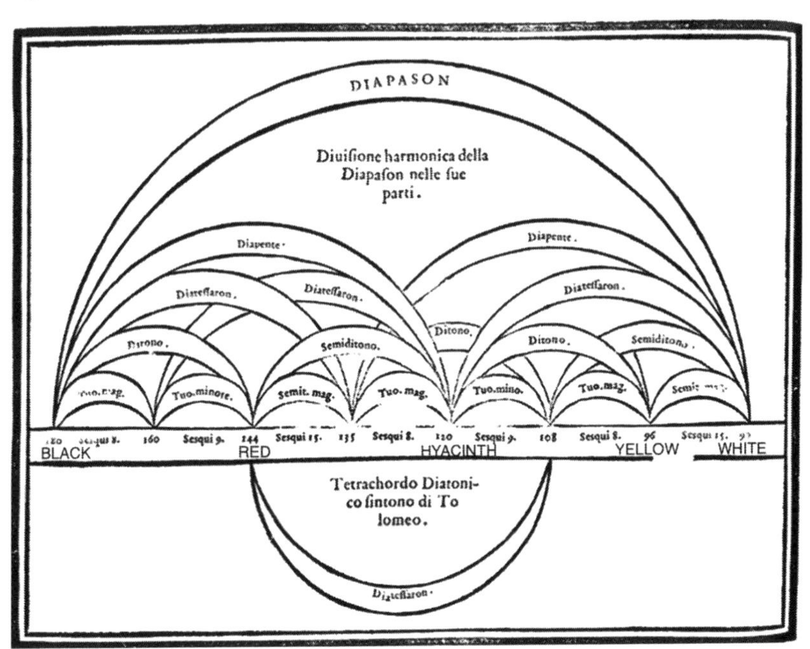

A small difference lies in the distance in this theoretical configuration between yellow and white, shown here as a major semitone, a unit Zarlino (1558) had laboriously described as a harmonious fraction of the diapente (121–22). Mocenigo (1581) had proposed instead that there would "also be the proportion of a [whole] tone, which is neither consonant nor dissonant," between these hues. He had further stipulated that the proportions between red and white, and between black and yellow would be dissonant (305).

In the most important contrast to Mainetti and other Aristotelian predecessors, Mocenigo (1581) insisted on the proximity of his system to actual artistic practices, privileging first the primary nature of three colors: "It is clear that painters can make neither red, nor hyacinth, nor yellow—any more than they can make white or black—from mixtures, unless a new concoction [that is, alteration by heat] is involved. However, all other hues can be produced from the admixture of any of these simple colors" (305). The schematic ordering of these secondary colors that followed would have impressed at least some of Mocenigo's initial audience less with its inaccuracy than with its apparent distance from the model of *On Sense and Sensible Objects*, and with its resemblance to lower genres such as the painter's manual and the book of secrets. "Blue arises from hyacinth and white," Mocenigo explained, "green from yellow to which some black has been added; crimson from blue and black; brown from black and white; ash-gray from white to which some black has been added" (ibid.).[16]

NOW PRINTED FOR THE FIRST TIME

Guido Antonio Scarmiglioni and Anselm de Boodt, two writers educated at the medical school of Padua in the late 1580s and eventual residents of Vienna and Prague, respectively, also offered modified versions of Mocenigo's system.[17] Scarmiglioni's *Two Books on Color* appeared only in 1601. Its breathless subtitle, *Now Printed for the First Time*, and its preface both portray it as a text composed years earlier, and its numerous references to other works include nothing published after 1590. Like de Boodt's eventual publication of 1609, it offered extensive reference to the practice of painters and dyers: Scarmiglioni gestured several times to various combinations of blue and yellow and of blue and red used to produce green and purple. Unlike the more pragmatic guide provided by de Boodt's *Natural History of Gems and Precious Stones*, however, *Two Books on Color* also drew upon the flexible resources of the musical argument.

In reviewing the notion that painters were unable to make the five so-called primaries "through mixtures, unless heating were involved," and that "by mingling these hues they easily obtain others," Scarmiglioni (1601, 112) objected first of all to the exclusion of green from this first rank of colors. The fact that one might observe, "as a quotidian experience," painters concocting this color from admixtures of yellow and blue did not persuade Scarmiglioni of its secondary status, but did justify its central place, equidistant from the two hues of which it was composed, in his array of seven primaries (120, 170). Matters were evidently more complicated for purple, whose confection he knew to involve "a small amount of red added to blue," for he located it between green and blue (117, 119, 169). Orange seems to have figured only briefly, as a substance produced when minium was moistened with water; it had no status as a separate hue (120). Though Scarmiglioni adopted the terms of Zarlino's harmonic intervals, comparing the relationship of white to black to the diapason, describing the position of green with respect to these two endpoints by referring to the diatessaron and the diapente, and defining the distance between white and yellow by a tone "neither consonant nor discordant," his spectrum does not conform to the diatonic scale as well as Mocenigo's does (Scarmiglioni 1601, 148, 174, 180, 187, 199). It is not surprising that he included no illustration of the arrangement.

De Boodt's *Natural History of Gems and Precious Stones* likewise emerged from the Paduan context, and also suffered a significant delay in publication. It mentioned the notion of primary colors only twice, and merely in passing, as if citing a truism rather than a point of contention. In his preface de Boodt (1609) stated that "from the colors white, black, red, blue, and yellow, painters can make a variety of hues," and subsequently he noted that "the principal colors, and those which are not made from the mixtures of others, are white, black, blue, yellow, red, and minium, which is made from calcined lead" (8, 25). In adding this last color—a bright orange pigment prized by painters—de Boodt provided a corrective to Scarmiglioni's recommendation of a water-based process, and more important, an instance of the use of heat, rather than mixing, to produce an unadulterated hue.[18]

Mocenigo, Scarmiglioni, and de Boodt do not seem to have offered a color system sufficiently robust to attract disciples. We might infer that Galileo Galilei's claim in May 1610, just after the publication of his *Sidereus Nuncius*, to have already written a short treatise "On Vision and Colors," involved an effort, or perhaps merely the intention to make such an effort, to improve upon an arrangement whose basis was the work of his father's great rival Zarlino.[19] The discussion of colors written by the prominent physician Epiphanio Ferdinando, which was published in early 1611 by the press from which Galileo's *Sidereus Nuncius* had emerged just eight months earlier, is notable for its tacit resistance to recent innovations. Its chartlike format, deployment of terms such as "diapason," "diapente," and "diatessaron," and ordering of the colors were nothing but a retreat to the proportional units of dark and light advocated sixty years earlier by Mainetti (Ferdinando 1611, 193).[20]

In these various efforts to reformulate and preserve the traditional connection of Pythagorean ratios with colors, we are far indeed from the antics of the pigment grinder, the brass merchant, and the die cutter, and yet Malaspina captures a striking common denominator. Published in 1609 by the same Venetian printing consortium used by Galileo in 1610 and Ferdinando in 1611, *Two Hundred Novellas* begins with a transparent fiction of anteriority much remarked by its original audience: even as it masquerades as a collection of tales told by speakers gathered in a villa to escape the plague of 1576, it blithely relates countless celebrated events of much more recent vintage. Malaspina lived in Venice from 1580 to 1591, but what is more crucial than the author's biographical particulars is the way in which his tale of the pigment grinder, opening with the formulaic "it is already many years ago," and treating the association of number with color as farce, mimics the familiar combination of prior discovery and deferred revelation. Just as Scarmiglioni and de Boodt alluded to a debate several decades old, and only "being printed for the first time" in 1601 and 1609, and Ferdinando and Galileo coupled bygone analyses of color with publications of 1611 or yet to come, so the mocking Malaspina presented the tale of the pigment grinder as an account of events long predating their moment of disclosure.

YELLOW, RED, AND BLUE

While Malaspina's gesture to this temporal lag suggests a kind of smug stasis in early modern color theory, a significant development soon followed. In 1613 the Jesuit François Aguilon offered a coherent discussion of the painter's primaries in his *Opticorum libri sex* (Six Books on Optics), published in Antwerp and accompanied by engravings designed by Peter Paul Rubens.[21] A crucial feature of visible phenomena, color emerges early as a topic in this seven-hundred-page treatise. Despite his ultimate rejection of the Aristotelian explanation of color, Aguilon invoked several of the arguments of *On Sense and Sensible Objects*, noting, for instance, that chromatic mixtures might occur through a layering of a translucent hue over a darker one, or as an optical impression of minute spots seen from a distance or, finally, as a genuine admixture of two different substances (Aguilon 1613, 39). He also distanced his discussion from the organic color changes addressed in pseudo-Aristotle's *On Colors*, and warned his readers that "we are not dealing here with concrete colors such as minium, dark purple, lake, cinnabar, indigo, ochre, orpiment, lead white, and the other things with which painters cover canvases, but rather with the visible qualities that inhere in them" (38).

His system had an elegant simplicity. "Yellow, red, and blue number, strictly speaking, as the three intermediate colors," Aguilon asserted. "Along with white and black they form a quintet of primary colors. Moreover, from these intermediate colors just as many secondary colors arise through three combinations. Orange is thus made of yellow and red, purple of red and blue, and from yellow and blue, finally, there is green. And from the mixture of all three of these intermediate colors a certain unpleasant hue is born, something livid and lurid, like a cadaver" (40).

The figure accompanying Aguilon's explanation is clearly a modified version of that traditionally deployed in discussions of consonance and dissonance. Aguilon wholly abandoned, however, the minute examination of various Pythagorean proportions: his system is characterized by symmetry, and stripped of terms imported from the discourse of harmony. Though he never alluded to the efforts of Mainetti, Werro, Mocenigo, or Scarmiglioni, Aguilon elsewhere suggested a certain resistance to efforts such as theirs. The preface of his work includes a passing condemnation of the obscurity of Pythagorean mysteries; more substantively, the discussion of colors is preceded by the censorship of those who insisted on commonalities between the senses (Aguilon 1613, "Lectori S[alutem"], second unnumbered page). "That which is perceived through color has only to do with sight; that which is discerned through sound, only with hearing; that which is known by scent, only with the sense of smell, and so forth," Aguilon warned (30). Even more mistaken than the erroneous comparison of sensible objects, he argued, was the belief that the difference between colors could be explained by reference to transparency, opacity, darkness, and shadow (ibid.). Worst of all, however, was the assumption that aesthetic judgments were other than matters of taste and opinion: "for beauty consists in harmonic division, which human reason barely recognizes; ugliness, in a certain obscure

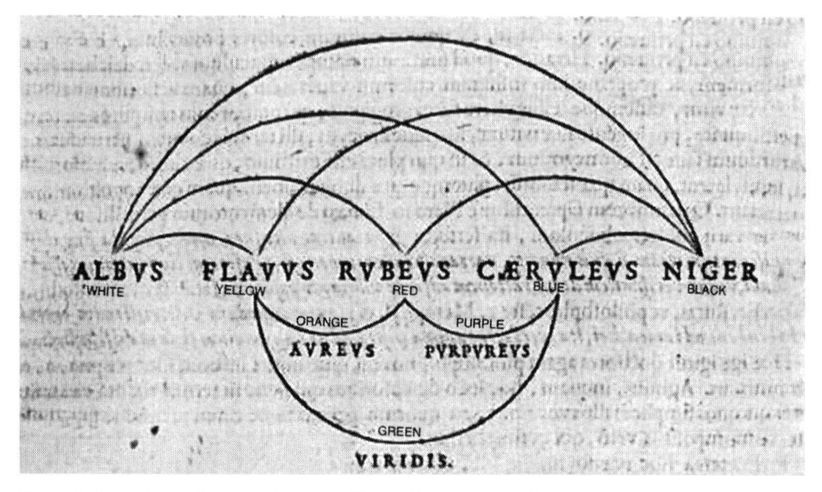

Figure 2. François Aguilon S. J., *Opticorum libri sex* (Antwerp: 1613), 40.

asymmetry of lines and qualities" (31). Put differently, consonance and dissonance could not be established by arithmetical means, and figured not at all in a discussion of the relationships between colors.

PAST ANTERIOR

Aguilon's theory of primary colors was parroted in the specialized ambit of the Jesuit thesis two years later (Felix and Denich 1615, 20). In general, however, his simultaneous rejection of the Aristotelian explanation and of the traditional association of colors with musical intervals seems to have gone unnoticed, unacknowledged, or unaccepted by natural philosophers.[22] As if in conformity with the pattern of tardy revelation of bygone findings, the Florentine physician Guido Guidi's work, written in the 1560s and based on Mainetti's adaptation of the Pythagorean ratios to color theory, was posthumously published in 1626; the Venetian physician Valerio Martini's *De colore libri duo sua aetate iuvenilia collecti* (Two Books on Color Composed in His Youth) emerged in 1638, looked back several decades to discussions at the University of Padua and concluded, "based on reason, experiment, and the authority of those who are expert in painting," that the six principal colors—white, gray, yellow, orange, blue-green, and black—were produced through admixtures of black and white (Martini 1638, 2:2). Without specifying its relationship to his prior and still unpublished "On Vision and Colors," in his *Assayer* of 1626 Galileo gestured in passing to his view that colors, like particular sounds, tastes, tactile sensations, and odors, were an artifact of the senses, and that ours is a monochrome world configured by indivisible quanta. In insisting that "a very long time would not be enough for me to explain, or rather shade in on paper, what little I understand of these matters, and thus I pass over them in silence," he avoided, to a degree, the problematic elaboration of an atomistic doctrine, and reverted to his pattern of infinitely deferred disclosure (Galilei 1967, 6:350).

Whether these debates were of any importance to the true protagonists of color mixing, the painters of early modernity, is by no means clear. Thus far, only a few works such as Peter Paul Rubens's *Juno and Argus* (ca. 1611) and Nicolas Poussin's *Christ Healing the Blind Man* (1650), where sight and light are overtly addressed, are considered direct responses to emergent color theory as formulated by Aguilon, though one might perhaps add to this meager list Guido Reni's *Union of Design and Color* (ca. 1620–25) and those self-portraits in which artists deliberately display a restricted palette (see Kemp 1990, 30–44). That said, it seems entirely possible that the most pronounced reactions to Aguilon's solution do not necessarily inhere in the expected genres or arise in conventional thematic treatments. Like Malaspina's narrative response to those prior attempts to mathematize color through Pythagorean ratios, they appear as a cluster of gratuitous details in a scenario whose ultimate referent is the original story of the forge.

By way of conclusion, then, I would like to consider the legacy of the debate over color in two works by Diego Velázquez, *Joseph's Bloodstained Coat Brought to Jacob*, and *Apollo at the Forge of Vulcan*, both completed in Rome around 1630 during his first Italian sojourn. I will argue that this issue is the crucial component of both paintings: the ostensible subjects, biblical and mythological, merely provide the pretexts. There are a number of contextual reasons to suspect that color theory would have been of interest to Velázquez in this period. The young and ambitious artist's first journey to Italy had been prompted by Rubens's visit to Madrid in 1628–29; before arriving in Rome he had spent a brief period in Venice; his travel throughout the country was facilitated by the Venetian ambassador to Spain, Alvise Mocenigo, cousin to Filippo Mocenigo; and when in Rome he resided in the Villa Medici, where Galileo was also staying (see Goldberg 1992, 453–56; Palomino 2007, 37–44, 76–86). Though these coincidences likely indicate no more than the relatively restricted number of participants in the cultural life of early modernity, it is also true that Velázquez acquired at some unknown point Aguilon's *Six Books on Optics*, as well as an unidentified work,

Figure 3. *Joseph's Bloodstained Coat*, 1630 (oil on canvas), Diego Rodriguez de Silva y Velázquez (1599–1660). Monasterio de El Escorial, El Escorial, Spain. Bridgeman Images.

possibly authored by Vincenzo Galilei, on music theory (Sánchez Cantón 1925, 3:389–91).[23] The best evidence for the importance of the debate over color, however, comes from the paintings themselves.

In his 1724 biography of Velázquez, Antonio Palomino (2007, 83–86) presented *Joseph's Bloodstained Coat* and *Apollo at the Forge* as companion pieces painted without commission but later offered to the Spanish king; apart from two landscape sketches of the villa where the artist may or may not have met Galileo, these are the only two canvases known to have been completed during the stay in Rome.[24] While scholars have emphasized their shared subject of deception, it must be noted that dishonesty enjoys very different handling in the two works. In *Joseph's Bloodstained Coat*, the stunned patriarch Jacob is deceived by his sons, who use the garment to convince him that his youngest and favorite child has perished. *Apollo at the Forge* is likewise a tawdry domestic drama in which Apollo tells Vulcan the unhappy truth about the infidelity of his consort Venus.

What the paintings do share is a set of formal resemblances and sustained attention to the medium itself.[25] That *Joseph's Bloodstained Coat* has something to do with color is not surprising, given that the *tunica polymita*

Figure 4. *Apollo at the Forge*, 1630 (oil on canvas), Diego Rodriguez de Silva y Velázquez (1599–1660). Prado, Madrid, Spain. Bridgeman Images.

in question was generally understood in the early modern period to have been woven "of diverse colors" (see Beyerlinck 1617, 216; de Mariana 1620, 28; 1617, 385; Cornelius a Lapide 1616, 258). Considering that this splendid garment was the catalyst of Joseph's quarrel with his brothers, and that it is the sole prop in their ruse, the small white item shown to the patriarch is much less impressive than one would expect. Its significance lies in this economy: flecked with faint red and yellow stains, and shadowed with blue and black, the coat recapitulates the emergent theory of primary colors. As if to reinforce the point, Velázquez distributed the primaries about the black and white tunics, the brightly lit, strongly modeled triad of blue, red, and yellow cloths on the left finding a subdued mirror image on the right. While white and black are clearly crucial to the different tonalities of the left and right sides of the canvas, their new status as something other than the source of all colors is indicated in two different ways. The rich gray cloak over the patriarch's robe, neatly posed against the juncture of light and dark walls, is the only hue that could be said to derive from those erstwhile primaries. And the black robe of the brother who bears the tunic "of diverse colors" is nothing other than the dark ground of the canvas itself, at once a representation, appropriately somber, of the liar's garment, and an entirely unworked section of the painting (see Brown and Garrido 1998, 40–45; Garrido 1992, 230–31).

Joseph's Bloodstained Coat incorporates a series of last efforts: after this painting, Velázquez never again used a dark ground, he abandoned the coarsely woven canvas for a finer fabric, and the brilliant Naples yellow next to the tunic "of diverse colors" does not reappear in later works (Brown and Garrido 1998, 43, 45). While any of these three developmental steps is plausibly associated with his artistic apprenticeship in Rome, they also appear crucial to Velázquez's insistence on the very nature of his medium in this painting. This emphasis reappears, albeit in a slightly different register, in the so-called companion piece of *Apollo at the Forge.* Here for the first time Velázquez prepared a luminous ground of an opaque lead white mixture. This modification, in tandem with the much denser weave of the canvas, contributes to the slightly more finished and even quality of this second work.

Despite these differences, the paintings share several features; both compositions include the device of the landscape in the upper left quadrant, and both involve a dramatic moment in which a group of five men confront a sixth character. While the focal point of *Joseph's Bloodstained Coat* was necessarily that pallid garment "of many colors," there is no such object in *Apollo at the Forge.* Rather than the double series of block-like primary colors, moreover, this work featured the secondary hues of orange, green, and purple, though only the first of these retains its initial intensity. The green of Apollo's crown, made of admixtures of azurite, iron oxide, and lead white, was reproduced, with varied tonality, in the garments of Vulcan and his centrally placed companions, while the clothing of the man working on the armor at the far right, originally a muted violet, was composed of lead white, iron oxide, vermilion, and a pale blue pigment, perhaps smalt, notoriously prone to discoloration (Brown and Garrido 1998, 46–56; Garrido 1992, 243).[26] Inevitably, the placement of these three secondary colors replicates the arrangement in Aguilon's diagram, a central arc of green falling between the orange and violet extremes. Except for the bluish sky beyond the forge, the primary colors do not appear in this painting, though black, white, and gray figure naturally in the metal objects produced by Vulcan and his assistants.

While the bright orange of both the metal on the anvil and of the fire recalls the occasional presentation of this color, when derived from calcined lead, as a primary hue, the context itself is puzzling. So, too, are Vulcan's four companions, traditionally identified as a trio of Cyclops, for they are neither giants nor one-eyed.[27] Nor is this Vulcan a deformed god, but merely a shocked cuckold. The easy translation of this episode to a vernacular idiom, though typical even of the young Velázquez, should not obscure the importance of that other forge where the fusion of color with consonance began. Simply put, the gratuitous details of this painting serve to evoke the moment when Pythagoras, routinely described as Apollo's son, entered a foundry where five hammers had been pounding molten metal; discarding one, he discovered the crucial ratios between the other four.[28]

Velázquez offers no way to assess the importance of this legend. As the background figure in *Apollo at the Forge* tends the bellows, and the man on the far right is using tongs, there are but three who wield hammers. These instruments differ noticeably in size, as in the Pythagorean story. They are complemented, however, not by one discordant and summarily discarded tool, but by at least five, and possibly more, scattered about the enclave. The pan of a balance, entirely unsuited to the oversized hammers, lies abandoned on the floor; a steelyard dangles unused next to the chimney. The apparent irrelevance of these weighing devices in *Apollo at the Forge* corresponds, roughly, to something like a visual pun in *Joseph's Bloodstained Coat*, the *baculus Jacobi* or "Jacob's staff" being a traditional instrument for calculating angles. This studied emphasis on a kind of inadequation between the physical world and our means of numbering and measuring its phenomena would seem the very antithesis of the Pythagorean episode at the forge, and a definitive break with that early attempt to measure and codify aesthetic production.

But we might just as easily conclude that the painting involves not a rejection of the neatness of Pythagoras's approach to sound and by extension to color but rather an uncanny prelude to that foundational moment. In such a reading, the emphasis is less on the evident disorder of the forge than on the very fact of its reduction to an image. Put differently, the scene at this forge, anterior to that of the origins of music, can be rendered *only* through visual means; like the banal motif of marital disharmony with which it is seemingly concerned, it cannot be captured through consonant ratios of sound. While clearly adhering to a vestigial or rather incipient version of the Pythagorean intervals in his treatment of color in both paintings, here Velázquez would have insisted on the absolute autonomy, priority, and permanence of his art. The work would have thus stood as a corrective to the long-standing subordination of color to sound and, by implication, of a crucial feature of painting to the strictures of musical harmony. If such was his intention, it is a matter of some irony that this aspect of the painting—its bid for priority and longevity—has been compromised by the instability of the medium.

In either guise, finally, *Apollo at the Forge* would appear remote from the tale of the tormented pigment grinder, and yet Malaspina's anecdote anticipates both readings. The decorous wreckage strewn about the forge, the disconcerting mismatch between the instruments depicted and the celebrated story of Pythagoras's discovery, and the suggestion of an irremediable incommensurability between the painter's medium and the harmonic ratios all figure as a sequel to the brass merchant's initial exposure of the strange place of number in the pigment grinder's shop. But as we might expect in a work whose focus is the antecedent to the Pythagorean moment in the forge, the canvas also articulates a corrective to the temporal feature parodied in *Two Hundred Novellas* and promoted without irony in early modern discussions of color. In the interest of undoing the priority claim of music, it refashions that insistence on a pronounced gap, typically on the order of a generation, between the discovery of something about the mathematization of nature under the aegis of sound and its disclosure in print. "Be like a father to me," the hapless pigment grinder asked the brass merchant, as if to avail himself of a previous generation's wisdom; "you will be like a son to me," Velázquez's Apollo might have observed to the as yet unborn Pythagoras.

NOTES

1. This and all subsequent translations are mine unless otherwise indicated.

2. More generally, see Fideler (1988); for another version of the story, see "Life of Pythagoras," in Guthrie (1988, 86–87); for background on Boethius's debt to Pythagorean and Ptolemaic arguments, see Goldberg (2011, 19–30). On the legend, its presuppositions, and consequences, see Heller-Roazen (2011, 11–59); on the place of Pythagorean mathematics in the world of Galileo Galilei, see Peterson (2011, 33–42, 57–65, 149–73, 257–58); on the importance of early modern music theory in the evolution of number theory, see Pesic (2010).

3. On the history of efforts to pair musical consonance with colors, see Gage (1993, 227–46) and Kuehni (2007).

4. On the relationship between artisanal knowledge of color mixing, particularly that of painters and dyers, and Isaac Newton's eventual treatment of white light, see Shapiro (1994).

5. On the traditional resistance to color mixing, and on the emergence of

the practice of layering in the early modern period, see Hall (1992, 15–16, 52–57, 71–73, 211–17).

6. Salviati's illustrations were for Guidi (1544). On Ghirlandaio's resistance to experimentation with color mixing, his reversion to the older mode of unbroken color established by Cennino Cennini, and on Salviati's imitations of various color modes, see Hall (1992, 57, 61, 67, 163–66).

7. See also Kirby, Nash, and Cannon (2010, 67, 147–48, 151, 244). As Gage notes, Alexander of Aphrodisias referred in passing, and dismissively, to the artificial production of purple and green around 200 AD; see Gage (1993, 31).

8. On the trade in early modern pigments and dyes in Venice, see Matthew (2002). On the prices of pigments in early modern Italy, and on their relation to genre, see Spear and Sohm (2010, 65–66, 101–4).

9. "Speech on Bringing in a Bill against Abuses in Weights and Measures," in Bacon (1868, 18).

10. Galilei's associate Ercole Bottrigari took up the argument in 1609 in his unpublished *Enigma of Pythagoras*.

11. On the commerce in colorants in Venice, see Matthew and Berrie (2010); Krischel (2010). For an overview of developments in the Venetian treatment of color, see Hall (1992, 199–235).

12. On the ongoing conflict between Zarlino and Galilei, see Heller-Roazen (2011, 61–69); Palisca (2006, 29–47, 142–44, 150–52); and Peterson (2011, 153–73); on Zarlino's senario in particular, see also Wienpahl (1959, 27–41). For a recent in-depth study of the entire dispute, see Goldberg (2011): on Zarlino's dedication to Alvise Mocenigo, see (2011, 44–45, 49), on his alleged attempt to delay Galilei's *Dialogue on Ancient and Modern Music* in 1581, see (2011, 220–21, 265–67).

13. On the emergence of the major and minor thirds, see Heller-Roazen (2011, 62, 80–81).

14. On the traditional discussions involving the status of the tone, see Heller-Roazen (2011, 28–29, 32–40, 53–54).

15. On the syntonic diatonic scale, and on Galilei's criticism of this choice, see Goldberg (2011, 57–64, 101–49, 240–48, 272–393).

16. For similar combinations, see Erizzo (1558, fol. 30v); Curaeus (1567, fol. 69 r–v); Caracciolo (1589, 257).

17. De Boodt's doctoral degree was awarded in 1586 or 1587; Scarmiglioni was awarded a degree in June 1589; see Zonta and Brotto (1969, 4:3; 1469–70; 4:4, 2354). On de Boodt and Scarmiglioni, see Parkhurst (1971); Shapiro (1994, 606–9); Gage (1993, 34–37, 93–96, 153–56, 165–68); Kemp (1990, 266, 275–76, 281–82).

18. On the confusion over this substance, see Falloppio (1564, 152–54, 164–66); Guidi (1626, 369); on the departure of minium and the paucity of orange in the quattrocento palette, see Hall (1992, 15, 208, 257); see further Kirby, Nash, and Cannon (2010, 84n40, 305, 457).

19. Galileo Galilei to Belisario Vinta, May 7, 1610, in Galilei (1967, 10:352). Curiously, the entry on Galileo written by Count Angelo de Gubernatis and published in an encyclopedia of 1901 refers to the work as "an essay, now lost, [establishing] the profound truth of the laws of consonance and dissonance, or the unity and variety of colors" (Adams and Rossiter 1901, 5:13).

20. The fact that Mainetti would be misidentified in the 1620s in Jacopo Soldani's poem "Contro gli aristotelici" as an exemplar of Paduan philosophy is perhaps an index of the particular impact of his arguments in the celebrated university of that city.

21. On Aguilon, see Parkhurst (1961); Kemp (1990); Shapiro (1994, 606–9).

22. On the resistance or indifference to Aguilon's argument, see Shapiro (1994, 615–18).

23. The work on music theory is generically described as "De arte música," and attributed to "Lipo Gailo," perhaps a misreading of "V^{zo} Galilei."

24. On the landscapes, see Brown and Garrido (1998, 57–61).

25. On the limited number of pigments favored by Velázquez, see Brown and Garrido (1998, 17–19).

26. On smalt, see Zahira Véliz, "In Quest of a Useful Blue in Early Modern Spain," and Nicola Costaras, "Early Modern Blues: The Smalt Patent in Context," in Kirby, Nash, and Cannon (2010, 389–414); on Velázquez's early use of smalt as a colorant (rather than as a siccative), see p. 393, as well as Brown and Garrido (1998, 39).

27. In his *Life of Velázquez*, Palomino notes that in the much later depiction of Vulcan painted by Juan Carreño and overseen by Velázquez, the Cyclops were three in number and named "Brontes, Steropes and Pyracmon" (Palomino 2007, 155).

28. Pythagoras's biographers generally ascribe belief in his divine nature and his descent from Apollo to reckless poets and to common people; occasionally he is said to be Apollo himself. For such references in the accounts of Iamblichus, Porphyry, and Diogenes Laertius, see Guthrie (1998, 57, 58, 59, 61, 80, 83, 97, 101, 109, 123, 128, 129, 144, 147). At least one early modern writer identified the workers encountered by Pythagoras in the forge as the Cyclops (Ringhieri 1551, fol. 144).

REFERENCES

Adams, C. K., and J. Rossiter, eds. 1901. "Angelo de Gubernatis, s. v. 'Galilei.'" In *Universal Cyclopedia and Atlas*. 12 vols. New York: D. Appleton and Company.

Aguilon, F. S. J. 1613. *Opticorum libri sex*. Antwerp: Plantin.

Aristotle. 1957. *On the Soul, Parva Naturalia, On Breath*. Trans. by W. S. Hett. Cambridge, Mass.: Harvard University Press.

Bacon, F. 1868. "Speech on Bringing in a Bill against Abuses in Weights and Measures." In *The Letters and the Life of Francis Bacon*, ed. J. Spedding and R. L. Ellis, vol. 3. London: Longmans, Green, Reader and Dyer.

Ball, P. 2001. *Bright Earth: Art and the Invention of Color*. Chicago: University of Chicago Press.

Beyerlinck, L. 1617. *Promptuarium morale*. Cologne: Antonii Hierati.

Boethius, A. M. S. 1989. *Fundamentals of Music*. Trans. with introduction and notes by C. Bower. Ed. C. Palisca. New Haven, Conn.: Yale University Press.

Bonora, E. 2011. "Filippo Mocenigo." In *Dizionario biografico degli italiani*, ed. A. M. Ghisalberti, vol. 75. Rome.

Brown, J., and C. Garrido. 1998. *Velázquez: The Technique of Genius*. New Haven, Conn.: Yale University Press.

Caracciolo, P. 1589. *La Gloria del Cavallo*. Venice: Niccolò Moretti.

Cornelius a Lapide, S. J. 1616. *Commentarius in Pentateuchum Mosis*. Antwerp: Haeredes Nutii et Meursium.

Curaeus, I. 1567. *Libellus physicus continens doctrinam de natura et differentiis colorum*. Wittenberg: Peter Seitz.

Dackerman, S. 2003. *Painted Prints: The Revelation of Color*. University Park: Pennsylvania State University Press.

De Boodt, A. 1609. *Gemmarum et lapidum historia*. Hanover: Wechelianus.

Drake, S. 1999. "Renaissance Music and Experimental Science." In *Essays on Galileo and the History of Science*, ed. N. M. Swerdlow and T. H. Levere, 3:190–207. Toronto: University of Toronto Press.

Erizzo, S. 1558. *Il dialogo di Platone intitolato il Timeo*. Venice: Comin da Trino.

Falloppio, G. 1564. *De metallis seu fossilibus tractatus*. Venice: Giordano Ziletti.

Felix, S., and S. Denich. 1615. *Praecipuae difficultates ex libris de Anima selectae*. Ingolstadt: Elisabeth Angerman.

Ferdinando, E. 1611. *Theoremata medica et philosophica*. Venice: Tomaso Baglioni.

Fideler, D. 1988. "The Monochord: The Mathematics of Harmonic Mediation."

In *The Pythagorean Sourcebook and Library*. Trans. and comp. K. S. Guthrie. Ed. D. Fideler, 24–28, 327–28. Grand Rapids, Mich.: Phanes Press.

Gage, J. 1993. *Color and Culture: Practice and Meaning from Antiquity to Abstraction*. Boston: Bulfinch.

Galilei, G. 1967. *Opere*. Ed. A. Favaro. 20 vols. Florence: Giunti Barbèra.

Garrido, C. 1992. *Velázquez: Tecnica y evolución*. Madrid: Museo del Prado.

Ghirlanda, D. 1960. "Orazio Celio Malespini." *Dizionario biografico degli Italiani*. Vol. 68.

Goldberg, E. 1992. "Velázquez in Italy: Painters, Spies, and Low Spaniards." *Art Bulletin* 74, no. 3: 453–56.

Goldberg, R. 2011. *Where Nature and Art Adjoin: Investigations into the Zarlino–Galilei Dispute, including an annotated translation of Vincenzo Galilei's Discorso intorno all'opere di Messer Gioseffo Zarlino*. PhD diss., Indiana University.

Guidi, G. 1544. *Chirurgia è Graeco in Latinum conversa*. Paris: Pierre Gaultier.

———. 1626. *Opera Omnia*. Frankfurt: Wechel.

Guthrie, K. S., trans. 1988. *The Pythagorean Sourcebook and Library*. Ed. and introduction D. Fideler. Grand Rapids, Mich.: Phanes Press.

Hall, M. 1992. *Color and Meaning: Practice and Theory in Renaissance Painting*. Cambridge: Cambridge University Press.

Heller-Roazen, D. 2011. *The Fifth Hammer: Pythagoras and the Disharmony of the World*. New York: Zone.

Kemp, M. 1990. *The Science of Art: Optical Themes from Brunelleschi to Seurat*. New Haven, Conn.: Yale University Press.

Kirby, J., S. Nash, and J. Cannon, eds. 2010. *Trade in Artists' Materials: Markets and Commerce in Europe to 1700*. London: Archetype.

Krischel, R. 2010. "The Inventory of the Venetian *Vendecolori* Jacopo de' Benedetti: The Non-Pigment Materials." In *Trade in Artists' Materials*, ed. J. Kirby, S. Nash, and J. Cannon, 253–66. London: Archetype.

Kuehni, R. 2007. "Development of the Idea of Simple Colors in the Sixteenth and Early Seventeenth Centuries." *Color Research and Application* 32, no. 2: 92–99.

Mainetti, M. 1555. *Commentarius mire perspicuous nec minus utilis in librum Aristotelis . . . De Sensu et Sensibilis*. Florence: Laurentius Torrentinus.

Malaspina, C. 1609. *Ducento Novelle*. 2 vols. Venice: Al Segno dell'Italia.

Mancosu, P. 2006. "Acoustics and Optics." In *Early Modern Science*, vol. 3 of *The Cambridge History of* Science, ed. K. Park and L. Daston, 596–631. Cambridge: Cambridge University Press.

de Mariana, J. 1617. *Biblia sacra cum glossa ordinaria*. Douai: Bellère.

———. 1620. *Scholia in vetus et novum testamentum*. Paris: Michel Sonnius.

Martini, V. 1638. *De colore libri duo sua aetate iuvenili collecti*. Venice: Ducali Pinelliana.

Matthew, L. C. 2002. "'Vendecolori a Venezia': The Reconstruction of a Profession." *Burlington Magazine* 144: 680–86.

Matthew, L. C., and B. H. Berrie. 2010. "'Memoria de colori che bisognino torre a vinetia': Venice as a Centre for the Purchase of Painters' Colours." In *Trade in Artists' Materials: Markets and Commerce in Europe to 1700*, ed. J. Kirby, S. Nash, and J. Cannon, 246–52. London: Archetype.

Merrifield, M. 1849. *Original Treatises dating from the XIIth to XVIIIth Centuries on the Arts of Painting*. 2 vols. London: John Murray.

Mocenigo, F. 1581. *Universales institutiones ad hominum perfectionem*. Venice: Aldus.

Palisca, C. 2006. *Music and Ideas in the Sixteenth and Seventeenth Centuries*. Urbana: University of Illinois Press.

Palomino, A. 2007. *Life of Velázquez*. Trans. N. A. Mallory, introduction M. Jacobs. London: Pallas Athene.

Parkhurst, C. 1961. "Aguilonius' Optics and Rubens' Color." *Nederlands Kunsthistorisch Jaarboek* 12: 35–49.

———. 1971. "A Color Theory from Prague: Anselm de Boodt, 1609." *Allen Memorial Art Museum Bulletin* 29: 3–10.

Pesic, P. 2010. "Hearing the Irrational: Music and the Development of the Modern Concept of Number." *Isis* 101: 501–30.

Peterson, M. 2011. *Galileo's Muse: Renaissance Mathematics and the Arts*. Cambridge, Mass.: Harvard University Press.

Ringhieri, I. 1551. *Cento giuochi liberali et d'ingenuo*. Bologna: Anselmo Giaccarelli.

Ruscelli, G. 1557. *De' secreti del donno Alessio Piemontese prima parte*. Milan: Fratelli da Meda.

Salazaro, D., ed. 1877. *L'arte della miniatura nel secolo XIV. Codice di un manoscritto della Biblioteca Nazionale di Napoli*. Naples: Tipografia editrice già del Fibreno.

Sánchez Cantón, F. J. 1925. "La libreria de Velázquez." In *Homenaje ofrecido a Menendez Pidal*, 3: 397–406. Madrid: Hernando.

Scarmiglioni, G. A. 1601. *De coloribus libri duo*. Marburg: Paul Egenolph.

Shapiro, A. E. 1994. "Artists' Colors and Newton's Colors." *Isis* 85: 600–30.

Spear, R., and P. Sohm, eds. 2010. *Painting for Profit: The Economic Lives of Seventeenth-Century Italian Painters.* New Haven, Conn.: Yale University Press.

Werro, S. 1581. *Physicorum libri decem.* Basil: Hervagiana.

Wienpahl, R. W. 1959. "Zarlino, the Senario, and Tonality." *Journal of the American Musicological Society* 12, no. 1: 27–41.

Zarlino, G. 1558. *Istitutioni Harmoniche.* Venice: n.p. [Pietro da Fino].

Zonta, G., and B. Giovanni, eds. 1969. *Acta graduum academicorum Gymnasii Patavini.* 5 vols. Padova: Antenore.

8

THE ROLE OF MATHEMATICAL PRACTITIONERS AND MATHEMATICAL PRACTICE IN DEVELOPING MATHEMATICS AS THE LANGUAGE OF NATURE

LESLEY B. CORMACK

THE SIXTEENTH AND SEVENTEENTH CENTURIES have long been seen as fundamentally important to an understanding of the changing study of nature. The changes in this period have been variously categorized by historians as philosophical, methodological, or mathematical, among other explanations. One of the profound transformations that took place was the introduction of mathematics into the language of description and explanation of nature. Many historians and philosophers have investigated this translation, which allowed the description of the occult and unseen forces of nature, introduced a new logical structure and language, and made natural knowledge increasingly useful for technological changes. Once adopted, the advantages of a mathematical lexicon were clear. But how and why did that adoption take place? In the sixteenth century, mathematics was a study separate from (and inferior to) natural philosophy; thus, the story of the mathematization of the worldview is also the story of how mathematics and mathematicians came to have a status previously afforded to philosophy alone. In order to understand how and why those who studied nature came to adopt the language of mathematics, we need to look at the men who were using mathematics in their everyday work—mathematical practitioners. Mathematics and mathematical practitioners played an essential role in the transformation of science in this early modern period, as we can see by examining the role of mathematics and mathematical practice, utility, commerce, and trade on the changing ideology and methodology of science.

THE MATHEMATIZATION OF NATURE

The great early and mid-twentieth-century historians of scientific ideas, scholars such as Edwin Burtt, Herbert Butterfield, and especially Alexandre Koyré and E. J. Dijksterhuis, argued that the move to mathematical language, and with this a move toward a mechanization of nature, was the defining characteristic of the age they called the scientific revolution (Burtt 1924; Butterfield 1949; Koyré 1957; Dijksterhuis 1961).[1] Indeed, the discipline of history of science really began in the twentieth century by focusing on the problem of the origin of modern science. The work of some of its great founders concentrated on what this important transformation was and how it took place. In the 145 years between Copernicus and Newton, people interested in the Book of Nature developed new methodologies including experimentation; new attitudes toward knowledge, God, and nature; a new ideology of utility and progress; and new institutional spaces and practices.[2] They began to view the world as quantifiable, investigable, and controllable. By the end of the period, the investigation of nature, still tied to theological concerns, but increasingly to practical ones as well, was carried out in completely new places, for different ends, and with quite different results.

Why was mathematics so powerful for these historians? First, and perhaps most obviously, modern science was heavily mathematical and so when they sought the origins of modern science, mathematics was a necessary prerequisite. Mathematics, especially Euclidean geometry, introduced a new logical rigor into argumentation. While there had been important disagreements about the status of mathematical truth, with Peter Ramus for example arguing that mathematics was a natural attribute of humans made more abstract and obscure by mathematicians such as Euclid, and Henry Saville insisting that abstraction demonstrated the perfection of mathematical knowledge (Goulding 2010), mathematics seemed to offer a clear language and to demonstrate underlying truths about nature. Those who argued for its importance, like John Dee or Isaac Newton, quoted the familiar biblical passage, "thou hast ordered all things in measure and number and weight."[3]

A problem with this grand historical narrative is that in the medieval and early modern periods mathematics was not part of natural philosophy. Mathematics was a separate area of investigation from natural philosophy and those interested in mathematical issues had usually tied such studies to practical applications, such as artillery, fortification, navigation, and surveying.[4] The mathematical quadrivium and natural philosophy were studied in different parts of the university curriculum and the status of mathematicians was substantially lower than that of philosophers or theologians (Feingold 1984; Biagioli 1989, 2007). Given this split, it is not self-evident that a natural philosopher would see the benefits of adding mathematics to his explanations of the world.

And yet, it is clear that many who studied and developed explanations of nature did employ mathematics. Studies of the work of Galileo, Huygens, Newton, and Leibniz, among others, point to the profound importance of mathematics to their explanations and worldviews. Such scholars asked different questions of nature—questions about measurement and prediction—because they had access to mathematics. They devised new ways of doing mathematics. So what were the origins of this interdisciplinary moment?

The key to understanding this development must be sought in the socioeconomic transformation of Europe, not simply in a metaphysical gestalt switch. A sociological change in who, where, and why the world was investigated was taking place.[5] A crucial category of scientifically inclined men downplayed by most historians of the period, the mathematical practitioners, was crucial to this transformation.[6] These mathematical practitioners became more important in the early modern period and provided a necessary ingredient in the transformation of nature studies to include measurement, experiment, and utility (Bennett 1991).[7] Their growing importance was a result of changing economic structures, developing technologies, and new politicized intellectual spaces such as courts and merchants' shops, and thus relates changes in 'science' to the development of mercantilism and the nation-state.

THE STUDY OF MATHEMATICS

In the early sixteenth century, few European scholars were interested in questions of mathematics. While Merton College in Oxford, for example, had been famous throughout Europe in the thirteenth and fourteenth centuries for its school of kinetics, this fame had dwindled by the sixteenth century and scientific study had largely been superseded by more humanistic pursuits. Most scholars and educational reformers in the first half of the sixteenth century were more concerned with the introduction of classical languages and literatures, and especially with the religious controversies swirling around them than with the structure of the natural world. This began to change in about the middle of the century and by 1600, natural philosophers and mathematicians were active and innovative. Why then did scholars, educational reformers, and practitioners change their minds?

One of the answers is that the goals of education altered dramatically during the century. Where few courtiers, politicians, or civil servants were university-trained at the beginning of the century, by 1600 a university education was practically a requirement (Stone 1964; see also Cormack 1997; McConica 1986). This changed the content of education as well as its delivery. The direction of education was also linked to patronage and as patronage patterns changed, natural philosophers and mathematicians were increasingly sought (Moran 1991; Biagioli 1993; Smith and Findlen 2002). These patrons, both merchants and courtiers, were responding to an evolving economic imperative, brought about by the rapid expansion of mercantile and trading opportunities. These activities, particularly by the trading companies, were in turn partly made possible by the increasing sophistication of the natural philosophers and mathematical practitioners.

The impetus for increased mathematical education and investigation came from outside the traditional scholarly community. Merchants and trading companies in northern Europe particularly were interested in applied mathematical practice, especially navigation, surveying, and accounting. This need became more pronounced as these merchants changed their focus to the Atlantic trade, putting them in competition with the much better informed Spanish and Portuguese. By the mid-sixteenth century, merchant companies believed that they needed a more theoretical grounding in mathematics, especially because they needed to navigate largely uncharted northern waters, and they began to patronize mathematical practitioners. Eventually, this led to important connections among these practitioners, skilled artisans, and natural philosophers.

The English Muscovy Company provides an interesting example of the new mercantile emphasis on mathematics. The merchants in this trading company recognized their need for mathematical knowledge in order to undertake significant ocean voyages and soon commissioned Robert Recorde to write mathematical books for the use of their navigators. In 1551 Recorde published *Pathway to Knowledge*, an explication of geometry through the first four books of Euclid's *Elements*, and in 1556 (reissued 1596), *The Castle of Knowledge, containing the explication of the Sphere*, dealing with spherical geometry, astronomy, and navigation (Recorde 1551; 1556). The latter was written and printed for the use of the Muscovy Company, and mentioned the Portuguese discoveries in order to illustrate the positions of the earth with respect to the sun. It was based on Ptolemy's astronomy and incorporated more recent astronomical work, including a brief, favorable mention of Copernican theory. Recorde's *Whetstone of Witte* (1557), his explication of algebra, was dedicated to the governors of the Muscovy Company and written, so Recorde claimed, to encourage the great exploration and trading enterprise on which they were embarked. He even promised to produce a future book (never written), in which "I also will shewe certain meanes how without great difficultie you mai saile to the North-Easte Indies. And so to Camul, Chinchital, and Balor" (Recorde 1557, fol. a3b). *Whetstone* was never reprinted, perhaps because it dealt with difficult mathematical concepts. It was based on German algebraic texts, including the treatment of the quadratic. Taken as a whole, Recorde's mathematical books contained a full course of mathematical study and many Elizabethan natural philosophers and mathematicians began their education with Recorde's books.[8] In this way, he was hugely influential in developing the English scientific endeavor, which therefore owed much to mercantile patronage.

The teaching and learning of mathematical knowledge in early modern Europe thus involved a complex interaction among scholars, practitioners, merchants, and gentry. Humanists and scholars at the university saw the value of mathematical knowledge for its intrinsic natural philosophical benefit as a way of understanding God's handiwork (see, for example, Elyot 1531, sig. 37a; Pace 1517, 109). Mathematics became an important part of both the formal and informal curricula at early modern universities (Feingold 1984). Equally, practitioners—instrument makers, entrepreneurial teachers, mathematical practitioners—were interested in mathematical knowledge for its application, its rhetorical power, and its value to potential patrons and employers (Recorde 1543, sig. A2a). Merchants and gentry needed secure knowledge for navigation, warfare, and investment.

Thus, the teaching and learning of mathematics took place at a variety of venues, some formal and some informal. The formal educational system was itself in a period of expansion and change, moving from an earlier church-based and -oriented institution, to one catering to a wider social demographic and to more political and mercantile career paths. At the same time, many young men (and some young women) had increasing access to printed texts, allowing them to teach themselves, and a new group of entrepreneurial teachers sprang up who could supply alternative instruction, either through personal tutoring, group lessons, or through the writing of self-help texts designed for the autodidact.

MATHEMATICAL PRACTITIONERS AND MATHEMATICAL LECTURES

London was a busy metropolis in the last decades of sixteenth century, both for the numerous and hard-working merchants and for those more interested in mathematical and natural philosophical pursuits (Harkness 2007).[9] The Inns of Court, Parliament, and the Royal Court all provided reasons for many young men and women to find their way to the city. Combined with a growing interest in trade, investment, and exploration, London was an increasingly attractive destination for young men from the country, fresh from university or their estates, eager to make their way in the world and to find communities of like-minded individuals. These new inhabitants of London, combined with skilled émigrés fleeing the religious troubles of the continent, ensured that there was both the expertise in mathematics and a ready market for this expertise. In the second half of the sixteenth century, a number of university-trained or self-taught men set themselves up as mathematics teachers and practitioners. These men, who we might call mathematical practitioners, sold their expertise as teachers through publishing textbooks, making instruments, and offering individual and small group tutoring. In the process, they argued for the necessity of practical knowledge of measurement, winds, surveying, and mapping, among others, rather than for a more philosophical and all-encompassing knowledge of the earth.

Most mathematical practitioners were university-trained, showing that the separation of academic and entrepreneurial teaching was one of venue and emphasis, rather than background. Mathematical practitioners claimed the utility of their knowledge, a rhetorical move that encouraged those seeking such information to regard it as useful.[10] It is impossible to know the complete audience for such expertise, but English mathematical practitioners seem to have aimed their books and lectures at an audience of London gentry, merchants, and occasionally artisans.[11] It is probably this choice of audience that most influenced their emphasis on utility, since London gentry and merchants were looking for practicality and means to improve themselves and their businesses.

Mathematical practitioners professed their expertise in a variety of areas, especially such mathematical applications as navigation, surveying, ballistics, and fortification. For example, Galileo's early works on projectile motion and his innovative work with the telescope were successful attempts to gain patronage in the mathematical realm.[12] Descartes advertised his abilities to teach mathematics and physics. Simon Stevin claimed the status of a mathematical practitioner, including an expertise in navigation and surveying.[13] William Gilbert argued that his larger philosophical arguments about the magnetic composition of the earth had practical applications for navigation.

In England, an early example of a mathematician using his expertise to improve the mathematical underpinning of these useful arts was Robert Recorde, employed by the Muscovy Company to give lectures and write a textbook in elementary mathematics in the 1550s (Cormack 2003; Johnston 2004). Recorde's early foray was to be repeated, especially in London, by mathematical practitioners, many of whom, such as Thomas Hood and Edward Wright, demonstrated an interest in mapping and navigation explicitly.

These mathematical practitioners offered lectures, individual tutelage, and the instruments to explicate the mathematical structure of the world. Sometimes this was done on a completely entrepreneurial model, that is, where the practitioner hung out his shingle and attracted clients through publishing and publicity. At other times, mathematics lectures were founded and supported by a small group of interested men, such as was the case with Thomas Hood.

THOMAS HOOD AS THE FIRST LONDON MATHEMATICAL LECTURER

Thomas Hood (1556?–1620) was the first mathematics lecturer paid by the city of London and thus fits a patronage model of mathematics lecturers. However, he also published and encouraged private pupils, and therefore was equally an entrepreneurial mathematics teacher. Hood attended Trinity College, Cambridge, where he received his bachelor's degree in 1578 and his master's in 1581.[14] In 1588, Hood petitioned William Cecil, Lord Burghley, to support a mathematics lectureship in London, to educate the "Capitanes of the trained bandes in the Citie of London."[15] This was a complex proposition because the Aldermen and Lord Mayor of London would be the ones paying the bills, but the Privy Council had to give its approval in order to allow the lectures to proceed.

Hood received the following positive response from the Privy Council: "The readinge of the Mathematicall Science and other necessarie matters for warlike service bothe by sea and lande, as allso the above saide traninge shalbe continued for the space of 2 yeares frome Michaelmas next to come and so muche longer as the L. Maior and the Citie will give the same alowance or more then at this present is graunted."[16] Hood's lectureship therefore went forward, held in the home of Sir Thomas Smith, merchant and later governor of the East India Company. The makeup of the audience is now unknown, although from the tone of his introductory remarks, published under the title *A Copie of the Speache made by the Mathematicall Lecturer, unto the Worshipfull Companye present . . . in Gracious Street: the 4 of November 1588*, Hood seemed to be talking to his mathematical colleagues and mercantile patrons, rather than to the mariners he insisted needed training (Hood n.d., 1588, sig. A2aff). The contents of Hood's lectures are also unknown, but the treatises bound with the British Library copy indicate that he stressed navigational techniques, instruments, astronomy, and geometry (Hood 1596, 1598).[17]

By 1590, Hood had been giving these mathematics lectures for almost two years, as he reported in his 1590 translation of Ramus: "so that the time limited unto me at the first is all most expired. . . . In this time I have binne diligent to profite, not onlie those yong Gentlemen, whom comonlie we call the captaines of this citie, for whose instruction the Lecture was first under taken, but allso all other whome it pleased to resorte unto the same" (Ramus 1590, sig. 2a). Hood identified himself on the title pages of all his books until 1596 as "mathematical lecturer to the city of London," sometimes advising interested readers to come to his house in Abchurch Lane for further instruction, or to buy his instruments.[18] His books explain the use of mathematical instruments such as globes, the cross-staffe, and the sector, suggesting that his lectures and personal instruction would have emphasized this sort of instrumental mathematical knowledge and understanding. While some historians have questioned what happened at Hood's lectures (or if indeed they did happen), this larger evidence indicates both that there were such lectures, and that a number of leaders of the community, as well as mathematical practitioners like Hood, thought they were important in creating mathematical literacy and conversation in the city of London.[19] This was the beginning of a recognition of the power of mathematics for understanding the answers to practical problems and with it a sense that mathematical answers were as legitimate as philosophical ones.

MATHEMATICS CHANGES THE GEOGRAPHICAL CONVERSATION

Given that the connection between mathematics and natural philosophy was a new interdisciplinary interaction, the best place to find such an interconnection would be in a study that blended measurement and larger philosophical theories. One such area of interest was to be found in the mathematical study of geography.

In sixteenth-century England, geography was a flourishing area of investigation. It was studied as part of the arts curriculum at both Oxford and Cambridge and therefore made up part of the worldview of most educated gentlemen and merchants.[20] The study of geography included a mathematical model of the earth, descriptions of its distant lands and inhabitants, and the local history of more immediate surroundings, what I have elsewhere labeled mathematical geography, descriptive geography, and chorography.[21] Because it relied on geographers of antiquity, such as Ptolemy and Strabo, to provide a backbone for modern investigation, geography was a discipline that used the methods of the humanists and the tradition of university scholars. Equally, geography was a study inspired by and reliant on new discoveries, voyages, and travels and so was integrally connected to the testimony and experience of practical men. Thus, geography existed as a point of contact for theoretical university scholars and practical men of affairs. Equally, it provides an excellent example of how mathematics could change the natural philosophical conversation, as well as the people conversing.

Geography embodied that dynamic tension between the world of the scholar, since geography was an academic subject legitimated by its classical, theoretical, and mathematical roots, and the world of the artisan, since it was inexorably linked with economic, nationalistic, and practical endeavors. It provided a synthesis that enabled its practitioners to move beyond the confines of natural philosophy to embrace a new ideal of science as a powerful tool for understanding and controlling nature. The usefulness of geographical study was of paramount importance to the new men attending the universities in ever greater numbers and it was this concept of utility to the state and to the individual that drove these new university men to investigate and appreciate geography.[22] The geographical community, then, was a wide-ranging group, with many different concerns and goals, but with a desire to be useful to the nation and to their own self-interest and a vision of England as an increasingly illustrious player on the world stage.

The English geographical community was complex, due in large part to its necessarily close connection between handwork and brainwork. Even the most theoretical geographer required the information and insight of navigators, instrument makers, cartographers, and surveyors in order to understand the terraequeous globe. This can be seen in the work of Richard Hakluyt, who used sailors' tales to construct a description of the world and

England's role in its discovery, and who in *Principal Navigations* created a predominantly practical document with important theoretical insights. Edward Wright, a serious mathematical geographer whose first-hand experience on voyages of discovery deeply affected his research program, also provides an important example of someone who mediated between theory and practice. Equally, the collaboration between John Dee, a university-trained mathematician and geographer, and Henry Billingsley, a London merchant, in the 1570 translation of Euclid indicates the fruitful exchange between the life of the mind and that of the marketplace (Dee 1570). Dee's career provides a particularly telling example of the importance of the theoretical-practical spectrum, and with it an interest in both mathematics and natural philosophy. Dee would probably have identified himself as a natural philosopher and certainly worked throughout his life to create a new theoretical worldview as well as to achieve a higher social status. Yet he was engaged much of the time in more practical mathematical pursuits, especially astronomical and geographical ones.[23] He advised most navigators setting out on northwest or northeast voyages, devised map projections and navigational instruments, and wrote position papers for the Privy Council on the political ramifications of English geographical emplacement.[24] Thus Dee, like many other mathematical practitioners, developed multiple and overlapping roles as scholar, craftsman, and statesman. This complex social world encouraged such men to integrate mathematics into their larger natural philosophical investigations and explanations.

WRIGHT AND HARRIOT AS MATHEMATICAL PRACTITIONERS

There are many examples of these interactive roles and disciplines. Two geographers who combined the life of the natural philosophical scholar with that of the mathematical practitioner were Edward Wright (Apt 2004) and Thomas Harriot (Roche 2004). Both were university-educated men, who had learned the classical foundations of their subject, as well as recent discoveries and theories. But these two were not isolated or traditional scholastics. Both went on prolonged voyages of discovery and learned navigation and its problems from the rude mechanicals and skilled navigators they encountered. They recognized the need to use mathematics to measure and understand the practical problems they encountered. They went beyond this practical knowledge, however, to try and formalize the structure of the globe and the understanding of the new world. Both were connected with important courts and patrons, and both used the cry of utility and imperialism to argue the need for geographic knowledge.

Edward Wright, the most famous English geographer of the period, was educated at Gonville and Caius College, Cambridge, receiving his bachelor's degree in 1581 and his master's in 1584. He remained at Cambridge until the end of the century, with a brief sojourn to the Azores with the Earl of Cumberland in 1589.[25]

In 1599 Edward Wright translated Simon Stevin's *The Haven-finding Arte* from the Dutch (Wright 1599a; Taylor 1954, #100). In this work Stevin claimed that magnetic variation could be used as an aid to navigation in lieu of the calculation of longitude (Wright 1599a, 3).[26] He set down tables of variation, means of finding harbors with known variations, and methods of determining variations. In his translation Wright called for systematic observations of compass variation to be conducted on a worldwide scale, "that at length we may come to the certaintie that they which take charge of ships may know in their navigations to what latitude and to what variation (which shal serve in stead of the longitude not yet found) they ought to bring themselves" (Wright 1599a, preface, B3a).[27]

Wright's work demonstrates a close connection between navigation and the promotion of a "proto-Baconian" tabulation of facts meant both for practical application and scientific advancement. Here appears the foundation of an experimental science, grounded in both practical application and theoretical mathematics, quite separate from any more traditional Aristotelian natural philosophy or Neoplatonic mathematics. Unfortunately, Wright's scheme was not entirely successful. By 1610, in his second edition of *Certaine Errors in Navigation*, Wright had constructed a detailed chart of compass variation—but he had also become more hesitant in his claims concerning the use of variation to determine longitude (Wright 1610a, sigs. 2P1a-8a; Waters 1958, 316).

Wright's greatest achievement was *Certaine Errors in Navigation* (1599), his appraisal of the problems of modern navigation and the need for a mathematical solution. In this book, Wright explained Mercator's map projection for the first time, providing an elegant Euclidean proof of the geometry involved. He also published a table of meridian parts for each degree, which enabled cartographers to construct accurate projections of the meridian network, and offered straightforward instructions on map construction (Wright 1599b, sigs. D3a-E4a; Taylor 1954, #99). He also constructed his own map using this method. Wright's work was the first truly mathematical rendering of Mercator's projection and placed English mathematicians, for a time, in the vanguard of European mathematical geography. It was equally significant for the close communication it claimed and required of theoretical mathematicians and practical navigators.

At about the turn of the century, Wright moved from Cambridge to London, where he established himself as a teacher of mathematics and geography, following in the footsteps of Robert Hood. At about the same time, he contributed to Gilbert's work on magnetism, providing a practical perspective to Gilbert's more natural philosophical outlook (Pumfrey 2002, 175–81). He created a world map using Mercator's techniques and probably aided in the construction of the Molyneux globes (Wallis 1952; 1989, 94–104). In the early seventeenth century, he is said to have become a tutor to Henry, Prince of Wales (elder son of James), a claim strengthened by Wright's dedication of his second edition of *Certaine Errors* to Henry in 1610 (Wright 1610a, sigs. *3a–8b, X1–4; Birch 1760, 389). Upon becoming tutor, Wright "caused a large sphere to be made for his Highness, by the help of some German workmen; which sphere by means of spring-work not only represented the motion of the whole celestial sphere, but shewed likewise the particular systems of the Sun and Moon, and their circular motions, together with their places, and possibilities of eclipsing each other. In it was a work by wheel and pinion, for a motion of 171000 years, if the sphere could be kept to long in motion" (Birch 1760, 389).[28]

Henry had a decided interest in such devices and rewarded those who could create them.[29] In addition, Wright designed and constructed a number of navigational instruments for the prince and prepared a plan to bring water down from Uxbridge for the use of the royal household (Strong 1986, 218; Wright 1610b, identified by Taylor 1934). In or around 1612, Wright was appointed librarian to Prince Henry, but Henry died before Wright could take up the post (Strong 1986, 212). In 1614, Wright was appointed by Sir Thomas Smith, governor of the East India Company, to lecture to the company on mathematics and navigation, for which he was paid £50 per annum (Waters 1958, 320–21). There is some speculation as to whether or not Wright actually gave these lectures, since he died the following year.

Wright thus provides a nice example of a mathematical practitioner who provided both intellectual and social connections between theory and practice. He was university-trained and worked as a teacher at various points in his career. He was interested in theoretical problems, including the mathematically sophisticated construction of map projections, and aided Gilbert in his philosophical enterprise. On the other hand, this was an academic who respected practical experience. He himself experienced the problems of ocean navigation, he built instruments, and he solicited the help and opinion of sailors and navigators. His motivation for this balancing of handwork and brainwork were many, probably including financial gain and social prestige as well as more intellectual concerns. He was certainly concerned with the usefulness of his investigations and, through the patronage support of aristocrats, Prince Henry, and the East India Company (somewhat latterly), was able to argue the utility of geographical knowledge both to imperial and mercantile causes. Mathematics provided a language for both his practical and theoretical pursuits, demonstrating a new integration of these different branches of knowledge.

Another preeminent figure in mathematical geography, also connected with Prince Henry, was Thomas Harriot (Shirley 1983). Harriot attended Oxford at the same time as Wright was at Cambridge. He matriculated from St. Mary's Hall in 1577 and received his bachelor's degree in 1580. By 1582 he was in the employ of Sir Walter Ralegh, who sent him to Virginia in 1585. Harriot, like Wright, was an academic and theoretical geographer whose sojourn into the practical realm of travel and exploration helped form his conception of the vast globe and of what innovations were necessary to travel it. Harriot's description of Virginia, seen in his *Brief report of . . . Virginia* (1588),[30] was "the first broad assessment of the potential resources of North America as seen by an educated Englishman who had been there" (Quinn 1974, 45).[31] Harriot compiled the first word list of any North American Indian language (probably Algonquin) (Shirley 1983, 133), a necessary first step of classifying in order to control, thus illustrating that inductive spirit never far from the heart of even the most mathematical geographer. He saw Virginia's great potential for English settlement, provided that the natives were treated with respect and that missionary zeal and English greed were kept to a minimum.[32] His advice concerning Virginian settlement was to prove important as the Virginia companies of the seventeenth century were established. This was the work of a man very aware of the practical and economic ramifications of the intellectual work of describing the larger world, as well as the imperial imperatives at work.

More important for Harriot were issues of the mathematical structure of the globe. Indeed his mathematics was bound up closely with his imperial attitude generally and the experience of his Virginian contacts in particular.[33] He was deeply concerned about astronomical and physical questions, including the imperfection of the moon and the refractive indexes of various materials (Shirley 1983, 381–416). Harriot was inspired by Galileo's telescopic observations of the moon and produced several fine sketches himself after *The Starry Messenger* appeared. He also investigated one of the most pressing problems of seventeenth-century mathematical geography—the problem of determining longitude at sea. Harriot worked long and hard on the longitude question and on other navigational problems, relating informally to many mathematical geographers his conviction that compass variation contained the key to unraveling the longitude knot (Harriot 1596).

Harriot was a mathematical tutor to Sir Walter Ralegh for much of the last two decades of the sixteenth century, advising his captains and navigators, as well as pursuing research interesting to Ralegh. As Richard Hakluyt said of Harriot, in a dedication to Ralegh: "By your experience in navigation you saw clearly that our highest glory as an insular kingdom would be built up to its greatest splendor on the firm foundation of the mathematical sciences, and so for a long time you have nourished in your household, with a most liberal salary, a young man well trained in those studies, Thomas Harriot, so that under his guidance you might in spare hours learn those noble sciences."[34]

As Ralegh fell from favor, eventually ending up in the Tower, Harriot began to move his patronage expectation to another aristocrat interested in mathematical and geographical pursuits, the ninth Earl of Northumberland (the so-called Wizard Earl). Although Harriot's relationship with Northumberland is somewhat obscure, he appears to have conducted research within Northumberland's circle and occasionally his household, as well as acting as a tutor as needed. Finally, Harriot was also connected with Henry, Prince of Wales, as a personal instructor in applied mathematics and geography, just as Wright had been (Shirley 1985, 81). It is likely that Wright and Harriot met at Henry's court. As two university-trained contemporaries, with very similar interests and experiences, they would have gained much from their association. Given their mutual interests, it would have made sense for them to discuss matters of mutual geographical and mathematical interest while at court together.

Harriot's career displays many of the same characteristics as Wright's. Harriot too was a man who drifted in and out of academic pursuits, from university, to Virginia, to positions as researcher and tutor for Ralegh and Northumberland. In some ways, he was less connected to practical pursuits than Wright, although his trip to Virginia and his work on longitude indicate his engagement with issues of practical significance. Harriot was also dependent on patronage, especially that of Ralegh and of Northumberland (poor choices as they turned out to be), and used this patronage to help create an intellectual community in which mathematical theory and imperial utility could be considered equally important.

Wright and Harriot, as well as a host of other geographers interested in this interconnection between theoretical and practical issues, combined an interest in the mathematical construction of the globe and a new, more wide-reaching understanding of basic geographical concepts with a desire for political and economic power on the part of princes, nobles, and merchants. This wide-ranging area of investigation encouraged associations to develop between academic geographers, instrument makers, navigators, and investors. The result was a negotiation between theoretical and practical issues, which helped introduce mathematics as a common language and rigorous means of analysis. This fruitful association between theory and practice helped to determine the kinds of questions these men asked, the kinds of answers that were acceptable, and the model of the world that would be developed. It was the work of mathematical practitioners such as these geographers that introduced mathematics to natural philosophical questions and to questions of natural knowledge more broadly. The utility, or at least the perceived utility, of such a language is part of the reason that mathematics became the language of nature in the years to come.

CONCLUSION

Hood, Wright, and Harriot provide good examples of the kind of investigators necessary for the introduction of the language of mathematics into the study of nature. These three men, and many other mathematical practitioners, represent the communication between theory and practice, both within their own careers and ideas, and between universities, courts, print shops, the shops of instruments makers, and many other liminal venues. Their lives and careers show that new locales were becoming important for the pursuit of natural knowledge, including urban shops and houses on the one hand, and the courts and stately homes of aristocratic and noble patrons on the other. Wright and Harriot also demonstrate within their scientific worldviews an interesting mixture of theory, inductive fact gathering, and quantification, which provided part of the changing view toward nature and its investigation so important for the changing emphasis on mathematization.

They were both concerned with practicality and utility, especially within the rhetoric they employed to argue their cause, and mathematics seemed to them to provide the useful answers necessary to their careers and their ambitions. But they were also convinced that mathematics would help them to understand the natural world more sufficiently. Their connections to mercantilism were important, but do not provide a complete answer to the changing emphasis of the study of nature (as Edgar Zilsel [1942] once suggested).[35] This was not science directed by the bottom line of mercantilist expenditure, but rather a more complex interaction among court, national and international intellectual communities, and mercantilist enterprise.

Thus, the mathematical practitioners provided an agent for the changing nature of the scientific enterprise in the early modern period. They created a fundamental step toward the introduction of the language of mathematics into the larger study of nature. They did this by combining theory and practice in a new and interesting way. They did so for reasons that included the economic and bourgeois changes that were directly affecting Europe. These men were also concerned with issues of nationalism, imperialism, cultural credit, and status, issues that do not fit easily into a more Marxist and materialist interpretation.

Did this change the enterprise of natural philosophy? Yes. Because these men were interested in mathematics, measurement and quantification became increasingly more significant. Their social circumstances ensured that the investigation of nature must be seen to be practical, using information from any available source, and science developed a rhetoric of utility and progress, as well as an inductive methodology, in response. Intimately connected to national pride and mercantile profit, the science that developed in this period reflected those concerns. In essence, in large part because of the work of mathematical practitioners like Hood, Wright, and Harriot, the investigation of nature began to take place away from the older university venue (though there remained important connections), with new methodologies, epistemologies, and ideologies of utility and progress. The scientific revolution had begun.

But there was still something missing. Hood, Wright, and Harriot did not make the transition to natural philosophers. Despite their best efforts, they remained mathematical practitioners. And by the end of the seventeenth century, mathematical practitioners had been reduced to technicians, whose presence became less and less visible.[36] Meanwhile, natural philosophers such as Robert Boyle and Isaac Newton removed themselves from the company of mathematical practitioners, even as they utilized the fruits of their labor. In other words, the final translation of mathematics as a tool and language of natural philosophy involved another social transformation, which devalued the very group that had made it possible.

NOTES

1. See Lindberg (1990, 1–26) for a discussion of early uses of the scientific revolution as a concept, and Lindberg (16) and Cohen (1994, 88–97) for a fuller treatment of Burtt.

2. Steven Shapin (1996), despite his opening caveat, does a good job of laying out some of the changes taking place that made up the scientific revolution, as more recently has John Henry (2001).

3. Wisdom of Solomon 11:20.

4. Kuhn (1977, 31–65) separates these two traditions. See also Cunningham and Williams (1993) and Dear (2001) for discussion of this separation.

5. Steven Shapin (1982) made a case for this new interpretation, and then, with Simon Schaffer, provided an extremely influential case study (1985).

6. With some modification, I take the important classification of the more practical men in Taylor (1954). For modern treatment of these crucial figures, see Bennett (1986) and Johnston (1991; 1994). Most recently, Pamela Long (2011) discusses these issues.

7. Kuhn (1977) provides an early attempt to claim a different history for mathematics and natural philosophy.

8. In Cormack (1997, 108, 110) I argue that Recorde's books were owned by many university students and college libraries in the late sixteenth and early seventeenth centuries.

9. For a more general discussion of early modern London, see Rappaport (1989) and Archer (1991).

10. Neal (1999) discusses some attempts to make mathematics appear useful.

11. Thomas Hood's lecture, (n.d. 1588) is a good example. See Harkness (2007) for a discussion of the complex interactions among London merchants, artisans, and scholars.

12. Of course, once Galileo successfully gained a patronage position, particularly with the Florentine Medici court, he left his mathematical practitioner roots behind and became a much higher status natural philosopher (Biagioli 1993).

13. Descartes was Jesuit-trained (Dear 1995).

14. Biographical material on Thomas Hood can be found in Taylor (1954, 40–41); Waters (1958, 186–89); Higton (2004).

15. British Library, *Lansdowne* 101, f. 56.

16. British Library, *Lansdowne* 101, f. 58.

17. Hood (1596; 1598) are bound together in BL 529, g. 6.

18. Thomas Hood lists himself as a mathematical lecturer on the frontispiece of the following books: Hood (n.d. [1588]; 1590); Ramus (1590); Hood (1592; 1596).

19. Further evidence of Hood's lectures is the fact that John Stow mentions them (1598, 57).

20. See Cormack (1997, 17–47) for a full treatment of the place of geography in the university curriculum.

21. See Cormack (1991) for a description of the different types of geography studied in sixteenth- and seventeenth-century England.

22. See Stone (1964) for an evaluation of the growing numbers of new men at the universities in this period.

23. For the natural philosophical work, see Clulee (1988). For his practical advising, see Sherman (1995).

24. In many ways, Dee is an English Galileo, providing a crossover from mathematical practitioner to court natural philosopher. It is no surprise, however, that Zilsel did not mention him, since his magical heritage, made famous by Frances Yates (1972), among others, discounted him in Zilsel's mind as a true scientist (Zilsel 1942; Harkness 1999).

25. As a result of this voyage, Wright wrote (1589), which was later printed by Richard Hakluyt, "written by the excellent Mathematician and Enginier master Edward Wright" (1598–1600). Hall (1962, 204), Waters (1958, 220), and Shirley (1985, 81) all cite this trip to the Azores as the turning point in Wright's career, his road to Damascus, since it convinced him in graphic terms of the need to revise completely the whole navigational theory and procedure.

26. Bennett (1991, 186) marks the relationship between magnetism and longitude as one of the important sites of the scientific revolution.

27. Waters (1958, 237).

28. "Mr. Sherburne's Appendix to his translation of Manilius, p. 86" in Birch (1760, 389).

29. Smuts (1987) especially mentions Salomon de Caus's *La perspective avec la raison des ombres et miroirs* (London, 1612), dedicated "Au Serenissime Prince Henry," 157.

30. Harriot, *Briefe and True Report*, reproduced verbatim in T. de Bry,

America. Pars I, published concurrently in English (Frankfurt, 1590) and in Hakluyt (1598–1600, 3:266–80).

31. See Alexander (2002) for an interesting interpretation of Harriot's mathematics.

32. The manuscript information concerning this expedition is gathered together in Quinn (1955, 36–53). Shirley (1983, 152ff) discusses Harriot's desire for noninterference. To see White's illustrations of this expedition, see White (2006).

33. Alexander argued that Harriot's work on the continuum was influenced by his view of geographical boundaries and the "other." "The geographical space of the foreign coastline and the geometrical space of the continuum were both structured by the Elizabethan narrative of exploration and discovery" (1995, 591). Alexander (2002) develops this further.

34. Richard Hakluyt, introduction to Peter Martyr, as quoted in Shirley (1985, 80). See Shirley (1983) for a fuller discussion.

35. See Raven and Krohn (2000, xx–xlvi) for an appraisal of the intellectual climate in which Zilsel worked.

36. Brotton (1997, 186) shows that cosmographers had become employees of the joint stock companies by the end of the seventeenth century, while Shapin (1994, 355–408) argues for the increasing invisibility of technicians. Sprat (1667, 392) celebrates the distance between gentlemen who create new knowledge and technicians who can only do as they are told. Jardine (2003) suggests that Hooke remained a technician.

REFERENCES

Alexander, A. R. 1995. "The Imperialist Space of Elizabethan Mathematics." *Studies in the History and Philosophy of Science* 26: 559–91.

———. 2002. *Geometrical Landscapes: The Voyages of Discovery and the Transformation of Mathematical Practice.* Stanford, Calif.: Stanford University Press.

Apt, A. J. 2004. "Wright, Edward (bap. 1561, d. 1615). In *Oxford Dictionary of National Biography*, ed. H. C. G. Matthew and B. Harrison; online edition, ed. L. Goldman. October 2006. http://www.oxforddnb.com/view/article/30029.

Archer, I. 1991. *The Pursuit of Stability: Social Relations in Elizabethan England.* Cambridge: Cambridge University Press.

Bennett, J. A. 1986. "The Mechanic's Philosophy and the Mechanical Philosophy." *History of Science* 24: 1–28.

———. 1991. "The Challenge of Practical Mathematics." In *Science, Belief, and*

Popular Culture in Renaissance Europe, ed. S. Pumfrey, P. Rossi, and M. Slawinski, 176–90. Manchester: Manchester University Press.

Biagioli, M. 1989. "The Social Status of Italian Mathematicians, 1450–1600." *History of Science* 27: 41–95.

———. 1993. *Galileo Courtier: The Practice of Science in the Culture of Absolutism*. Chicago: University of Chicago Press.

———. 2007. *Galileo's Instruments of Credit: Telescopes, Images, Secrecy*. Chicago: University of Chicago Press.

Birch, T. 1760. *Life of Henry, Prince of Wales, Eldest Son of King James I*. London.

British Library. *Lansdowne MS*. 101.

Brotton, J. 1997. *Trading Territories: Mapping the Early Modern World*. London: Reaktion.

Burtt, E. A. 1924. *The Metaphysical Foundations of Modern Science*. London: Dover.

Butterfield, H. 1949. *The Origins of Modern Science, 1300–1800*. London: Macmillan.

Clulee, N. 1988. *John Dee's Natural Philosophy: Between Science and Religion*. London: Routledge.

Cohen, H. F. 1994. *The Scientific Revolution: An Historiographical Enquiry*. Chicago: Chicago University Press.

Cormack, L. B. 1991. "'Good Fences Make Good Neighbors': Geography as Self-Definition in Early Modern England." *Isis* 82: 639–61.

———. 1997. *Charting an Empire: Geography at the English Universities, 1580–1620*. Chicago: University of Chicago Press.

———. 2003. "The Grounde of Artes: Robert Recorde and the Role of the Muscovy Company in an English Mathematical Renaissance." *Proceedings of the Canadian Society for the History and Philosophy of Mathematics* 16: 132–38.

Cunningham, A., and P. Williams. 1993. "De-centring the 'Big Picture': The Origins of Modern Science and the Modern Origins of Science." *British Journal for the History of Science* 26: 407–32.

Dear, P. 1995. *Discipline and Experience: The Mathematical Way in the Scientific Revolution*. Chicago: University of Chicago Press.

———. 2001. *Revolutionizing the Sciences: European Knowledge and Its Ambitions, 1500–1700*. Princeton, N.J.: Princeton University Press.

Dee, J. 1570. *The Mathematical Preface to the Elements of Geometrie of Euclid of Megara*. London: John Daye.

Dijksterhuis, E. J. 1961. *The Mechanization of the World Picture*. Oxford: Oxford University Press.

Elyot, T. 1531. *The Boke Named the Governor.* London: Thomas Berthelet.

Feingold, M. 1984. *The Mathematicians' Apprenticeship. Science, Universities, and Society in England, 1560–1640.* Cambridge: Cambridge University Press.

Goulding, R. 2010. *Defending Hypatia: Ramus, Savile and the Renaissance Rediscovery of Mathematical History.* New York: Springer.

Hakluyt, R. 1598–1600. *Principal Navigations, Voiages, Traffiques and Discoveries of the English Nation.* 3 vols. London: G. Bishop, R. Newberie, and R. Barker.

Hall, M. B. 1962. *The Scientific Renaissance 1450–1630.* London: Dover.

Harkness, D. E. 1999. *John Dee's Conversations with Angels: Cabala, Alchemy and the End of Nature.* Cambridge: Cambridge University Press.

———. 2007. *The Jewel House. Elizabethan London and the Scientific Revolution.* New Haven, Conn.: Yale University Press.

Harriot, T. 1596. "Of the Manner to observe the Variation of the Compasse, or of the wires of the same, by the sonne's rising and setting." British Museum manuscript Add. MS 6788.

Henry, J. 2001. *The Scientific Revolution and the Origins of Modern Science.* Basingstoke: Houndmills.

Higton, H. K. 2004. "Hood, Thomas (bap. 1556, d. 1620)." In *Oxford Dictionary of National Biography,* ed. H. C. G. Matthew and B. Harrison, n.p. Oxford: Oxford University Press. http://www.oxforddnb.com/view/article/13680.

Hood, T. (n.d. [1588]). "A Copie of the Speache made by the Mathematicall Lecturer, unto the Worshipfull Companye present . . . in Gracious Street: the 4 of November 1588." London.

———. 1590. *The Use of the Celestial Globe in Plano, set foorth in two Hemispheres.* London: [by John Windet] for Tobie Cooke.

———. 1592. *The Use of Both the Globes, Celestiall, and Terrestriall, most plainely delivered in forme of a Dialogue.* London: Thomas Dawson.

———. 1596. *The Use of the Two Mathematicall Instruments, the Crosse Staff, . . . And the Iacobs Staffe.* London: by Richard Field for Robert Dexter.

———. 1598. *The Making and Use of the Geometricall Instrument, called a Sector.* London: John Windet.

Jardine, L. 2003. *The Curious Life of Robert Hooke: The Man Who Measured London.* New York: Harper.

Johnston, S. 1991. "Mathematical Practitioners and Instruments in Elizabethan England." *Annals of Science* 48: 319–44.

———. 1994. *Making Mathematical Practice: Gentlemen, Practitioners, and Artisans in Elizabethan England.* PhD diss., University of Cambridge.

———. 2004. "Recorde, Robert (c. 1512–1558)." In *Oxford Dictionary of National*

Biography, ed. H. C. G. Matthew and B. Harrison, n.p. Oxford: Oxford University Press; online edition, ed. Lawrence Goldman. http://www.oxforddnb.com/view/article/23241.

Koyré, A. 1957. *From a Closed World to an Infinite Universe.* Baltimore: Johns Hopkins University Press.

Kuhn, T. S. 1977. "Mathematical versus Experimental Tradition in the Development of Physical Science." In *The Essential Tension: Selected Studies in Scientific Tradition and Change,* 31–65. Chicago: Chicago University Press.

Lindberg, D. 1990. Introduction to *Reappraisals of the Scientific Revolution.* Ed. Lindberg and R. Westman, 1–26. Cambridge: Cambridge University Press.

Long, P. 2011. *Artisan/Practitioners and the Rise of the New Sciences, 1400–1600.* Corvallis: Oregon State University Press.

McConica, J. K. 1986. "Elizabethan Oxford: The Collegiate Society." In *The Collegiate University,* ed. J. K. McConica, 645–732. Vol. 3 of *The History of the University of Oxford,* ed. T. H. Aston. Oxford: Clarendon Press.

Moran, B. T., ed. 1991. *Patronage and Institutions. Science, Technology, and Medicine at the European Court.* Rochester, N.Y.: Boydell Press.

Neal, K. 1999. "The Rhetoric of Utility: Avoiding Occult Associations for Mathematics through Profitability and Pleasure." *History of Science* 37: 151–78.

Pace, R. 1517. *De Fructu qui ex Doctrina Percipitur.* Basel: Froben.

Pumfrey, S. 2002. *Latitude and the Magnetic Earth.* Duxford, UK: Icon.

Quinn, D. B. 1955. *The Roanoke Voyages, 1584–1589: Documents to Illustrate the English Voyages to North America.* London: Ashgate.

———. "Thomas Harriot and the New World." In *Thomas Harriot: Renaissance Scientist,* ed. J. W. Shirley. Oxford: Clarendon Press.

Ramus, P. 1590. *The Elementes of Geometrie.* Trans. T. Hood. London: Printed by John Windet for Thomas Hood.

Rappaport, S. 1989. *Worlds within Worlds: Structures of Life in Sixteenth Century London.* Cambridge: Cambridge University Press.

Raven, D., and W. Krohn. 2000. Introduction to *Edgar Zilsel: The Social Origins of Modern Science,* ed. D. Raven, W. Krohn, and R. S. Cohen, xx–xlvi. Dordrecht: Kluwer Academic Publishers.

Recorde, R. 1543. *The Grounde of Artes.* London: Reynold Wolfe.

———. 1551. *The Pathway to Knowledge.* London: Reynold Wolfe.

———. 1556. *The Castle of Knowledge, containing the explication of the Sphere.* London: Reynold Wolfe.

———. 1557. *Whetsone of Witte.* London: John Kingston.

Roche, J. J. 2004. "Harriot, Thomas (c.1560–1621)." In *Oxford Dictionary of National Biography*, ed. H. C. G. Matthew and B. Harrison; online edition, ed. L. Goldman. October 2006, http://www.oxforddnb.com/view/article/12379.

Shapin, S. 1982. "History of Science and its Sociological Reconstructions." *History of Science* 20: 157–211.

———. 1994. *A Social History of Truth: Civility and Science in Seventeenth-Century England.* Chicago: University of Chicago Press.

———. 1996. *The Scientific Revolution.* Chicago: University of Chicago Press.

Shapin, S., and S. Schaffer. 1985. *The Leviathan and the Air Pump: Hobbes, Boyle, and the Experimental Life.* Princeton, N.J.: Princeton University Press.

Sherman, W. H. 1995. *John Dee: The Politics of Reading and Writing in the English Renaissance.* Amherst: University of Massachusetts Press.

Shirley, J. W. 1983. *Thomas Harriot: A Biography.* Oxford: Oxford University Press.

———. 1985. "Science and Navigation in Renaissance England." In *Science and the Arts in the Renaissance*, ed. J. W. Shirley and F. D. Hoeniger, 74–93. Washington, D.C.: Folger Shakespeare Library.

Smith, P. H., and P. Findlen. 2002. *Merchants and Marvels: Commerce, Science and Art in Early Modern Europe.* New York: Routledge.

Smuts, R. M. 1987. *Court Culture and the Origins of a Royalist Tradition in Early Stuart England.* Philadelphia: University of Pennsylvania Press.

Sprat, T. 1667. *The History of the Royal Society of London.* London.

Stone, L. 1964. "The Educational Revolution in England, 1560–1640." *Past and Present* 28: 41–80.

Stow, J. 1598. *Survey of London.* London: [John Windet for] John Wolfe.

Strong, R. 1986. *Henry, Prince of Wales and England's Lost Renaissance.* New York: Thames and Hudson.

Taylor, E. G. R. 1934. *Late Tudor and Early Stuart Geography 1583–1650.* London: Octagon.

———. 1954. *Mathematical Practitioners of Tudor and Stuart England.* Cambridge: Cambridge University Press.

Wallis, H. M. 1952. "The Molyneux Globes." *B.M. Quarterly* 16: 89–90.

———. 1989. "'*Opera Mundi*': Emery Molyneux, Jodocus Hondius and the First English Globes." In *Theatrum Orbis Librorum*, ed. T. Croiset van Uchelen, K. van der Horst, and G. Schilder, 94–104. Utrecht: HES Publishers.

Waters, D. W. 1958. *The Art of Navigation in England in Elizabethan and Early Stuart Time.* Chicago: University of Chicago Press.

White, Thomas. 2006. "Picturing the New World. The Hand-Coloured De Bry Engravings of 1590." University Library, University of North Carolina at Chapel Hill. http://www.lib.unc.edu/dc/debry/about.html.

Wright, E. 1589. "The Voiage of the right honorable George Erl of Cumberland to the Azores." In *Principal Navigations, Voiages, Traffiques and Discoveries of the English Nation*, ed. R. Hakluyt, vol. 2, part 2, 155 [misnumbered as 143]–168. London: G. Bishop, R. Newberie, and R. Barker.

———. 1599a. *Haven Finding Art.* London: G. Bishop, R. Newberie, and R. Barker.

———. 1599b. *Certaine Errors in Navigation.* London: Valentine Sims [and W. White].

———. 1610a. *Certaine Errors in Navigation.* 2nd ed. London: Felix Kingsto[n].

———. 1610b. *Plat of Part of the Way Whereby a Newe River May be Brought from Uxbridge to St. James, Whitehall, Westminster, the Strand, St. Giles, Holbourne and London.* MS.

Yates, F. 1972. *The Rosicrucian Enlightenment.* London: Routledge.

Zilsel, E. 1942. "The Sociological Roots of Science." *American Journal of Sociology* 47: 552–55.

9

LEIBNIZ ON ORDER, HARMONY, AND THE NOTION OF SUBSTANCE

Mathematizing the Sciences of Metaphysics and Physics

KURT SMITH

THE MARCH 1694 edition of *Acta Eruditorum* included a short article by Leibniz titled "On the Correction of First Philosophy and the Notion of Substance."[1] In it Leibniz complains about an emerging trend of obscurity in metaphysics, attributing it in part to a widening rift between work in metaphysics and work in mathematics. In July of that year Leibniz wrote to Jacques-Bénigne Bossuet, including with the letter a copy of a manuscript titled "Reflections on the Advancement of the True Metaphysics and Particularly on the Nature of Substance Explained by Force."[2] This was an expanded version of the *Acta Eruditorum* article. As in the article, Leibniz complains about the growing rift between metaphysics and mathematics. His remedy for philosophers is implicit though clear: the growing obscurity in metaphysics could be wiped clean by the co-opting of mathematics. By November of that same year, he looks to have adopted his remedy, writing to the Marquis de l'Hospital, "My metaphysics is all mathematics, so to speak, or it can become so" (1846–60, GM 2:255–62).[3] This suggests that Leibniz believed that he had been able to wipe clean any obscurity that may have existed with respect to the most important components of his metaphysics, by clearly articulating them by way of certain concepts taken from mathematics.

In both the *Acta Eruditorum* article and the manuscript sent to Bossuet, Leibniz claims that the centerpiece of his new philosophical system is his *notion* (*notione*) of a substance. For not only does it yield "important truths about God, the soul, and the nature of body" (Leibniz 1998 WF, 141), but it makes a connection between the concepts of substance and force, a connection that turns out to be of great importance to his physics. One of his earliest attempts at discussing the notion of a substance in light of certain mathematical concepts appears in a 1686 manuscript not published during his lifetime, but published later (in 1846) and then officially titled *Discourse on Metaphysics*.[4] So, he had been hard at work for some time (at least eight years) before the *Acta Eruditorum* article using concepts taken from mathematics to more clearly formulate certain components of his metaphysics. A look at other texts, especially those written around the time of the *Acta Eruditorum* article, so around 1695, show that after the *Discourse* he had made advances on his use of mathematics as a means of making clearer his metaphysics, which in turn, as emphasized in this chapter, helped to clarify his physics. Specifically, the advances can be found in the *New System* and the *Specimen Dynamicum*. For lack of a better term, I shall call this use of mathematics in metaphysics and physics the "mathematization" of these two sciences. I recognize that this is not the only way this term can be used. I propose to look carefully at the mathematization as it emerges in Leibniz's work, focusing specifically on how it clarifies his metaphysical concepts of *order* and *harmony*, concepts vital to the physics.

The *Discourse* is among the earliest texts in which we find Leibniz discussing the notion of a substance and what he calls universal order (*l'ordre universel*) in light of certain concepts taken from mathematics. The universal order, he says, is regulated by the most general of God's laws (*le plus generale des loix de Dieu*), which, he is clear to assert, are exceptionless (*sans exception*; *Discourse*, §7; Leibniz 1879 [GP] 4:432). Like everything in the cosmos, the notions of individual substances, which we shall see are themselves laws, are also subject to the most general of God's laws. Although the particular examples taken from mathematics do not help make clearer the sense in which this order is *universal*, they do help to make clearer Leibniz's metaphysical conception of order, on which the notion of an individual substance is built. About order he writes:

> Suppose, for example, that someone puts a number of completely haphazard points on paper, as do people who practice the ridiculous art of geomancy. I say that it is possible to find a geometrical line whose notion is constant and uniform according to a certain rule, such that the line passes through all the points, and in the same order as they were drawn. And if someone drew a continuous line which is sometimes straight, sometimes follows a circle, and sometimes of some other kind, it would be possible to find a notion (*notion*) or rule (*regle*) or equation (*equation*) common to all the points on this line in virtue of which these same changes would occur. (*Discourse* §6; GP 4 431)

Order is inherent in the cosmos. "For regards the universal order," he says, "everything conforms to it. So much is this true that not only does nothing happen in the world which is absolutely irregular, but also that we can't even imagine such a thing" (ibid.). Even in the case where points are drawn haphazardly on a piece of paper, an equation nevertheless could be constructed that would express a rule that would describe a line that not only connects the points but connects them in exactly the order in which they were drawn. Of course, the line represents only *an* order discoverable in the data (here, the data are the points), its notion or equation understood as that which expresses that order as represented by that line.

The principal aim of the discussion of the metaphysics of order is to make *intelligible* the concept of order, to make clear its very *possibility*. This concept is in turn important to the physics, since one of the aims of the physics is to provide an account of *the* order of things (as opposed to just *an* order). The order dealt with in the physics presupposes the metaphysics of order. About his own use of mathematics in the *Discourse*, Leibniz writes, "I make use of these comparisons in order to sketch some imperfect picture of the divine wisdom, and to say something which might at least raise our minds to some sort of conception of what cannot be adequately expressed" (GP 4 431). In a letter to the Count Ernst von Hessen-Rheinfels, which dealt specifically with issues in the *Discourse*, Leibniz says that when conceiving the notions of individual substances (the notions understood as "final species"), in line with Aquinas, we should not conceive them "physically, but metaphysically or mathematically" (GP 2 131).[5] These statements align with what he will say later in the *Acta Eruditorum* article, the manuscript sent to Bossuet, the *New System*, the *Specimen Dynamicum*, and the letter to l'Hospital.

First let us consider some specific instances in which he appeals to mathematics in making clearer certain metaphysical concepts. As noted, the mathematical concepts of *equation* and *geometrical line* were applied in section 6 of the *Discourse*, where he mentions the order discoverable among the points haphazardly drawn on a piece of paper. These mathematical concepts are again used in section 8, in the discussion of the notion of Alexander the Great.[6] In a 1690 letter to Arnauld, we get a slightly different picture. There, Leibniz says of each substance that it "contains in its nature the law (*legem*) of the continuous progression (*seriei*) of its own workings and all that has happened to it and all that will happen to it" (GP 2 136).[7] Scholars have referred to this law, the law that determines the series of changes in an individual substance over time, as the "law-of-the-series."[8] Leibniz in fact likens the law-of-the-series to what occurs in mathematics, where a rule or equation determines a series in numbers (Cover and O'Leary-Hawthorne 1999).[9] We have, then, two competing mathematical "pictures" of an individual substance and the notion that determines (or expresses) it. The first is the picture of a geometrical line and the equation that determines (or expresses) it; the second is the picture of a numerical series and the equation that determines (or expresses) it. They are related but distinct pictures. To be sure, both are metaphors in this context, but they are importantly instrumental in Leibniz's attempts at making clearer his metaphysics. They are part of what I earlier called the mathematization of his metaphysics. The first picture is again used in section 13, where this time Leibniz offers up a circle in his conception of the notion of Caesar (*Discourse*, §13; GP 4 437). The second picture is used in section 30. Leibniz says there that God "continually conserves and continually produces our being in such a way that our thoughts occur spontaneously and freely in the order laid down by the notion of our individual substance" (*Discourse*, §30; GP 4 454). Predicate-talk has been replaced with thought-talk, the notion of an individual substance now cast as that which orders a substance's series of thoughts. And, in yet another letter to Arnauld, Leibniz puts both mathematical pictures to use, casting the notion as that which orders the unfolding sequence or series of events that constitutes the individual substance while casting this substance's duration over time, the unfolding of that history of events, as a geometrical line (GP 2 43).[10] Although much can be said about the second picture—of individual substance as numerical series[11]—I shall focus on the first, which depicts an individual substance, and the notion that determines it, in terms of a geometrical line and the equation that organizes items to "produce" it.

EQUATIONS AND ORDER

In an early manuscript, dated July 11, 1677—so almost a decade before the *Discourse*—Leibniz instructs us on how to construct an equation of a geometrical line.[12] The construction is not the point of his discussion, but is required for the more sophisticated mathematical analysis to come. For our purpose we need to focus only on the equation's construction. First, he constructs a curve *DC* on paper, and then constructs, relative to *DC*, two straight lines, *AS*, which he calls "*y*," and *AB*, perpendicular to *AS*, which he calls "*x*."[13] In other letters he will refer to *AS* as the *ordinate* and to *AB* as the *abscissa*.

The equation of *DC*, he says, is constructed by thinking of it as expressing a relation between *AB* and *AS*. In other words, the equation of *DC* will produce ordered pairs $<x_n, y_n>$, which will be points located on *DC*, where x_n is located on *AB* and y_n is located on *AS*. Thus, referring now to Figure 1, the equation of *DC* will produce the ordered pairs $<x_1, y_1>$, $<x_2, y_2>$, and so on, which, using *AB* and *AS* as referents, are points lying on *DC*. There are a few philosophical issues that should be brought to light, so let me pause briefly to do that, after which I shall return to the specifics concerning equations and geometrical lines.

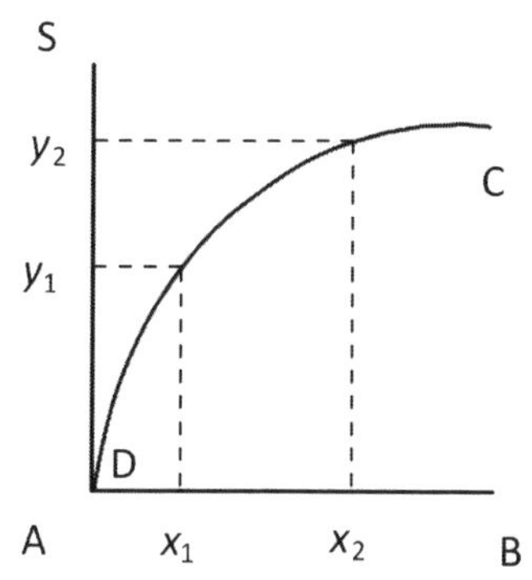

Figure 1. Algebraic construction of DC from AS and AB.

In 1695, Leibniz tells Foucher, in response to Foucher's remarks concerning the *New System*, that there is an important distinction between ideal things such as lines and points, and real or actual things such as corporeal substances (bodies), specifically offering the example of a sheep's body. One difference between the two kinds of entity is that the sheep's body, for instance, is a "concrete thing" or "a mass," and so is an aggregate of an infinity of bodies (though in being a unified living thing is an infinity of bodies organized by the notion of this specific animal, and is presumably a genuine corporeal substance), where its "parts," those bodies constituting the aggregate, are prior to the whole. By contrast, the geometrical line, an ideal thing, is not an aggregate of its "parts." Here, the "parts" that he has in mind are in fact points. In speaking about how others have confused ideal and real things, Leibniz says that they have mistakenly thought that "lines are made up out of points" (WF 185). So, the "parts" here are not line segments but are, as he puts it, "the primary elements in ideal things," or points. For ideal things, the whole is prior to its parts or its elements.[14] Even though a line is not an aggregate of points like the sheep's body is an aggregate of smaller bodies, we can nevertheless conceive a line in terms of "relations which involve eternal truths" (WF 185). If so, what are the *relata*? According to the above account, *DC* can be understood as a relation of at least two other lines, *AB* and *AS*. Specifically, what the equation of *DC* shows is precisely how each *point* x_n on *AB* is related to a point y_n on *AS*. Now, there are an infinite number of possible points or "locations" x_n on *AB*. Likewise for *AS*—there are an infinite number of possible points or "locations" y_n on *AS*. Despite this, the equation of *DC* can take any x_n on *AB* and pair it with a unique y_n on *AS*. Here is a sense, then, in which we can understand how it is that an organizing principle orders or organizes infinitely many elements or, as he calls them in the remark to Foucher, "primary elements," which results in the "production" of a geometrical line. Likewise, I think, the *notion* of an individual substance *S* can be understood as ordering or as organizing infinitely many simple substances, which results in the "production" of the individual substance *S* over time.

This early metaphysical view of an individual substance and the notion responsible for organizing it is, certainly by 1695, extended in some form to all *bodies*. In the *New System*, for example, Leibniz makes clear the difference between the souls or forms of a "superior order" and those "sunk in matter which," he says, "in my view are to be found everywhere" (WF 146).[15] The former are those souls that express the notions of individual substances, the latter are simpler in some sense and of an inferior order. Even so, the latter are simple substances, or what in the *New System* he refers to as "*real and animated point(s)*" (Leibniz 1695, 145) and later as "*metaphysical points*" (149). In the *Specimen Dynamicum*, he casts this animation, or the activity of any informed material locus (i.e., animated or metaphysical point), in terms of *conatus* (WF 154–79; GM 6 234–54).[16] "Since only force and its resultant effort exist at any moment," he says, "and since every effort tends in a straight line, it follows that *all motion is rectilinear or composed of rectilinear motions*" (WF 173; GM 6 252) The idea, I think, is that each metaphysical point expresses its own motion, where its motion is rectilinear. This is one sense in which metaphysical points differ from mathematical points. Such motions are smaller than can be measured. In fact, they are smaller than can be perceived.[17] Yet out of them arises much of the motion we observe in the cosmos. Though not clear, perhaps these will become the *petites perceptions* of the *New Essays* (1703–5)—simple, active, and yet insensible (Leibniz 1991, 49–67; GP 5 41–61).[18]

Even as late as in the *Principles of Nature and Grace* (1714), Leibniz writes:

> In nature, everything is full. There are simple substances everywhere, genuinely separated one from another by their own actions, which continually change their relationships. Every simple substance, or individual monad, which forms the centre of a composite substance (an animal, for example)

> and the principle of its unity, is surrounded by a *mass* made up of an infinity of other monads which constitute the body of that central monad; and in accordance with the ways in which that body is affected, the central monad represents, as in a kind of *centre*, things which are outside it. This *body* is *organic*, when it forms a kind of natural automaton or machine, which is a machine not only in its entirety, but also in its smallest noticeable parts. Because of the plenitude of the world everything is linked, and every body acts to a greater or lesser extent on every other body in proportion to distance, and is affected by it in return. It therefore follows that every monad is a living mirror, or a mirror endowed with internal activity, representing the universe in accordance with its own point of view, and as orderly as the universe itself.[19] (WF 258–66; GP 6 598–606)

A "central" monad is the one responsible for organizing a "surrounding mass" of an infinity of monads, this central monad functioning as a *principle of unity.* "Each monad," he says, "together with its own body, makes up a living substance. Thus not only is there life everywhere, together with limbs or organs, but there are infinite levels of life among monads, some of which are dominant over others to greater or lesser extents" (WF 260). This, I think, harkens back to the notion of an individual substance as introduced in the *Discourse.* The dominant or central monad expresses itself as the organizing principle of an individual substance. And although I cannot argue for it here, allow me to at least suggest that Leibniz seems to believe, certainly by the time he writes the *Specimen Dynamicum*, that this is so whether that substance is Alexander the Great, a red blood cell, the sun, or a piece of iron.

There appear to be earlier versions of the idea of the central monad and the mass of monads surrounding it, as it was introduced in the *Principles of Nature and Grace.* One such version can be found in the *Specimen Dynamicum.* This earlier version does not make anything of the distinction between the living body of Alexander, say, and the corpse of Alexander. Rather, it treats of *bodies* generally, or what Leibniz will refer to as corporeal substances. This is, I think, compatible with the view in the *Discourse.*

In a 1686 letter to Arnauld, Leibniz had said that the human body "or the *corpse*, considered in isolation from the soul, can only improperly be called a *substance*, like a machine, or a heap of stones, which are only beings by aggregation" (WF 117).[20] So, postmortem, the corpse of Alexander, supposing that "it" is no longer organized by the notion of Alexander, is now a heap or aggregate of bodies. This aggregate is not an individual substance, for it is now simply a heap of bodies (or corporeal substances) with nothing functionally relating or organizing them into a unified being with a singular aim. When the notion of Alexander *did* organize these bodies, it organized them so that every body (that now constitutes the heap) was then directed by the notion toward some specific end and no other. This body, the ante mortem "living" body of Alexander, was an individual substance—it was Alexander the Great.[21] And, it may even have counted as a genuine *corporeal* substance. This is suggested when Leibniz writes, "if there are corporeal substances, man is not the only one" (WF 147; GP 4 474–75). He has philosophical reasons for thinking that bodies other than human bodies are corporeal substances. During the period he writes the letter to Arnauld (1686), Leibniz holds that a soul is a purely active being.[22] Matter is a purely passive medium through which a soul acts. A *body*, then, is the result of a soul's *acting through* matter, and in this sense, as a *union*, can be understood to be what he calls a corporeal substance. If this is right, then it would seem that all *bodies* must be understood as being animate or organic. So, although the corpse postmortem is simply an aggregate of an infinite number of bodies, given that each member of the aggregate is a *body*, then each is "governed" by some organizing principle, which if not a soul of the superior order is a soul of the inferior order, the kind of soul that God had apparently "sunk" into matter.

In the *Specimen Dynamicum*, Leibniz tells us that a body will possess two kinds of *active force*, namely *primitive active force* and *derivative active force*. Primitive active force, he says, "is inherent in all corporeal substances as such." It is "none other than the first entelechy—[which] corresponds to the *soul* or *substantial form*." Derivative active force "is as it were the limitation of primitive force brought about by the collision of bodies with each other" (WF 155–56; GM 6 234–54). This latter kind of force is "the force by which bodies actually act and are acted upon by each other" (WF 157). If we think of the primitive active force as the corporeal substance's internal principle of organization, we might in turn think of the derivative active force in terms of how this organized body relates to or "harmonizes with" all other bodies. The derivative force, in other words, requires us to consider a body as it *relates* to all others. Here are a few other details. The velocity (*velocitatem*) of a body, he says, which in this context seems to be the *speed* of a body (i.e., the body's change of place over time), "taken together with direction is called *conatus*, while *impetus* is the product of the mass of a body and its velocity (speed)" (WF 157). Here, we begin to see an application of mathematics that is *not* metaphorical. Impetus is the *product* of mass and speed (where, if by *velocitatem* he means *speed in a direction*—though it is not clear that he means this—then impetus is the product of mass and what we would today recognize as velocity).

Leibniz introduces a number of distinct though related conceptions of force—*dead force, living force, respective force, individual force, total force, partial force, directive* or *common force*, to name several. Here, I shall focus on only a few. First, consider two bodies, *A* and *B*, where both are in motion. Even though *A* and *B* can be understood as bodies themselves, each considered in itself a *single* body, Leibniz also allows them to be understood as the aggregates of smaller bodies (and each of those smaller bodies the aggregates of even smaller bodies, and so on *ad infinitum*). Call those bodies constituting *A*, $a_1, a_2, \ldots a_n$, and those constituting *B*, $b_1, b_2, \ldots, b_m$. *Respective* or *individual force*, he says, is the force by which the bodies constituting *A*, for example, namely $a_1, a_2, \ldots, a_n$, can act on one another; likewise, it is the force by which the bodies constituting *B*, namely $b_1, b_2, \ldots, b_m$, can act on one another. The interactions occur *internal to* the aggregates *A* and *B*, respectively. By contrast, *directive* or *common force* is that force by which the *aggregate itself*, namely *A*, can act on something else, for example on *B*; and it is that force by which the *aggregate itself*, namely *B*, can act on something else, for example on *A* (WF 158–59). Clearly, when we appeal to directive or common force, we treat *A* and *B* as singular, or we might even say *unified*, entities, where no reference to constituent bodies is made. (Of course, *A* and *B* could be constituent bodies of some larger body, in which case, understood as constituents the force by which they act on one another will be individual force. But I will ignore this here.) This is consistent with what Leibniz says, for example, in a response to Abbé Gouye, where he notes that depending on what we choose to explain, we can regard the earth, for example, as a point, as it stands to its orbit around the sun, or a ball held in one's hand as a point, as it stands, say, to the circumference of the earth, knowing full well that they are spheres and not points.[23] Likewise, I think, in the *Specimen Dynamicum*, Leibniz allows himself a way to treat *A* and *B* as points, or rather, as points in motion, where the motions of *A* and *B* are represented as geometrical lines.

He explains this within the context of introducing the idea of a body's *center of gravity*, where this center can be conceived as a (metaphysical) point (WF 173). So, when we think of *A* as a singular body in motion, we conceive of *A* in terms of its center of gravity. This comes within the context of introducing two kinds of motion, which not coincidentally are related to the two kinds of force introduced earlier, individual force and common force. Consider again bodies *A* and *B*. *Individual motion*, he says, is the individual motions of the bodies constituting *A*, namely $a_1, a_2, \ldots, a_n$; and, if considering *B*, the individual motions of the bodies constituting *B*, namely $b_1, b_2, \ldots, b_m$. As was the case with individual force, individual motion occurs *internal to* the aggregates *A* and *B*, respectively. *Common motion* is the motion of *A* or the motion of *B*, each considered as a center of gravity, these centers, as I said, being understood as (metaphysical) points. Now, we know that the notion of *A*, for example, will be an organizing principle, the central monad to use the language of the *Principles of Nature and Grace*, that organizes an infinity of a "surrounding mass" of monads. This surrounding mass will no doubt include bodies $a_1, a_2, \ldots, a_n$. Of course, the same holds for the notion of *B* and $b_1, b_2, \ldots, b_m$. It is worth noting that the notion of a_1, for example, also has a center of gravity, and each body constituting it also has its own center of gravity, and so on *ad infinitum*. There is much to juggle here. So, how are we to conceive all of this? It is one thing to understand the notion of *A* as an equation that determines a geometrical line, where the line represents *A*'s motion, but it is quite another thing to understand how the notion of *A* works to organize an infinity of rectilinear motions, which include the motions of $a_1, a_2, \ldots, a_n$, and ultimately the distinct *conata* of an infinity of metaphysical points.

With this sort of challenge in mind, I think, Leibniz develops a mathematical procedure that in fact allows him to solve a system of equations, which makes intelligible a *harmony* among distinct rectilinear motions, so that the latter can be shown to converge on a singular point, "forming" a body's center of gravity. The procedure deals with how to find what we today call a *determinant*. This will have wider application, for it will also help to make Leibniz's conception of universal order clearer. But I shall focus only on how the determinant makes clearer how the bodies constituting *A*, for example, might be conceived as being ordered or organized to form *A*'s center of gravity.

MATRICES AND HARMONY

Leibniz's development of the determinant is, as far as I am aware, found in only a few texts. The first is a 1693 letter to l'Hospital, the second is an untitled manuscript written sometime before 1693 (GM 2 238–40).[24] He does not use the word "determinant." The term came later. Leibniz begins the description of his new procedure by telling l'Hospital how to understand his new notation, starting with a system of linear equations:

$$10 + 11x + 12y = 0$$
$$20 + 21x + 22y = 0$$
$$30 + 31x + 32y = 0$$

Here, he has set each equal to zero. We are given three equations with two unknowns. Today, mathematicians would say that this system is overdetermined. This means that the number of equations is greater than the number of variables (unknowns). The numbers here are placeholders, and so are not really numbers, but are, as he calls them, pseudo-numbers (*nombre feint*; S 268; GM 2 239).[25] Each is a two-digit pseudo-number, each digit representing important information. For instance, the first digit tells us which linear equation a term belongs to. So, the first digit 1 in the pseudo-numbers **1**0, **1**1, and **1**2 tells us that these terms belong to the first linear equation; the first digit 2 in the pseudo-numbers **2**0, **2**1, and **2**2 tells us that these terms belong to the second linear equation, and so on. The second digit tells us which variable, if any, the term is the coefficient of. So, the second digit 0 in the pseudo-number 1**0** tells us that this term is the coefficient of no variable; the second digit 1 in the pseudo-number 1**1** tells us that this term is the coefficient of the *x* variable; the second digit 2 in the pseudo-number 2**2** tells us that this term is the coefficient of the *y* variable. Thus, the pseudo-number 32 tells us that this term belongs to the third linear equation and is the coefficient of the *y* variable. To emphasize this visually, later in the letter Leibniz writes the first digit larger than the second, the second digit written to look a bit like a subscript. So, 10, 11, 12 become 1_0, 1_1, 1_2.

About the pseudo-numbers, he is clear to say that since they are not really numbers, when he instructs us to multiply 10 and 22, for example, which he writes "10.22," we should not multiply these as though *they* were numbers. So, we should not arrive at 220 (ibid.).[26] Rather, 10 and 22 are placeholders, 10 standing for the term in the first linear equation that is coefficient of no variable, 22 standing for the term in the second linear equation that is coefficient of the *y* variable. So, supposing that **3** is in the 10-place and **5** in the 22-place, then 10.22 tells us to multiply **3** and **5**. Here, **3** and **5** *are* numbers. So, the product is **15**. Now, the procedure is geared to eliminate and subsequently reduce the number of variables. Ultimately, "the final equation [is] freed from the two unknowns that we wished to eliminate" (S 269; GM 2 239–40).[27]

Today we do not usually call the terms 10, 20, and 30, that is, terms that are the coefficients of *no* variable, *coefficients*. They are referred to as *constants*. But, Leibniz refers to them as coefficients. Although he appears to be laying out a single procedure, there are, I think, several related procedures. The first can be understood as a method of eliminating unknowns. Working the procedure on two of the above linear equations, we can make out a rule that yields the same results as Cramer's rule.[28] Gabriel Cramer (1704–52) was a student of Johann Bernoulli, and was no stranger to Leibniz's work. The rule appears in his *Introduction à l'analyse des lignes courbes algébraique* (1750).[29] Let's look at what Leibniz says, considering two of the above linear equations:

$$10 + 11x + 12y = 0$$
$$20 + 21x + 22y = 0$$

First, let's move the constants (coefficients with no variables) over to the right side of their respective equations. So:

$$11x + 12y = -10$$
$$21x + 22y = -20$$

Leibniz instructs us to multiply the first equation by the y coefficient of the second equation (so, we multiply the terms of the first equation by 22) and multiply the second equation by the negative of the coefficient of the y variable of the first equation (so, we multiply the terms of the second equation by –12) (S 268; GM 2 239).[30] So:

$$11x.22 + 12y.22 = -10.22$$
$$-12.21x + (-12.22y) = -12.-20$$

We can make this more manageable by cleaning up the terms:

$$11.22x + 12.22y = -10.22$$
$$-12.21x - 12.22y = 12.20$$

Next, we add the two equations together. (12.22y) and (–12.22y) cancel each other out, leaving:

$$11.22x - 12.21x = 12.20 - 10.22$$

Now we solve for x:

$$x = \frac{12.20 - 10.22}{11.22 - 12.21}$$

This quotient is precisely what Cramer's rule yields.

The rule requires that we first construct the coefficient matrix. This is a matrix containing only those coefficients attached to variables. Looking back at our original two linear equations, these are 11, 12, 21, and 22, the coefficients of the variables x and y. So:

$$\begin{matrix} 11 & 12 \\ 21 & 22 \end{matrix}$$

Following Leibniz with what I believe is a slightly different procedure (to be considered shortly), we cross-multiply these terms and take the difference. Specifically, we cross-multiply 11 and 22 (so, 11.22) and cross-multiply 12 and 21 (so, 12.21) and subtract the latter from the former. 11.22 – 12.22 is the determinant of this coefficient matrix. Now, to compute the x value, we substitute in the coefficient matrix the x coefficients with the constants, which are Leibniz's coefficients attached to no variables. In the two linear equations, these are –10 and –20. So:

$$\begin{matrix} -10 & 12 \\ -20 & 22 \end{matrix}$$

We now do the same cross-multiplication and take the difference: –10.22 – (–12.20), which is –10.22 + 12.20, or 12.20 – 10.22. This is the determinant of this new matrix. We compute x by dividing the determinant of this new matrix by the determinant of our original coefficient matrix:

$$x = \frac{12.20 - 10.22}{11.22 - 12.21}$$

which is the quotient that Leibniz's procedure produces. To be crystal clear, I will apply this to a real case. Consider this system of linear equations.

$$\begin{aligned} -6 + 2x + 3y &= 0 \\ -15 + 4x + 9y &= 0 \end{aligned}$$

Now, to make this simpler, following what we did above, move the constants over to the right-hand side of their respective equations. So:

$$\begin{aligned} 2x + 3y &= 6 \\ 4x + 9y &= 15 \end{aligned}$$

We can construct the coefficient matrix (which, recall, includes only those terms that are coefficients of variables) and apply Leibniz's cross-multiplication procedure.

We cross-multiply 2 and 9, and 3 and 4, and take the difference (I will use contemporary notation to make this simpler).

$$(2)(9) - (3)(4) = 18 - 12 = 6$$

The determinant of the coefficient matrix is 6. Now, let's substitute the constants 6 and 15 for the x coefficients 2 and 4 in the coefficient matrix:

63159

Cross-multiply and take the difference:

$$(6)(9) - (3)(15) = 54 - 45 = 9$$

So, the determinant of this new matrix is 9. Now, we compute x by dividing the determinant of this new matrix by the determinant of the original coefficient matrix.

$$x = \frac{9}{6} \text{ or } \frac{3}{2}$$

The procedure for computing y is similar: substitute the constants for the y coefficients of the original coefficient matrix, calculate the determinant of this new matrix (the determinant is 6), and divide this determinant by the determinant of the original matrix (which is 6). Here, $y = 6/6$ or 1. The solution to this system of equations is $< 3/2 \, , 1 >$, which tells us that the lines of this system intersect at this and no other point.

A second procedure allows Leibniz to deal with what I earlier noted was an overdetermined system (more equations than unknowns). Recall Leibniz's original three linear equations. Remove the variables and addition signs (following Leibniz, I now adopt his alternate notation. So, 10 is now 1_0, 11 is now 1_1, and so on.):

$$\begin{matrix} 1_0 & 1_1 & 1_2 \\ 2_0 & 2_1 & 2_2 \\ 3_0 & 3_1 & 3_2 \end{matrix}$$

This is an ancestor of a 3 × 3 *matrix*. Leibniz does not call it this, of course. It must be again stressed before getting into the details of the procedure that 1_0, 2_0, 3_0, which we today call constants, are included in this matrix. This is acceptable. But usually when they are included along with coefficients proper (terms that are coefficients of variables) mathematicians refer to the matrix as an *augmented matrix*. Again, this is a distinction that Leibniz does not seem to recognize. He treats the constants, as I said, as coefficients.[31]

Here is how the procedure goes. Begin with the matrix-like structure:

$$\begin{array}{lll} 1_0 & 1_1 & 1_2 \\ 2_0 & 2_1 & 2_2 = 0 \\ 3_0 & 3_1 & 3_2 \end{array}$$

A study of the outcome of his procedure suggests that we should multiply diagonally in both directions. Sum the products when moving right, subtract the products when moving left. Always begin with terms in the first equation. So, moving diagonally (to the right), start with 1_0, multiply that to 2_1, and multiply that to 3_2. So, $1_0 . 2_1 . 3_2$. Go to the next term in the first equation, 1_1, and moving diagonally multiply that to 2_2, and because we run out of terms, move to the first term of the third equation, 3_0. So, $1_1 . 2_2 . 3_0$. Go to the last term in the first equation, 1_2, and since we again run out of terms, multiply 1_2 to the first term of the second equation, 2_0, moving diagonally multiply that to 3_1. So, $1_2 . 2_0 . 3_1$. Since we got them by moving right, we sum these three products: so, $(1_0 . 2_1 . 3_2) + (1_1 . 2_2 . 3_0) + (1_2 . 2_0 . 3_1)$. So far, so good. But now we must move diagonally in the other direction. So, again begin with 1_0 and move diagonally this time to the left. Of course, we run out of terms, so move to the *last* term of the second equation, 2_2, and multiply by this, then move diagonally and multiply 3_1. So, $1_0 . 2_2 . 3_1$. Continue with the second term of the first equation, 1_1 (still moving diagonally to the left), multiply that to 2_0, and since we run out of terms, move to the *last* term of the third equation, 3_2. So, $1_1 . 2_0 . 3_2$. Begin now with the last term of the first equation, 1_2, moving diagonally multiply that to 2_1, and multiply that to 3_0. Because we got these by moving left, instead of sums we take the difference. Putting this together with the prior sums we get:

$$(1_0 . 2_1 . 3_2) + (1_1 . 2_2 . 3_0) + (1_2 . 2_0 . 3_1) - (1_0 . 2_2 . 3_1) - (1_1 . 2_0 . 3_2) - (1_2 . 2_1 . 3_0) = 0$$

Ignoring for the moment that Leibniz is including the constants here, if this were a system of three equations in three unknowns, his procedure in fact produces the determinant of the 3 × 3 matrix. What Leibniz notes, however, is that his procedure will work for "eliminating the unknowns in any number of equations of the first degree, provided that the number of equations exceeds by one the number of unknowns" (S 269; GM 3 5). So, here, insofar as we have *three* equations with *two* unknowns, the procedure, he says, should work.[32]

Given what he arrives at (I will state this shortly), he looks to move the last three terms (the differences) to the right-hand side of the equation, getting:

$$(1_0 . 2_1 . 3_2) + (1_1 . 2_2 . 3_0) + (1_2 . 2_0 . 3_1) = (1_0 . 2_2 . 3_1) + (1_1 . 2_0 . 3_2) + (1_2 . 2_1 . 3_0)$$

What Leibniz in fact writes is:

$$\begin{matrix} 1_0 . 2_1 . 3_2 & & 1_0 . 2_2 . 3_1 \\ 1_1 . 2_2 . 3_0 & = & 1_1 . 2_0 . 3_2 \\ 1_2 . 2_0 . 3_1 & & 1_2 . 2_1 . 3_0 \end{matrix}$$

Here, he has ordered the left and right-hand sides of the equation using his matrix-like structures, where the first term of the equation on the left-hand side ($1_0 . 2_1 . 3_2$) is located on top, the second term ($1_1 . 2_2 . 3_0$) is located underneath that, the third underneath that; likewise the first term of the equation on the right-hand side ($1_0 . 2_2 . 3_1$) is located on top, the second term is located underneath that, and so on.

What does this tell us? For starters, if the lines determined by the various equations of the system intersect, then there will be a unique, nontrivial, solution. This solution is what Leibniz's determinant procedure is able to find. By *nontrivial* it is meant that the equations in the system are not simply equations of the same line, which if so, would entail that there are an *infinite* number of solutions. And, on the flipside, if all of the lines did not intersect, there would be *no* solution. What does this mean? Leibniz's procedure guarantees that if there is a solution it will be a unique $<x_n, y_n>$ pair that solves each equation in the system. This will be the point at which *all* of the lines intersect. And this is precisely what we wanted to understand when trying to conceive a body's center of gravity as a system of rectilinear motions converging on a single point. In terms of body *A*, and the bodies $a_1, a_2, \ldots, a_n$ that constitute it, given that we possessed the linear equation of a_1, the linear equation of a_2, and so on, and given that they form the center of gravity of *A*, which is a single (metaphysical) point, Leibniz shows that we can go some way toward understanding, and even calculating, this convergence. To be sure, when *A* is conceived as a point in motion, the notion of *A* can be taken to be a linear equation. But conceived as an organized collection of bodies, the notion of *A* can be taken to be a system of linear equations, where the determinant now helps to make clearer the possibility of *A*'s center of gravity.

Leibniz employs the above concepts taken from mathematics to make clearer his metaphysical conceptions of order, harmony, and the notion of an individual substance. We saw that he uses the concepts of geometrical line, equation, and what is now called the determinant of a matrix. The geometrical line and equation found a place in his metaphysics, the two used to make clearer his conceptions of order, individual substance, and the notion of a substance, and they also look to have found a place in his physics, the two used to represent the motion of a body. The determinant also looks to have found a place in his metaphysics, though admittedly not as prominently as the line and equation, this apparatus used to make clearer his conception of harmony among individual substances. And it also looks to have found a place in his physics, though again admittedly not as prominently as the line and equation, used to understand how an aggregate of moving bodies might harmonize so as to converge on a single (metaphysical) point, forming a body's center of gravity. Using mathematics to this end, this is one sense in which Leibniz can be seen as "mathematizing" the sciences of metaphysics and physics.

ABBREVIATIONS

GM Leibniz, G. W. 1846–60. *Leibnizens Mathematische Scriften*

GP Leibniz, G. W. 1879. *Die Philosophischen Schriften von Gottfried Wilhelm Leibniz*

L Leibniz, G. W. 1976. *Philosophical Papers and Letters* (followed by volume and page numbers)

S Leibniz, G. W. 1959. *A Source Book In Mathematics* (followed by page numbers)

WF Leibniz, G. W. 1998. *G. W. Leibniz: Philosophical Texts* (followed by page numbers)

NOTES

I am greatly indebted to the editors for their suggestions on how to improve this chapter, especially Benjamin Hill who officially provided comments at the workshop that produced this volume, and, of course, to the participants of the workshop. I also owe thanks to Douglas Marshall for his advice. Last, I thank Roger Ariew, Richard Arthur, Dan Garber, Geoff Gorham, and Doug Jesseph for extended discussions at the workshop that helped me to better understand Leibniz's view.

1. The original Latin title was: "*De prima philosophiae Emendatione, et de Notione Substantiae*," which can be found reprinted in Leibniz (1879, 4:468–70). Hereafter, I shall refer to this collection, the philosophical writings, as "GP," fol-

lowed by the volume and page number. I am using an English translation found in Leibniz (1976, 2:432–34).

2. An English translation of this manuscript can be found in Leibniz (1998, 139–42). The original manuscript and correspondence can be found in Bossuet (1912).

3. This letter is No. 1. It is dated November or December 27, 1694. The quote is taken from p. 258: "Ma metaphysique est toute mathematique pour dire ainsi ou la pourroit devenir."

4. The manuscript is found as an untitled work in GP 4 427–63. The section headings are found in a letter to the Count Ernst von Hessen-Rheinfels, GP 2 12–14. An English translation can be found in Leibniz (1931, 68–72). When not my own, I am using English translations included in WF, 54–93. I am also relying on the translation found in Leibniz (1991).

5. Leibniz to Count Ernst von Hessen-Rheinfels (1687 or 1688). This is letter 24, in Leibniz (1931, 238).

6. *Discourse*, §8; GP 4 433. Leibniz had in fact sent the *Discourse* to the Count and to Arnauld, both of whom corresponded with Leibniz and with one another about it.

7. Leibniz to Arnauld, March 23, 1690. This was part of Arnauld's response to the *Discourse*. This is letter 26, in Leibniz (1931, 244).

8. See, for example, Cover and O'Leary-Hawthorne (1999). I am indebted to Richard Arthur for bringing this to my attention. I am borrowing the phrase "law-of-the-series" from Cover and O'Leary-Hawthorne.

9. Cover and O'Leary-Hawthorne (1999) quote Leibniz as follows: "the essence of substances consists in . . . the law of the sequence of changes, as in the nature of the series in numbers" (220). This comes from Leibniz (1923).

10. Leibniz to Arnauld, May 1686. This is letter 8.

11. Cover and O'Leary-Hawthorne (1999) offer an excellent discussion of this picture. It is worth noting, I think, that they actually never take Leibniz up on his use of mathematics in this context. So, their analysis is solely conceptual or logical.

12. The manuscript is included in Leibniz (2005).

13. Here I alter the orientation of his drawing (I spin it 90 degrees counterclockwise) so that *AS* and *AB* align with the more familiar *y*-axis and *x*-axis as we position them today. Leibniz actually includes two other lines: parallel to *AS* he puts *BC* and parallel to *AB* he puts *SC*, making a rectangle with the four lines. I only require the simpler drawing to make my point. The inclusion of x_1, x_2, y_1, y_2 along with the dashed lines is my own.

14. Here, Leibniz casts points as "extremities" of lines. This is tricky, for it is easy to think that he means that they are simply the "ends" or the "tips," so to speak, of a finite line segment. But this would be wrong. Rather, in the discussion he speaks of ratios, for example 1/2 and 1/4, where, I take it, he means the following: Assume line *PQ*. Divide *PQ* at *R*. The *ratio* of *PRPQ* is a point on *PQ*. Call this point a_1. Now divide *PQ* at *S*, where $PS \neq PR$. The ratio of *PSPQ* is a point on *PQ*. Call this point a_2. Since $PR \neq PS$, $a_1 \neq a_2$. We construct points, the elements, out of the whole, the line. In this sense, the whole is prior to its "parts." As I note in the body of the paper, I do not read "parts" as denoting smaller line segments.

15. I cite the English translation in WF 143–52. The quote is from WF 146. GP 4 477–87 has a revised draft. The brackets reflect the additions.

16. From *Specimen Dynamicum: An Essay in Dynamics, Showing the Wonderful Laws of Nature Concerning Bodily Forces and Their Interactions, and Tracing Them to Their Causes* (1695).

17. For an excellent discussion of the issues surrounding infinitesimals, incomparably small elements, and conatus, which I here suggest by the appeal to *metaphysical points*, see Jesseph (1998, 6–40).

18. From the preface to *New Essays*. The remark about insensible perceptions is made in AG 56.

19. From *Principles of Nature and Grace Based on Reason* (1714), a version of which forms sections 61–62 of the *Monadology*.

20. Leibniz to Arnauld, November 28/December 8, 1686. This, it seems to me, is aligned with what Aristotle says in the *Categories*. There, he speaks of a corpse as being a *man*, but only homonymously so.

21. Alexander's soul may appear to perish, since it is no longer perceived as organizing bodies existing at a level of normal human perception. This may be what human beings typically call "death." His notion, however, could continue to operate, but now over bodies that are insensible to human beings. So, Alexander's visible corpse does not count as evidence of the death (or annihilation) of Alexander's soul or its activity in the cosmos. About this, see what Leibniz says in the *New System* (WF 147; GP 4 474–75).

22. See, for example, Mercer (2001). For detailed studies of Leibniz's views on substance as they evolve over his career, see Leibniz (1994; 2009).

23. I am indebted to Douglas Jesseph's paper for this passage (1998, 30). There, Jesseph quotes the passage from the *Journal*, printed in GM 6 95–96.

24. Leibniz to l'Hospital, April 28, 1693. An English translation of a portion of the letter is in Leibniz (1959, 267–69). The untitled manuscript can be found

in GM 3 5–6. An English translation of a portion of the manuscript can be found in S (269–70). I rely on Smith's dating of the untitled manuscript.

25. 1693 letter to l'Hospital.

26. That is, $10.22 \neq 220$. Ibid.

27. Ibid.

28. I show some of this in Smith (2010, 172–73).

29. Carl Boyer (1985) notes that this rule appears earlier in Colin Maclaurin's 1748 *Treatise of Algebra*, published two years before Cramer's book.

30. Letter to l'Hospital.

31. In Smith (2010, 170–75), I discuss some of this material, though I ignore the present concern over coefficients proper, constants, matrices, and augmented matrices. What I say there is essentially correct, though the procedure I work out is mostly only suggested by what little Leibniz says, though it is consistent with it. I also show how Leibniz's letter to l'Hospital prefigures Cramer's rule. This seems to me still to be essentially correct. Finally, since I am confessing things, I overstate in Smith (2010, 175) what Leibniz actually shows in this letter—specifically, I say that *he* arrives at a particular quotient (which is a quotient that Cramer's rule yields). While it is true that the procedure Leibniz describes will yield the quotient, *he* does not actually produce that quotient in the letter.

32. Some of what Leibniz says in these texts suggests that he was toying with what today we think of as cofactors and minors. But this is murky. Even so, given Leibniz's 3×3 matrix we could eliminate the column containing the constants, or coefficients with no variable, by way of an expansion by cofactors. This may be what he means when he says, "Make all combinations of the coefficients of the letters, in such a way that more than one coefficient of the same unknown and of the same equation never appear together. These combinations, which are to be given signs [plus or minus] in accordance with the law which will soon be stated, are placed together, and the result set equal to zero will give an equation lacking all the unknowns" (S 269; GM 3 5). But restrictions on length of chapters in this volume prohibit me from pursuing that here.

REFERENCES

Adams, Robert. 1994. *Leibniz: Determinist, Theist, Idealist.* Oxford: Oxford University Press.

Bossuet, J. B. 1912. *Correspondance de Bossuet.* Ed. C. Ubain and E. Levesque. 15 vols. Paris: Hachette.

Boyer, C. 1985. *A History of Mathematics*. Princeton, N.J.: Princeton University Press.

Cover, J. A., and J. O'Leary-Hawthorne. 1999. *Substance and Individuation in Leibniz*. Cambridge: Cambridge University Press.

Garber, Daniel. 2009. *Leibniz: Body, Substance, Monad*. Oxford: Oxford University Press.

Jesseph, D. M. 1998. "Leibniz on the Foundations of the Calculus: The Question of the Reality of Infinitesimal Magnitudes." *Perspectives on Science* 6, nos. 1–2.

Leibniz, G. W. 1695. *New System of the Nature of Substances and Their Communication, and of the Union which Exists between the Soul and the Body*. Originally published in *Journal des savants*, no. 23 (June 27, 1695): 294–300; no. 24 (July 4, 1695): 301–6.

———. 1846–60. *Leibnizens Mathematische Scriften*. Ed. C. I. Gerhardt. 7 vols. Berlin: Schmitt.

———. 1879. *Die Philosophischen Schriften von Gottfried Wilhelm Leibniz*. Ed. C. Gerhardt. 7 vols. Berlin: Weidmann.

———. 1923. *Gottfried Wilhelm Leibniz: Sämtliche Schriften und Briefe*. Deutsche Akademie der Wissenschaften. Series 6, vol. 3. Berlin: Akademie Verlag.

———. 1931. *Leibniz*. Trans. G. Montgomery. Chicago: Open Court.

———. 1959. *A Source Book in Mathematics*. Trans. T. Cope. Ed. D. E. Smith, 267–69. New York: Dover Publications.

———. 1976. *Philosophical Papers and Letters*. Trans. and ed. L. E. Loemker. The Netherlands: Springer.

———. 1991. *Discourse on Metaphysics and Other Essays*. Trans. R. Ariew and D. Garber. Indianapolis: Hackett Publishing.

———. 1998. *G. W. Leibniz: Philosophical Texts*. Trans. and ed. R. S. Woolhouse and R. Francks. Oxford: Oxford University Press.

———. 2005. *The Early Mathematical Manuscripts of Leibniz*. Trans. J. M. Child. New York: Dover Publications.

Mercer, C. 2001. *Leibniz's Metaphysics: Its Origin and Development*. Cambridge: Cambridge University Press.

Smith, K. 2010. *Matter Matters: Metaphysics and Methodology in the Early Modern Period*. Oxford: Oxford University Press.

10

LEIBNIZ'S HARLEQUINADE

Nature, Infinity, and the Limits of Mathematization

JUSTIN E. H. SMITH

DESCARTES: "IN THE SAME STYLE AS THE REST"

René Descartes writes to Marin Mersenne in a letter of 1639 that if his account of the circulation of blood, among other things, "turns out to be false, then the rest of my philosophy is entirely worthless" (1964–76, AT 2 501). He does not say that if his account is wrong then his circulation theory, or his medicine, or his physiology will be worthless. He says that his *philosophy* depends on the correctness of his explanation of cardiac motion. This is, indeed, putting quite a lot of weight on a question that appears to be of only the remotest concern to philosophers today—there is, for example, no specialization in the philosophy of cardiology—and at the very least it should motivate any historian of early modern philosophy to reconsider the way he or she conceptualizes the philosophical project, and to strive to study this project in its early modern expression in a way that is adequate to the self-conception of its leading exponents. For Descartes, the medical philosophy aimed to prolong the human body's life, and at the base of any human or animal body's life is the continuation of certain vital processes, most particularly respiration, digestion, and circulation. These, in turn, must be understood if they are to be influenced by human art and intervention. The project of understanding, in turn, for Descartes, means first and foremost, as he put it in the 1637 *Discourse on Method*, explaining a given domain of nature "in the same style as the rest"[1] (1964–76, AT 6 45). But what style is this?

Many commentators would be tempted to call it 'the mechanical style,' and to see this as synonymous with 'the mathematical,' but these categories are rightly being subjected to a certain degree of scholarly revision in recent scholarship, including the present volume. Roger Ariew, in chapter 4, emphasizes the polysemy of the notion of mathematization in the early modern period: different thinkers meant different things when they asserted of their own projects that these were mathematical, or that they involved mathematization of domains of nature that previously had not been given such a treatment. According to Ariew, the historiographical approach to early modern natural philosophy that prevailed in the twentieth century, in particular in the work of Burtt (1925), Dijksterhuis (1961), and Koyré (1957) (hereafter, collectively, BDK), is fundamentally misguided. On the new, revisionist approach, there has been a basic conflation of *mechanization* and *mathematization*. Certainly, if we understand the latter notion as involving the substitution of "mathematical idealities for the concrete things of the intuitively given surrounding world,"[2] then there is indeed little primary-textual evidence that Descartes and his successors were intent on mathematizing the natural world in this way. The twentieth-century consensus on this point is now rightly in the course of being displaced. But we may also understand by 'mathematization' another sort of endeavor: that of accounting quantitatively for natural processes that had previously been dealt with in strictly qualitative terms. Typically, such a quantitative approach would be elaborated in terms of the mass, figure, and motion of the particles involved in the process: particles that could be described in terms of their size, weight, shape, and speed. This description would then, it was hoped, render large-scale natural processes comprehensible.

It is plain that such a description was part of what Descartes had in mind when he expressed the desire to explain physiological processes "in the same manner as the rest." And while it may be the case that Descartes himself did not think of this sort of explanation as part of a mathematization of nature, there can be no doubt that subsequent physiologists, many of them working under the banner of Cartesianism, did so conceive it. Lorenzo Bellini and Niels Stensen, certainly, to cite two prominent examples, thought of their own iatromechanical projects as a sort of mathematization of medicine, and moreover the model of such mathematizing was geometry. This much is clear from the very title of a work such as Steno's *Elementorum myologiae specimen seu Musculi descriptio geometrica* [*Specimen of the Elements of Myology, or, A Geometrical Description of the Muscle*] of 1667 (see Kardel 1994; also, e.g., Bellini 1662). And Leibniz, for his part, who early in his career praised both Bellini and Stensen for "mathematizing in medicine" (Smith 2011, 84) evidently takes 'mechanical' and 'mathematical' as synonyms, at least in their application to the study of physiological processes. Thus he writes to Arnauld: "One must always explain nature mathematically and mechanically, provided that one understands that the very principles or laws of mechanics of force do not depend on mathematics alone, but on certain metaphysical reasons (Leibniz 1875–90, G 2 58).[3] In sum, Ariew is certainly right to insist that we must not conflate mathematization and mechanization in Descartes himself, and it is also certainly wrong to suppose that mathematizing iatromechanists would think of mathematics as an idealizing abstraction away from bodies. And yet, the supposition that the approach to mechanical nature as something mathematically tractable, the approach *more geometrico*, lay at the heart of the mechanical project is by no means an invention of mid-twentieth-century historiography. It goes back to the very self-understanding of early modern mechanical philosophers themselves. This is, at least, the conclusion we must draw if we limit our consideration of early modern mechanical philosophy to iatromechanism, or to the medical-scientific and physiological study of living bodies. In this respect, as in no doubt many others, it is crucial for the new, post-BDK historiography to consider the life sciences alongside the other domains of natural philosophy, in order to arrive at a sufficiently clear picture of just how much from the BDK thesis needs to be thrown out, and how much deserves to be retained.

Stensen, Bellini, and others were hoping to extend what they saw as the recent successes of 'mathematizing' in the study of the nonliving world. But it would be a mistake to suppose that this order of operations, this attempt at an extension of mathematical methods (again, not necessarily in the BDK sense, but simply in the sense of quantitative tractability) from celestial mechanics and ballistics to medicine, amounted to a movement from the most pressing to the less pressing of matters, or that the foundational sciences were taken care of first, and then the less important sciences were followed up by the *arrière-garde*. It would be more correct to say, in fact, that from the perspective of early modern thinkers themselves, physics was mathematized first only because physics is relatively easy compared to physiology and medicine. The fact that its objects more readily submit to a mathematical treatment than the parts of living bodies may have meant that lessons could be extended from physics to the life sciences, but this did not entail that physics was seen as somehow more fundamental. Quite the contrary, its mathematization appears to have been seen as a sort of preliminary skirmish in the build-up to the great battle, whose victory would have been the crowning achievement of the new philosophy, namely, the mathematization of medicine. It is not that the mathematization of medicine would provide a peripheral confirmation of the robustness of an explanatory approach already thoroughly established in a more foundational science, but rather that the *failure* to mathematize medicine, the inability to account for it "in the same style as the rest," would amount to a falsification of the entire endeavor, indeed of 'philosophy' as Descartes understands it.

But alas, by the end of the century, physics will have its *Principia mathematica*, while the philosophical treatment of living bodies, by contrast, will have fragmented into a sort of small-'e' empiricist agnosticism, on the one hand, and a fairly reactionary vitalism on the other, which took the phenomena of life as in principle unsusceptible to a treatment "in the same style as the rest." By the end of the following century, we would find Kant declaring that in principle there could never be, as he put it, "a Newton of the blade of grass" (G 5 400, 18ff). What happened? Why did this program fail? And what can its failure show us about the project of modern natural philosophy as a whole? In this chapter, I would like to argue that the philosophy of G. W. Leibniz provides significant insight into the fate of the early modern project of mathematizing living nature. While early on he had hoped to see mathematical methods extended to the analysis of the composition of organic bodies, and while in his mature analysis of these bodies the idea of an actual infinity plays a central role, nonetheless, for reasons I will proceed to spell out, Leibniz's mature theory of living bodies, according to which they consist in infinitely many bodies in hierarchical relations to one another ad infinitum, cannot be considered a victory for mathematization. Leibniz was no Newton for the blade of grass.

LEIBNIZ: "C'EST TOUT COMME ICI"

There is an interesting, if at first not obvious, connection between this last desideratum of Descartes' philosophical program, on the one hand, and, on the other, the so-called Harlequin principle, as stated a few decades later by Leibniz: *c'est tout comme ici*, "it's all as it is here." For both Descartes and Leibniz, it is a core conviction that different domains of nature must not be seen as requiring different sorts of explanation. Concretely, this means both a collapse of the Aristotelian separation between the superlunar and the sublunar, as well as of that between the living and the nonliving. Planets, projectiles, and muscles must all be subjected to the same sort of treatment as the others. The implications of the Harlequin principle for Leibniz are however rather different from those of Descartes' desire to explain everything "in the same style as the rest." Leibniz, in fact, appears to be drawing this principle from a few, likely unexpected, sources, and it is worthwhile to trace his version of the principle back to them, in order to gain a clearer picture of what he himself intends.

Harlequin—'Arlequin' or 'Arlecchino'—is a stock character of the Italian and French *commedia dell'arte* traditions. He is known, in some of his multiple iterations, for wearing a many-layered costume that makes it impossible to disrobe him. Thus Leibniz describes his conception of the infinite folds of organic bodies in the *New Essays* of 1704 as follows: "It is . . . like Harlequin, whom they wanted to disrobe on stage, but could never arrive at the end, because he had I don't know how many clothes the ones on top of the others: although these replications of organic bodies to infinity, which are in animals, are neither so similar to one another nor as layered upon one another as the clothes, the artifice of nature being of a completely different subtility" (G 5 309).[4]

In a letter to Damaris Masham of the same year, Leibniz makes oblique reference to Anne Mauduit Nolant de Fatouville's *Arlequin, Empereur dans la lune*, a comedy that first appeared in 1683 or 1684 (G 3 343).[5] He recalls the expression of a view according to which "everywhere and all the time, everything's the same as here." In fact, in this play, the phrase, "c'est tout comme ici" is repeated several times by different characters (the Doctor, Colombine, Isabelle), listening to Arlequin's description of life on the moon and affirming that this is "just like" life on Earth. The conclusion of the play consists in a resounding repetition of this phrase by the entire cast, yet this is in response to Arlequin's description of the daily habits of lunar women—they wake up past noon, take three hours to get dressed, travel to the opera in carriages—and it has nothing at all to do with the structure of the matter making up the lunar world or anything of the sort.[6]

What, then, is the connection between the two references to Harlequin in Leibniz—the reference to the character's onionlike costume, on the one hand, and the reference to the "tout comme ici" principle on the other? In order to answer this question, we need to take stock of the full range of Leibniz's interest in his era's science fiction. In the *New Essays*, in addition to referencing Nolant de Fatouville's fantasy about the emperor of the moon, Leibniz also cites another work describing travel through the solar system, and indeed one that shares many familiar themes from Nolant de Fatouville's work: Cyrano de Bergerac's *Histoire comique des États et Empires du Soleil*, first published in 1662 as a sequel to his *Histoire comique des États et Empires de la Lune*, published posthumously in 1655.[7] "I am also of the opinion," Leibniz writes, "that *genii* apperceive things in a way that has some relationship with ours, even if they had that curious gift that the imaginative Cyrano attributes to some animated natures in the Sun, composed of an infinity of small birds that, moving according to the command of the dominant soul, form bodies of every kind" (G 5 204).[8] In the tale of the voyage to the sun, Cyrano's narrator describes several encounters with composite beings. For example, he describes an encounter with a miniature man who emerges out of a pomegranate that has fallen from a tree. This man identifies himself as "the king of all the people who constitute that tree" from which the fruit has just fallen (de Bergerac 1858, 195). Soon after this, "all the fruits, all the flowers, all the leaves, all the branches, and finally the entire tree, fragmented into little men: seeing, sensing, and walking" (196). And soon the men begin to dance: "As the dance grew tighter, the dancers blurred into a much more rapid and indiscriminate stampede: it seemed that the purpose of the Ballet was to represent an enormous Giant; for by dint of coming together and augmenting the speed of their movements, they mixed so closely that I now perceived only one great Colossus . . . This human mass, previously boundless, reduced itself little by little so as to form a young Man, of an average size" (198–99).

The little king proceeds to jump into the composite man's mouth, and, in the role of 'dominant monad,' proceeds to give the composite golem its principle of unity: "All this mass of little men did not, before now, give any sign of life; but as soon as it had swallowed its little King, it no longer felt itself to be anything but one" (ibid.).[9]

In his earlier work, describing a no less delirious lunar voyage, Cyrano's narrator defends heliocentrism, but does so on the grounds that "all bodies that are in Nature need this radical fire," and therefore it must be "at the heart of this Kingdom, so as to be able to promptly satisfy the needs of each part" (de Bergerac 1858, 35). He compares this placement to the location of an animal's genital organs at the center of the body, or to "the pits at the center of their fruit; and just as the onion conserves, under the protection of a hundred skins that surround it, the precious germ, where ten million others will draw their essence; for this apple is a little universe of its own" (ibid.).

Assuming that Leibniz in fact read and was influenced by both Cyrano de Bergerac and Nolant de Fatouville, as he reports in the *New Essays*, we are now in a position to see the connection between the two versions of the Harlequin principle: everything is as it is here, which is to say, first of all, that there is no need to invoke a different sort of explanation for superlunar bodies as for sublunar bodies. Second, the sort of explanations proffered for macroscopic bodies will be the same as that for microscopic ones; scale is of no relevance in determining the appropriate explanation for the structure and motion of a given portion of the natural world. Finally, the structure that characterizes every body in the natural world is one of bodily individuals that are in turn capable of constituting greater composite bodily individuals. Individual bodily beings are simultaneously worlds apart, but are also, at the same time and no less, implicated in the constitution of greater bodily beings. A given bodily being can 'fragment,' but no being can pass from a bodily state into a nonbodily one, nor indeed from bodily existence to nonexistence *tout court*, just as the king of the people formerly constituting the tree falls away in the form of a fruit and then emerges from the fruit as a little man.

For Leibniz, the need to account for everything "in the same manner as the rest" is articulated in terms of the *tout comme ici* principle. In turn, the way the world is structured *ici*, in the sublunar sphere, and more particularly in what we would call the 'biosphere'—and indeed even more particularly in the plant or animal body—is unlike anything Descartes was prepared to imagine. For Leibniz, the world is conceptualized on the model of the animal body, which is in turn conceptualized as an infinitely structured assemblage of infinitely many constituents—corporeal substances, all of which stand in hierarchical relations of domination and subordination relative to all others. Leibniz does not look at the animal body and hope that *it* might be explained in the same manner as the planets and projectiles; rather, he looks at the planets and projectiles and asserts that *they* are to be explained in the same manner as the animal body. What remains constant from Descartes to Leibniz, then, is the desire to explain the entire world in a unified way, and to do so, in some broad sense, 'mathematically'; what changes, though, are both the domain of nature from which the explanation is to be drawn, as well as the particular branch of mathematics that is looked to as a potential source of answers. The evolution is from the explanation of everything on the model of inanimate bodies, and by appeal to geometry, to the explanation of everything on the model of animate bodies, and by appeal to infinity.

MATHEMATICAL AND BODILY INFINITY

On Leibniz's mature view, already clearly in evidence by the 1704 *New Essays*, everything is as it is here, and moreover here it consists in infinitely structured or 'folded' body. For Leibniz, this infinite structure is synonymous with 'organism,' understood not as a count noun (there is no talk of 'organisms'), but as an abstract noun characterizing all natural bodies. To be characterized by organism is to be infinitely structured, which, to follow out Leibniz's trail of synonyms still further, is to be a 'divine machine.' Thus infinity lies at the very heart of Leibniz's mature account of the natural world, and in this respect he could not differ more sharply from Descartes. What it is for each philosopher to offer an account of any given segment of nature "in the same style as the rest," then, will be very different in each case.

As Ohad Nachtomy rightly notes, "Descartes delineated an irreconcilable gap between the infinite creator and its finite creatures, suggesting that it would be not only cognitively impossible but also morally and theologically wrong for us to investigate the infinite" (Nachtomy 2014, 9–28). In this connection, Descartes is hewing far more closely to the traditional understanding of infinity, among the vast majority of philosophers up until his time. Aristotle effectively curtailed serious commitment to an actual infinity with his well-known observation that "nature avoids what is infinite, because the infinity lacks completion and finality, whereas this is what nature always seeks" (Aristotle 1943, 7; 1.1.715b15). Leibniz was emboldened in his embrace of an actual infinity in large part by the innovative, imaginative, and fairly radical philosophical speculation of predecessors such as Henry More and Giordano Bruno. But, more important still, he was able to incorporate infinity into his philosophy in a very concrete, and not merely speculative way, as a result of important attainments of his in pure mathematics. As Nachtomy well explains: "Leibniz discovered a rational method to treat infinity in mathematics. By translating infinitesimal quantities into finite ones, arguing that they can be regarded as variables, smaller or larger than any assignable quantity, he showed that infinitesimals could in fact be used in calculations. Leibniz's sophisticated approach (evident in his early work in mathematics) certainly contributed to his applying infinity in other domains of his philosophy as well. For, given this approach, one could feel free using infinity without falling into paradox" (Nachtomy 2014, 12). Liberated from fear of paradox, Leibniz is also freed up to develop what might be considered a highly counterintuitive account of the structure of the natural world, as consisting in infinitely structured bodies, which result from the conspiracy of infinitely many subordinate bodies, yet cannot be said to be made up out of these subordinate bodies, since these subordinate bodies have exactly the same sort of structure as the bodies they in turn serve to constitute, and so on ad infinitum. Indeed in the end there simply is no rock-bottom level of the physical world that serves to make up composite entities in the way that bricks make up a house. Divine machines—and in the end such machines are all there is—are divine precisely to the extent that they cannot be analyzed into fundamental constituent parts; which is in the end just another way of saying that they cannot be analyzed away, and are therefore immortal.

Leibniz repeats this account of the structure of natural bodies numerous times, along with its various corollaries such as that of the immortality of corporeal substance. Thus in a letter to Malebranche of 1679, Leibniz writes: "There is even room to fear that there are no elements at all, everything being effectively divided to infinity in organic bodies. For if these microscopic animals are in turn composed of animals or plants or other heterogeneous bodies, and so on to infinity, it is apparent, that there would not be any elements" (Smith 2011, 235).[10] In a note on a letter of Michelangelo Fardella from 1690, amply discussed in a recent study by Dan Garber (2009), we find a similar expression of the commitment to the infinite structure of matter, though now expressed by explicit analogy to geometry: "There are substances everywhere in matter, just as points are everywhere in a line . . . Just as there is no portion of a line in which there is not an infinite number of points, there is no portion of matter which does not contain an infinite number of substances" (AG 1989, 105). Consider, finally, this passage from a text entitled "On Body and Force, against the Cartesians," written just two years before the *New Essays*, in which Leibniz nicely brings together the infinite structure of the organic body with the divinity and immortality of the corporeal substance: "Moreover, a natural machine has the great advantage over an artificial machine, that, displaying the mark of an infinite creator, it is made up of an infinity of entangled organs. And thus, a natural machine can never be absolutely destroyed just as it can never absolutely begin, but it only decreases or increases, enfolds or unfolds, always preserving in itself some degree of life or, if you prefer, some degree of primitive activity" (AG 253). This is Leibniz's mature account of the structure and activity of natural bodies in general, and it is at the same time his account of living bodies, since again, for Leibniz, there is no body that is not living: to be alive just is to have organic structure, which is to say to consist in individual corporeal substances standing in relations toward one another of nestedness ad infinitum, and there simply are no bodies that are not like this. This means that in the final analysis it is the study of the living body, and in actual practice the study of the human body, that will serve as the model and guide for the study of nature in general. Sometimes, the discipline that sees to this study is described as 'physiology,' sometimes as 'animal economy.' But most often it is called, simply, 'medicine.'

"MATHEMATIZING IN MEDICINE"

We have seen that Leibniz, like Descartes, wishes to explain everything in nature in the same way, and we have seen that for Leibniz this way will come to involve the application of infinity to the account of the structure of natural bodies.[11] For Leibniz, moreover, on the final analysis all natural bodies are living bodies, in the sense that there is nothing that is not a divine machine, nothing that is not characterized by organism. We have also seen that Leibniz and many of his contemporaries suppose—notwithstanding any revisions that need to be made to the BDK thesis as applied to Descartes—that it is a desirable thing to 'mathematize nature,' that is, to render all natural bodies tractable by subjecting them to quantitative methods of analysis, and to suppose that what is causing the qualitative features of living bodies are in the end the mass, shape, and motion of those bodies' microanatomical constituents. The question now arises, though, whether Leibniz would himself consider his analysis of living bodies in terms of their infinite structure a case of successful "mathematizing in medicine."

The young Leibniz is very optimistic about the project of mathematizing medicine, and, as we have seen, he cites Stensen and Bellini as his models for this undertaking. In the *Directiones ad rem medicam pertinentes* of 1671, he anticipates that medicine will eventually be exhaustively explicable in terms of the mass, figure, and motion of bodily particles. Yet later on Leibniz will come to see the project as only partially realizable. As he writes to Michelotti in a letter on animal secretion of 1715: "There may be many mechanical causes that explain secretion. I suspect however that one should sooner explain the thing in terms of physical causes. Even if in the final analysis all physical causes lead back to mechanical causes, nonetheless I am in the habit of calling 'physical' those causes of which the mechanism is hidden" (Leibniz 1768). Here, then, 'physical' contrasts with 'mechanical' to the extent that the latter lends itself to immediate mathematization, given the state of our knowledge and our capacity for observation, whereas physical explanation remains avowedly hypothetical.

Early on, for Leibniz, it had been the work of Steno that served as a model for the possible mathematization of medicine and related fields of investigation. Steno had argued, most importantly in his *Elementorum myologiae specimen seu Musculi descriptio geometrica* of 1667, that the nerves and muscles alike contract and expand without the influx or efflux of any new material. For the Danish physician, this argument was of particular importance for the broader argument against the animal spirits playing a role in animal motion. Steno is effectively attempting to demonstrate the mechanism of contraction *more geometrico*, namely, by showing how the shortening of the fibers that constitute the muscles is alone sufficient to account for muscular contraction.

Ultimately, Leibniz will find the Stenonian account inadequate, and will come to believe that the sort of mechanical explanation that should be hypothesized in accounting for the motion of the muscles will be one that attributes an important role to elasticity. The elasticity of the inner parts is conceived as a sort of force (*vis elastica*) that keeps the body in motion through countless imperceptible vibrations in a manner analogous to the "vibrations" of perceptions that endure in the soul as memories. As Leibniz writes in a letter to Bernoulli of May 6, 1712: "In organic beings many things seem to consist in perpetual, imperceptible vibrations, which, when we perceive them to be at rest, are in fact being held back by contrary vibrations. Thus in truth we are led back to an elastic force. I suspect that memory itself consists in the endurance of vibrations. Thus there does not appear to be any use for a fluid that goes by the name of animal spirits, unless it is traced back to the reason itself of the elastic force" (G 3.2 884–85). This strategy of explaining the dilation and contraction of the parts of the body in terms of an effervescence that brings about a sort of vibration far antedates Leibniz's correspondence with Bernoulli and seems to be traceable most directly to the influence of Boyle's *New Experiments Physico-Mechanicall, touching the Spring of the Air, and its Effects* of 1660. Leibniz writes as early as the *Corpus hominis* of the mid-1680s:

> While it is granted that the seat of effervescence is in the heart, it nonetheless is easily communicated to the whole body by the blood vessels, just as

> [when] we attempt to heat an enormous cask of wine with a small fire, if the fire be applied through a small copper utensil, connected with the vessel through a tube. Seeing moreover that in any ebullition there is an excessive dilatation, the vapor is nevertheless not expelled, but rather it is necessary that it in turn be pushed along, whence arises respiration, indeed in all exceedingly great efficient [causes] there is a certain reciprocation of restitutions such as we note in oscillating pendula, or in vibrating chords. (Smith 2011, 295)

What will be new by the time of the 1712 letter to Bernoulli is Leibniz's interest in describing a mental process such as memory as parallel to the bodily vibration that is brought about by the elastic force and that keeps the body in perpetual motion. Bernoulli would hold that "in the whole machine of the human body, every smallest particle involved in a movement is moved either directly by an order of the soul or by muscles. All these muscles follow strictly and steadily the laws of mechanics." Leibniz was very impressed with Bernoulli's work, yet there is no way, given Leibniz's conception of body and soul as parallel automata, that he could have agreed with his physician friend as to the dual sources of motion in the body. Leibniz would certainly agree that the muscles follow the laws of mechanics, and that the origin of motion in the muscles is a mechanically explicable pyrotechnical event, yet for him no "order of the soul" could make a difference in the succession of a corporeal substance's states. The reason for this is spelled out at length in Leibniz's arguments against Stahl's account of how the soul moves the body.

Leibniz agrees that the body contains the principles of its own motion, and this will be the major point of contention around which his debate with Stahl circles. The debate, at least as both of its participants understood it, was not about 'vitalism,' a notion that would not even come to be meaningful until well after the deaths of both Leibniz and his opponent. Yet if we must categorize Leibniz anachronistically in terms of this doctrine, we may say with firm conviction that he is an antivitalist: for him, the growth, motion, and preservation of a living body can be exhaustively accounted for without appeal to the soul. The soul is not responsible for life.

In many respects, Leibniz's development with respect to the question of mechanism reflects the development of his philosophy as a whole: we see an early, fervent commitment to the explanatory promise of the mechanical philosophy, followed by a gradual reintroduction of a role for teleological explanation, one that sees it as coexisting with mechanical explanation, and that sees each of the two types of explanation as accounting for one and the same world at different metaphysical levels and in view of different epistemological exigencies. Leibniz's rediscovery of teleology, however, and his mature view that natural beings are not 'mere' machines but rather corporeal substances, should not be seen entirely as an abandonment of the project of mathematizing nature. Rather, the conception of both the structure and complexity of nature on the one hand, and on the other the sort of mathematical model that could be useful in the study of nature, changed in tandem. For Leibniz, in both mathematics and in nature, the key concept would be one that, as we have already seen, had been utterly excluded in first-wave mechanical philosophy such as that of Descartes: infinity.

DIVINE MACHINES

Leibniz's mature antivitalism is far from a further development of approaching anatomy and physiology *more geometrico*. If there is any extent to which mathematics continues to provide a model for the study of living bodies, then the domain of mathematics that suggests itself is not geometry, but rather the infinitesimal calculus. As we have already seen, Nachtomy perceives a close relationship in Leibniz between the mathematics of infinity on the one hand and his analysis of the structure of organic bodies on the other. However, there are some grounds for caution in perceiving Leibniz's engagement with the mathematical problem of infinitesimals as shining any sort of light on his metaphysics of body. There are no smoking-gun texts that lead us directly from the one to the other, though Nachtomy does make a rather compelling case that the infinite nestedness of parts within parts that defines Leibniz's mature conception of the organic body could not have taken shape in the way it did if Leibniz had not been working through the problem in many other areas of his philosophical reflection. This includes his reflection on the composition of the continuum and his contributions to the development of the infinitesimal calculus, as well as his reflections on the nature of freedom and the idea of infinite analysis of complete concepts. In a very broad sense, then, we may cautiously say that for Leibniz infinity is a central concern in many aspects of his thought, and that this concern is reflected equally, with interesting parallels across both domains, in Leibniz's mathematics and in his account of the structure and motion of so-called living bodies. Leibniz's ultimate account of these bodies, we might say, differs from the account of respiration, circulation, generation, and so on, that had been sought after by first-wave mechanists in roughly the same way the infinitesimal calculus differs from geometry.

As Nachtomy has well noted, there is no domain of the natural world that does not involve infinity. Rather than rejecting infinity, as both Aristotle and Descartes had recommended, Leibniz is insistent that any adequate explanation of nature must involve the notion of infinity. As he writes to Foucher: "I am so much in favor of actual infinity that, instead of admitting that nature rejects it, as it is vulgarly said, I hold that it affects it everywhere, for better marking the perfections of its author" (G 1 416). Nachtomy rightly stresses however that Leibniz employs different notions of infinity in different contexts, and that he is particularly careful to distinguish between infinity in a mathematical context, "which concerns abstract and ideal entities," and in a metaphysical context, "which concerns concrete and real beings" (Nachtomy 2014).

The principal difference between the two contexts, as Nachtomy's distinction suggests, is that mathematics concerns itself with *ideal* entities, which thus have no real divisions in them but rather are literally continuous, that is, are such that they are not constituted out of real parts, but rather any part or section can be taken out of any given part or section at will. Thus ideal continua are actually infinitely divisible, for Leibniz, whereas concrete and real beings are actually infinitely *divided.* Recall, in this connection, Leibniz's notes on a letter to Fardella, from 1690. There, he drew an analogy between the composition of the organic body on the one hand and the relationship between points and lines on the other, arguing that bodies are no more built up out of fundamental parts than lines are built up out of points. But there is a crucial difference, namely, that there is no conceivable need to account for how lines are built up at all, since on the final analysis there simply are no such things as real lines. Bodies, however, demand to be accounted for; they are real, and therefore really constituted in some way or other, even if they are not constituted out of physical atoms. The answer, again, is that bodies are constituted as divine machines, which is to say that they result from the conspiracy of infinitely many other organic bodies, and so on without end, ad infinitum, no matter how far down you may wish to go in your analysis of the organic bodies constituting other organic bodies.

From the mid-1690s until his death, Leibniz gives several explicit accounts of what he means by this technical term, all of which amount to variations on the same core idea.[12] He says to Stahl in 1709, for example, that "organic machines are nothing other than machines in which divine invention and intention are expressed to a greater extent" (Leibniz 1720, 135). And five years earlier, in a letter to Damaris Masham, he writes: "I define *organism* or a natural machine, as a machine each of whose parts is a machine, and consequently the subtlety of its artifice extends to infinity, nothing being so small as to be neglected, whereas the parts of our artificial machines are not machines. This is the essential difference between nature and art, which our moderns have not considered sufficiently" (G 3 356).

"Organic," as an adjective, in the seventeenth century had no particular biological connotation (and of course "biological" had no connotation whatsoever). Rather, it was first and foremost a description of anything that has interrelated, working parts, whether physical or conceptual; anything, that is, that the Greeks would have recognized as an *organon*, a term any serviceable Greek-English lexicon would translate as "instrument" or "tool." Working with this minimal definition, we arrive already at the surprising conclusion that if we wish to avoid anachronism we must stop reading early modern occurrences of the term "organic" as antonyms of "mechanical," and instead interpret them as *synonyms.*

Anne Conway illustrates this original synonymy of "organic" and "mechanical" very clearly in her *Principles of the Most Ancient and Modern Philosophy*, published posthumously in 1690, when she writes that an animal is *not* "a mere Organical body like a Clock, wherein there is not a vital Principle of Motion" (Conway 1996). Similarly, in his *Lexicon Philosophicum* of 1662, Johannes Micraelius (1996) defines "organic parts" as "composite heterogeneous parts . . . They are members of the body, which nature exploits for uses that are necessary for life." The "inorganic" in turn is the "intellect, for it does not have its own organ of the body of which it makes use" (Micraelius 1662). The organic is whatever has working parts—machines and animals alike—and the inorganic is that which lacks parts, which is to say, that which is mental or intellectual. But Micraelius gives no possibility for distinguishing among different kinds of organicity, just as some efforts to describe animals as simple machines offered no criteria for distinguishing among different varieties of mechanicity. This is what Leibniz would provide: for him, organicity is a special variety of mechanicity; for Leibniz, in contrast with Conway, the horse's body *is* "organical." A body is organic, Leibniz explains, "when it forms a kind of automaton or natural machine." For Leibniz, unlike Conway, the horse's body, even though it is organic, is not simply like a watch, since to be an organic body is not to be a "mere" organic body. Leibniz defends the organicity of the horse by denying it to the watch. This distinction might seem obvious today, but until Leibniz made it, it went against the very meanings of the words involved.

Leibniz would agree with Aristotle's general line of reasoning, according to which it is the function, and not the material constitution, of an organ or an animal that makes it the sort of organ or animal that it is. For him it is impious to argue that eyes see simply because they are so structured as to be able to see, rather than that they were structured in order to see. Leibniz will favor the function of the organ by tracing its existence to a divine creator and to its, so to speak, "intelligent design." Such a consideration certainly could not have interested Aristotle, yet in his as in Leibniz's case the organ exists for the execution of a function, rather than that it happens to fulfill that function simply because it exists. Leibniz would certainly also agree with Aristotle that just as a blind eye is an eye in name only, so, too, a cadaver is a man only by convention. What makes the blind eye merely a nonfunctioning organ, rather than a dead animal, is that the seeing eye that it once was, was what it was only insofar as it contributed to the telos of the creature as a whole.

While Leibniz's understanding of "organic" does mark a new turn in the history of the concept, it is still not the antonym of "mechanical" that many commentators have taken it to be. In Leibniz's view, an organic body is distinct from a clock with respect to the complexity of its constitution, but Leibniz continues to agree with Conway that an organic body, considered in itself, lacks a single, dominant, vital principle. For Leibniz, an organic body is distinct from a mere mechanical body in that it is infinitely complex, but this does not mean that the organic body per se is something the explanation of which requires the introduction of an immaterial vital principle. It is true that metaphysically speaking an organic body is always dominated by the soul or form of the animal or corporeal substance to which it belongs, but physically speaking, the difference between an organic body and an inorganic body is found in the complexity of the organic body: it and all of its parts and the parts of the parts, ad infinitum, are machines of nature.

The organic body of the fish, then, insofar as it is the body *of* the fish, will in fact never be without a dominant monad or unifying entelechy. Yet the block of marble, at least as a whole, is always without one, even though every part of the organic matter making up the block of marble is part of some corporeal substance. An organic body can at most be conceptually distinct from a corporeal substance, while in fact there is never an organic body that is not the organic body of a corporeal substance.

Any arbitrarily chosen parcel of matter is extremely unlikely to constitute in itself one organic body, even if there is no part of it that is not so constituted. As Leibniz writes in 1702, the organic body, taken separately, is just a special kind of aggregate, while the union of this organic body with an entelechy is one per se, and not a mere aggregate of many substances, for there is a great difference between an animal, for example, and a flock. And further, this entelechy is either a soul or something analogous to a soul, and always naturally activates some organic body. Which, taken separately, indeed, set apart or removed from the soul, is not one substance but an aggregate of many, in a word, a machine of nature.

The organic body, then, is a machine of nature, even if, taken together with the soul rather than separately, the whole thing is not a machine at all, but a corporeal substance. Insofar as we are considering the organic body of the fish, as distinct from its soul, we are considering something on an ontological par with a pile of sawdust, even though the fish, which consists in this organic body and an ichthyoid soul, is of an ontologically higher rank than the pile. The block of marble is made up entirely of organic matter, but is only an aggregate, insofar as it is not, as a whole, unified by a dominant monad or entelechy. The fish's body is also made up entirely of organic matter, but the fish itself is a corporeal substance and not an aggregate, insofar as there is a dominant monad, the fish's soul, uniting the organic body. While it is true that souls and bodies are not *really* separable, their conceptual separation is of central importance.

We find the same point also in the *New Essays*. Animated bodies, Leibniz says there, can be picked out by their interior structures. Body and soul can each be taken separately, and each suffices for the determination of the identity of the thing in question. Neither influences the other, but each expresses the other perfectly, the one being the concentration in a unity of what the other disperses throughout a multitude. Leibniz emphasizes that the organic body may be taken separately (*pris à part*), which is to say that organic bodies just are the machines of nature, or that which remains mechanical in its least parts, and which does not require the introduction of the capacity for perception that would be required in the exhaustive account of a corporeal substance.

In *Divine Machines*, I have developed at greater length this distinction between the organic body and the corporeal substance. There is no need to dwell on it any further here. It is enough to be clear that, for Leibniz, the corporeal substance is to be understood in relation to its ends, which it has in virtue of the domination relation of its soul or entelechy to the infinitely many other monads implicated in it; the natural or divine machine, by contrast, which is to say the organic body, is to be understood without regard for its ends or for its unification under the domination of an entelechy, but only with regard to its infinite structure.

It is precisely this infinite difference between the natural and the artificial machine that, for the mature Leibniz, will come to be coextensive with, and also come to replace, the more familiar distinction between the living and the nonliving. As Leibniz makes particularly clear already in the *Protogaea* of the early 1690s, the formation of crystals, by contrast with that of animals and plants, can be exhaustively analyzed in terms of "external contiguity," that is to say in terms of the regular repetition of radial and polygonal shapes. This is important, because it sharply delineates anything formed geologically, including crystals, from the realm of the organic. Crystals and organic bodies are in fundamentally different ontological categories, yet this difference cannot be accounted for by the fact that the former are 'nonliving' while the latter are 'living,' since, strictly speaking, for Leibniz organic bodies are *not* living. Rather, organic bodies can be explained exhaustively in terms of their vegetative structure, while *life*, in turn, is simply a capacity of immaterial perceiving monads. As Leibniz writes to Stahl, in response to the Halle physician's account of supposedly irreducibly vital processes: "I do not wish to quarrel over words. It is the author's wish to call 'life' what others call 'vegetation'" (Leibniz 1720, 11). Animal bodies vegetate but are not alive, for the mature Leibniz, and vegetative structure consists precisely in this: that whereas crystals, for example, are generated out of the finite repetition of regular geometric forms, vegetative bodies are ungenerable, to the extent that they consist in an infinite structure with no lower limit to its composition, and thus no possibility of ever being decomposed into its elementary constituents, or of ever having been built up in time from such constituents.

At this point, we have considered just about every aspect of the structure and nature of organic bodies, or divine machines, and of corporeal substance that Leibniz was willing or able to impart to us. We have seen that they are real, infinitely structured entities, whose composition Leibniz on occasion wishes to describe on analogy to the relationship between points and lines, even though he knows that this analogy cannot be terribly helpful, to the extent that lines and bodies belong to entirely different ontological categories, and lines, as ideal entities, do not really need to be constituted at all. But how, if bodies are real, composite entities, can they fail to be built up out of fundamental parts, as houses are from bricks, rather than simply resulting from requisites? What is it that yields a big body if not several smaller bodies? But if there is no lower limit to the analysis of smaller bodies into smaller bodies still, then how can a body of any size ever be yielded by composition?

Nachtomy, as we have seen, believes that Leibniz is emboldened in accounting for the composition of real bodies by appeal to infinity as a result of his parallel success in the mathematical treatment of infinite quantities. Having banished paradox from the mathematics of the infinite, Leibniz was now ready to explain the bodily world as well by appeal to infinity. On Nachtomy's view, this was a great coup de grâce of Leibniz's philosophy: to give us a novel, postmechanist philosophy of nature by extending his successes in the mathematics of the infinite to the study of the natural world. Yet as I have been arguing here, the not-merely-mechanical, end-directed corporeal substance is something quite distinct from the infinitely structured organic body. To the extent that Leibniz invokes infinity in his analysis of natural bodies, in other words, he does so precisely in order to *preserve* a variety of mechanism, even if this amounts to a mechanism with a very considerable twist. This is precisely the point of Leibniz's insistence that, as he writes to Stahl, "organism is in truth mechanism, but more exquisite" (Leibniz 1720, 9). 'Exquisiteness,' for Leibniz, does not reach back to a pre-Cartesian understanding of nature, but rather radically modifies the mechanical philosophy in order to give an account of nature that is adequate to its complexity, and that is also almost wholly original in Leibniz (with, obviously, a complex prehistory in figures such as Giordano Bruno, Nicholas of Cusa, Henry More, and many others).

A final problem with the suggestion that Leibniz is offering us a postmechanical nature by extending his successes in the study of the mathematics of the infinite to his account of the natural world is precisely that we may doubt that his invocation of 'infinity' in his account of the structure of bodies has much to do with mathematical infinity at all. It is certainly true, as Nachtomy suggests, that Leibniz was able to some extent to banish paradox from the mathematical treatment of infinity by treating infinitesimal quantities as variables. But again, the freedom to treat them in this way in the end rested for Leibniz on the fact that mathematics is only concerned with ideal entities, yet the very challenge that the natural world poses for Leibniz is that it requires us to account for something real, actually existing, and resistant to fictions. Leibniz determines that the matter making up this world is actually infinitely divided, and that the organic bodies, in which all existent matter is wrapped up, consist in bodies nested within one another ad infinitum. It is not clear that in elaborating this remarkable account of the natural world, which is to say of the living world (to say that the two are coextensive is in the end exactly the same as to say that all matter is wrapped up in organic bodies or divine machines), Leibniz successfully steers clear of paradox.

CONCLUSION

Leibniz's theory of divine machines may, in sum, be seen as a continuation of the project of the mathematization of nature, even if it amounts to a continuation in such different terms as to be almost unrecognizable. From the treatment of nature *more geometrico* that inspired early mechanical philosophy, we witness a shift to an approach to nature inspired more by the mathematics of the infinite and of infinitesimals. But this shift is based, ultimately, on an untenable analogy between two different realms, the ideal and the real, and in the end it does not seem that Leibniz's success in advancing a paradox-free treatment of the mathematical infinite enabled him to provide a fully compelling account of the infinite structure of natural bodies.

We may ask whether an early mechanical philosopher such as Descartes would have seen the introduction of the concept of the organic, which is to say, again, the concept of the infinite structure of bodies, as a failure of the mechanical program, or rather as a final perfection of it. But one thing that is certain is that Descartes and Leibniz both saw the need to account for the structure and origins of what are commonly called 'living beings' as likely *the* most pressing task for the new natural philosophy to fulfill. Leibniz's account of living bodies would come to serve for him as the explanation of body in general, and in this respect we may say, with Garber (1985), that for Leibniz it is what we would call 'biology,' rather than physics, that is the foundational science of nature. Leibniz hoped early on to 'mathematize' biology, or what he called 'medicine,' but seems to have grown skeptical of the possibility of doing so as his mature philosophy developed. At the same time as his mature philosophy developed, however, he became increasingly enthusiastic about an account of living nature that rested on the notion of infinity. As I have attempted to argue, however, against Nachtomy's provocative and in many respects compelling account, this employment of the notion of infinity in the mature account of nature does not amount to a successful instance of the mathematization of nature, nor even a conscious attempt to mathematize it.

There are, of course, many sources feeding into Leibniz's use of infinity in his philosophy, and in particular into his account of the infinitely nested structure of natural bodies. His work in the mathematics of infinity no doubt played a role. Although the full case has yet to be made, however, a no less significant source for Leibniz's conception of worlds within worlds, and of the constitution of composite beings out of countless other such beings, appears to have come from the science fiction of which he was an avid reader: the imaginative flights of fancy from the likes of Cyrano de Bergerac, for example, who dreamed up a trip to the sun, and an encounter there with deliriously strange beings. Leibniz, the genius eclectic, did not dismiss Cyrano's pomegranate men as the products of a mere *rêverie*, but instead saw in them a reflection of the very earnest account of the natural world he spent much of his intellectual energy, drawing on sundry and often surprising sources, to elaborate.[13]

ABBREVIATIONS

AG Ariew, R., and D. Garber, eds. 1989. *G. W. Leibniz: Philosophical Essays*

AT Descartes, R. 1964–76. *Oeuvres de Descartes*

G Leibniz, G. W. 1875–90. *Die Philosophische Schriften von G. W. Leibniz*

NOTES

1. "De la description des cors inanimez & des plantes, ie passay a celle des animaux & particulierement a celle des hommes. Mais, pourceque ie n'en auois pas encore assez de connoissance, pour en parler du mesme style que du reste, c'est a dire, en demonstrant les effets par les causes, & faisant voir de quelles semences, & en quelle faôn, la Nature les doit produire, ie me contentay de supposer que Dieu formast le cors d'un homme, entierement semblable a l'un des nostres, tant en la figure exterieure de ses membres qu'en la conformation interieure de ses organes."

2. Citing Ariew's citation of Sophie Roux summarizing Husserl on mathematization (see Ariew, chapter 4, this volume).

3. "Il faut tousjours expliquer la nature mathematiquement et mecanique-

ment, pourveu qu'on sçache que les principes mêmes ou loix de mecanique ou de la force ne dependent pas de la seule étendue mathematique, mais de quelques raisons metaphysiques."

4. *Nouveaux essais* 2, ch. 7, §42. "[C]'est . . . comme Arlequin qu'on voulait dépouiller sur le théâtre, mais on n'en put venir à bout, parce qu'il avait je ne sais combien d'habits les uns sur les autres: quoique ces réplications des corps organiques à l'infini, qui sont dans un animal, ne soient pas si semblables ni si appliqués les unes aux autres, comme des habits, l'artifice de la nature étant d'une tout autre subtilité."

5. Leibniz to Masham, May 8, 1704.

6. Anne Mauduit Nolant de Fatouville (credited anonymously as 'Monsieur D***'), *Arlequin Empereur dans la Lune* (de Fatouville between 1765 and 1814).

7. Antonio Nunziante has offered an excellent analysis of the relevance of this part of Cyrano's work for our understanding of Leibniz's theory of corporeal substance, and it is Nunziante who first brought this connection to my attention (see Nunziante 2011).

8. *Nouveaux essais* 2, ch. 23, §43. "Au reste je suis aussi d'avis que les Genies appercoivent les choses d'une maniere qui ait quelque rapport à la nostre, quand même ils auroient le plaisant avantage, que l'imaginatif Cyrano attribue à quelques Natures animées dans le Soleil, composées d'une infinité de petits volatiles, qui en se transportant selon le commendement de l'ame dominante forment toutes sortes de corps."

9. "Tout cet amas de petits hommes n'avoit point encore, avant cela, donné aucune marque de vie; mais, sitot qu'il eut avalé son petit Roi, il ne se sentit plus être qu'un."

10. Aristotle 1, 2:719.

11. Portions of this section were developed previously in Smith (2011) though in the course of making a very different argument about the place of medicine and physiology in Leibniz's philosophy.

12. Portions of the present treatment of the distinction between "organism," "organic body," "mechanism," and "corporeal substance" were previously developed in Smith (2011) though, again, in the course of making a very different argument about the significance of these concepts in Leibniz's philosophy.

13. Elsewhere, I have argued for the important role of the empirical discoveries of microscopy in the development of Leibniz's theory of composite substance, and indeed here I am not at all seeking to subvert that account, nor to

refute Nachtomy's, but only to recommend that here, as in so many other areas of Leibniz's thought, no single monocausal account will do. See, in particular, Smith (2011).

REFERENCES

Ariew, R., and D. Garber, eds. 1989. *G. W. Leibniz: Philosophical Essays*. Indianapolis, Ind.: Hackett.

Aristotle. 1943. *On the Generation of Animals*. Ed. and trans. A. L. Peck. Cambridge, Mass.: Harvard University Press.

Bellini, L. 1662. *Exercitatio anatomica Laurentii Bellini Florentini de structura et usu renum*. Florence: Ex Typographia sub signo Stellae.

de Bergerac, C. 1858. *Histoire comique des États et Empires de la Lune et du Soleil*. Ed. P. L. Jacob. Paris: Adolphe Delahays.

Burtt, E. A. 1925. *The Metaphysical Foundations of Modern Physical Science: A Historical and Critical Essay*. London: Kegan Paul.

Conway, A. 1996. *Principles of the Most Ancient and Modern Philosophy*. Ed. A. P. Coudert. Cambridge: Cambridge University Press.

Descartes, R. 1964–76. *Oeuvres de Descartes*. Ed. C. Adam and P. Tannery. 2nd ed. Paris: Vrin.

Dijksterhuis, E. J. 1961. *The Mechnization of the World Picture*. Oxford: Oxford University Press.

de Fatouville, A. M. N. (1765 and 1814). *Arlequin Empereur dans la Lune*. Troyes: Garnier.

Garber, D. 1985. "Leibniz and the Foundations of Physics: The Middle Years." In *The Natural Philosophy of Leibniz*, ed. K. Okruhlik and J. R. Brown, 27–130. Dordrecht: Reidel.

———. 2009. *Leibniz: Body, Substance Monad*. Oxford: Oxford University Press.

Kardel, T., ed. and trans. 1994. *Steno on Muscles*. Philadelphia: American Philosophical Society.

Koyré, A. 1957. *From the Closed World to the Infinite Universe*. Baltimore: Johns Hopkins University Press.

Leibniz, G. W. 1720. In G. E. Stahl, *Negotium otiosum, seu Skiamachia*. 135. Halle: Orphanotrophei.

———. 1768. *G. G. Leibnitii Opera omnia*. Ed. Louis Dutens. 6 vols. Geneva.

———. 1875–90. *Die Philosophische Schriften von G. W. Leibniz*. Ed. C. Gerhardt. 7 vols. Berlin: Weidmannsche Buchhandlung.

Micraelius, J. 1996 [1662]. *Lexicon Philosophicum terminorum philosophis usitatorum*. Düsseldorf: Stern-Verlag Janssen.

Nachtomy, O. 2014. "Infinity and Life: Infinity in Leibniz's View of Living Beings." In *The Life Sciences in Early Modern Philosophy*, ed. O. Nachtomy and J. E. H. Smith, 9–28. Oxford: Oxford University Press.

Nunziante, A. 2011. "Continuity of Discontinuity? Some Remarks on Leibniz's Concepts of 'Substantia Vivens' and 'Organism.'" In *Machines of Nature and Corporeal Substances in Leibniz*, ed. J. E. H. Smith and O. Nachtomy, 131–44. New York: Springer.

Smith, J. E. H. 2011. *Divine Machines: Leibniz and the Sciences of Life*. Appendix 1. Princeton, N.J.: Princeton University Press.

11

THE GEOMETRICAL METHOD AS A NEW STANDARD OF TRUTH, BASED ON THE MATHEMATIZATION OF NATURE

URSULA GOLDENBAUM

> Mais ie n'ay resolu de quitter que la Geometrie abstracte, c'est a dire la recherche des questions qui ne seruent qu'a exercer l'esprit; & ce affin d'auoir d'autant plus de loysir de cultiuer vne autre sorte de Geometrie, qui se propose pour questions l'explication des phainomenes de la nature.
> —DESCARTES TO MERSENNE, JULY 27, 1638

THE CONNECTION BETWEEN GEOMETRICAL METHOD AND MATHEMATIZATION OF NATURE

Use of the geometrical method has long been criticized, even before Kant, for being inappropriate in the field of philosophy. There is above all the general reluctance to accept the ponderous method of geometrical demonstration in philosophy. This method is considered to require definitions and demonstrations of propositions,[1] rarely commenting and explaining, thus providing little communication with the audience, while eschewing irony and rhetoric altogether. Many philosophers characterized this geometrical method as a mere external means of presentation, without contributing anything philosophical.[2] In the case of Spinoza, his use of the geometrical method was even considered as a major hindrance to the reader's getting into his precious inner doctrine.[3]

My aim in this chapter is to argue that the geometrical method is not just a form of presentation or demonstration. Rather, this method was seen as a *new standard* of knowing natural things. It was embraced as a new epistemological approach to the external world, laying the ground for a new type of science and philosophy. The geometrical method of early modern times was no longer just geometry but was clearly connected with the new *science* of mechanics. It was the wedding of mathematics and the *art* of mechanics, that is, the art of building machines that brought about Galileo's new *science* of mechanics. Turned into a science, mechanics was no longer the art of building machines and "tricking nature." Rather, the entire world was now considered as a machine, constructed by a divine mechanic.

While this new mechanical science owed much to geometry, it in turn had the greatest impact on geometry and mathematical development in general.[4] While Euclid had defined a line by the motion of a point rather accidentally, early modern mathematicians and then Hobbes made the generation of geometrical figures the starting point for understanding geometrical figures through their causes. Hobbes, moreover, used this new explanatory device far beyond mathematics and formulated the general epistemological principle that we can only know what we can generate (OL I, 9; De corpore I, 1, #8). Alan Gabbey (1995) first pointed to the often overlooked achievement of Spinoza who first used the geometrical method to understand the functioning of human beings, especially of their emotions and of the consequences for their morals and politics (142–91). Of course, Spinoza also uses the geometrical method to *present* his argument by compelling demonstrations. What is of greater significance though, is the fact that he uses this method to lay ground for a scientific ethics: "In this respect the Ethics was more radical than (say) Newton's *Philosophiae naturalis principia mathematica* (1687), where at least the *mathematica* and the *philosophia naturalis* were both parts of *philosophia speculativa*" (Gabbey 1995, 147–48).

Spinoza's *entire metaphysics*, the notorious parallelism of his philosophical system, rests on the new geometrical method, namely on the conviction that ideas follow one another in the same order and connection as things in the world, causing one another. According to the analytical geometry of Descartes and Viète, the figure of a circle parallels its equation if considered in the framework of the Cartesian coordinates with a defined unit. There the figure of the circle was substantially identical with its equation without any visible similarity of the figure with its equation.[5] What gave rise to the geometrical method as a new epistemological standard in philosophy, embraced by all rationalists, was Galileo's mathematization and thereby mechanization of nature, the wedding of mathematics to mechanics.

These philosophers saw mathematics and the geometrical method as the high road to certainty in human knowledge while raising caution about mere sense experience. The mathematical method could not only provide certainty but also freedom from authorities and from biased interpretations of mere empirical facts: “For there is not one of them,” Hobbes complains about past philosophers, “that begins his ratiocination from the definitions, or explications of the names they are to use; which is a method that hath been used only in geometry, whose conclusions have thereby been made indisputable” (Hobbes 1994, 24; Leviathan v, 7). And Spinoza states that the belief alone “that the judgments of the Gods far surpass man’s grasp . . . would have caused the truth to be hidden from the human race to eternity, if mathematics, which is concerned not with ends, but only with the essences and properties of figures, had not shown men *another standard of truth*” (Ethics I App, C 441; emphasis added).

This enthusiasm for the new geometrical method in close connection with the mathematization of nature stirred up resistance. The aversion of theologians—as well as of Christian philosophers such as Henry More, Locke, Kant, and the German idealists—against the geometrical method was not a result of their deep mathematical insights. Rather it was due to their fear of necessitarianism as well as of hubris. They feared that the geometrical method, with its claim to provide knowledge as certain as that of God (adequate ideas), would lead to the claim of human “omniscience.”[6] The mathematization of nature would introduce the mathematical necessitarianism into nature and take away God’s will and humans’ free will.[7] Fighting Wolffianism, Joachim Lange, the chief pietist theologian, and Valentin Ernst Löscher, the leader of orthodox Lutheran theologians (Löscher 1735, 126–29), both saw the geometrical method, especially the genetic definitions, in clear connection with the mathematization of nature.[8] They saw the aim of certainty of human knowledge as delivering divine certainty to human beings. They considered the view that genetic definitions provide the highest standard of truth as a threat to throw theology from the throne in order to install philosophy as the highest knowledge.[9] Lange and Löscher both blamed the geometrical method for dismissing the truth of the Christian religion[10]—because the latter rested on historical knowledge alone, thus lacking adequacy.[11]

What made theologians so vigilant against this method is just the assumption that we can use mathematical and further mechanical methods to study the *inner* structure of nature, God’s creation—and that we can come up with a kind of knowledge that is superior to mere sense perception and mere empirical investigation, nay, which is even like that of God himself.

In contrast, the Lutheran theologians praise empiricist philosophers and scientists.

As a matter of fact, such fundamental theological suspicion against mathematics as a tool of natural science had been expressed by Catholic theologians before, against Galileo—that is, right from the outset of the new mathematical science of mechanics. In the documentation of the trial one can find, among other listed accusations against Galileo, "to badly state and declare that there is a certain equality, in understanding geometrical things, between the human and divine intellect" (Galilei 1907, 19:326–27).[12] Thus, from the very beginning, the enthusiasm to mathematize the science of nature and thereby gain access to the inner structure of nature through discovering mathematics within nature stirred up strong theological resistance.

This critical approach to the mathematization of nature was perfectly in agreement with the tradition of Aristotle and even of Plato (Cassirer 1946, 277–97). With all respect and enthusiasm for mathematics, they would never have thought of nature, that is, of pebbles, mountains, rivers, plants, or animals, as being mathematically structured. These natural things could be known only by observation and classification. In contrast, Galileo, and thereafter Descartes, Spinoza, and others considered nature to be constructed mathematically. Therefore, we needed mathematical science to learn about the functioning of nature. Cassirer emphasizes this difference: "For what does the term 'science' mean in Galileo's system? It never means mere probability, it means necessity. It means no mere aggregate of empirical facts or haphazard observations; it implies a deductive theory. Such a theory must be capable of demonstration; it cannot be based on mere opinion or probability. If it is not possible to attain such a deductive truth about physical phenomena, then Galileo's scientific ideal, the ideal of modern dynamics, breaks down" (Cassirer 1946, 281). And indeed, the term "necessary demonstrations" as Galileo's tool of mechanical science occurs again and again in his writings.[13]

In early modern times, it was first the theologians who defended the view that nature, as God's creation, is incomprehensible to human beings, knowable only by observation, externally, and by chance. This defensive position was soon taken up by (Christian) philosophers, who tried to escape the mathematization of nature and thereby escape the comprehensibility of nature by mathematics (Goldenbaum, forthcoming, "How Theological Concerns Favor Empiricism over Rationalism"). I see mathematizing nature and the rejection thereof as the root of the well-known opposition of the two philosophical camps of rationalism and empiricism, an opposition that continued into the battles between Wolffianism and Pietism during the eighteenth century in Germany, and which provided the background for Kant's entrance into philosophy. Kant's well-known rejection of the geometrical method as inappropriate did not result from Kant's work on mathematics either (Kant 1998, 630–48). He had argued against the mathematization of nature and the geometrical method in his very first book on the estimation of forces (Goldenbaum, forthcoming) when he barely knew recent mathematics.[14] He simply shared the German Lutheran theologians' fear of necessitarianism and tried to save free will by allowing a mutual influence of body and soul (Goldenbaum, forthcoming).

In addition to the traditional objection against human hubris when claiming a human knowledge of nature like God's own, there lingered another threat for Christian religion that arose out of the mathematization of nature—namely necessitarianism. What constitutes the great advantage of mathematical cognition, namely necessary knowledge, would allow for necessary knowledge about nature as well—if nature were mathematically structured. Natural processes were necessary and could be comprehended by means of causal connections. As a result, if mathematics were to structure God's creation it would mean the end of free will, that of God as well as of human beings. Neither would God be capable of acting arbitrarily according to His good will, nor could human beings have a free will to act according to the good or bad. It was these theological concerns that produced a deep suspicion among Christian philosophers against using the geometrical method beyond mathematics. Exemplary of this philosophical suspicion against the mathematization of nature and the geometrical method is the development of Henry More's relation to Descartes and his philosophy. While More had never been a partisan of Descartes, he admired his work in the beginning and appreciated the special status of the soul in this philosopher's system. However, from More's first comments one can grasp his sensitivity regarding Descartes' emphasis on the geometrical method, on mathematics as the high avenue to knowledge of nature, and on the *necessity* of the knowledge we are able to obtain in this way (More 1711, esp. 3, 8, 40). But More would turn against Cartesian*ism* as atheism as soon as its necessitarianism became notorious, through Descartes' partisans, as Alan Gabbey has shown (1982, 173–74). Although More would never accuse Descartes himself of being an atheist, he recognized that the mathematization of nature would inevitably result in necessitarianism, leading to atheism. This became manifest with Hobbes's *Leviathan*, Ludewijk Meijer's *Philosophia Scripturae interpres*, and 'Cartesian' writings, not to mention Spinoza (ibid., 233–50). Of course, the critics of the mathematization of nature well agreed that we can produce mathematics within our minds. But they strongly detested mathematics *in* nature, outside of ourselves; or, if there were mathematics in nature, they held that we could not have any access to it due to our limited capacities. As a result, John Locke even stated that there could never be such a thing as a science of natural bodies (Locke 1975, 560; Essay IV, 3, #29).

In this chapter I first explain this *new* geometrical method and its new use in philosophy and science in general, thereby refuting the abovementioned prejudices against its appropriateness in these fields (see next section). I then focus on the concept of definitions as the cornerstone of the new geometrical method and discuss the concept of adequate ideas, showing how it is closely related to the concept of causal definitions, which was first introduced by Thomas Hobbes. I then describe how the Christian philosopher Leibniz embraced the new geometrical method and the concept of causal definitions that led to his logical containment theory and how he struggled to avoid necessitarianism while retaining the geometrical method and the mathematization of nature. Leibniz was aware that it was the threat of necessitarianism that caused theologians and especially British philosophers to reject the geometrical method in metaphysics. He assured the Cambridge Platonists that, based on his approach, mathematics and the geometrical method would no longer threaten free will and could be successfully applied (GP 3 363–67, 401–3).

WHAT IS THE GEOMETRICAL METHOD?

Although ancient mathematics used the method of deducing demonstrations from axioms and definitions, surprisingly, the term "geometrical method" came into use only during early modern times. Since Zabarella, it has been described as involving two aspects, both belonging to the method, namely the resolutive and compositive methods, also known as the analytic and synthetic methods (Cassirer 1974, 1 136–44). While the former is considered to be helpful for discovery and invention (but difficult to bring under rules), the latter is appreciated for *ensuring* the certainty of the results due to a complete deduction of propositions; that is, demonstration. Thus it was the synthetic method in particular that provided the *compelling force* to convince others of the correctness of a solution, while scientists and especially mathematicians did not care much about a *gapless* deduction if they were working to solve a problem (Breger 2008, 191–92). It was above all the *compelling* force of the method that made it highly attractive to early modern philosophers and scientists. The familiar anecdote about Hobbes's wondering about the Pythagorean theorem and being struck by the compelling power of its demonstration may illuminate this enthusiasm (Aubrey 1898, 1 332–33; Hobbes 1994, lxvii). But it was this compelling power of the geometrical demonstration as well that caused theological concerns about the method, and, moreover, about mathematics in general, as soon as this geometrical method was applied beyond pure mathematics.

Usually, when we think of geometrical method today we associate it with what we see when we open a book of Euclid, or (if we are looking for its use in philosophy) what we see in Spinoza's *Ethics*. Instead of a coherent flow of text, the lines are broken up into different types of text—definitions, axioms, postulates, propositions, and demonstrations. Although geometrical method is often associated with such complicated mathematical procedures, the great mathematician Pascal broke it down to only two essential rules in his fragment on the geometrical spirit that came down to us as a part of the *Logic of Port-Royal*: (1) not to employ any term whose meaning is not defined and (2) not to advance any proposition that is not demonstrated by known truths (Pascal 2000, 155–56). These simple rules do not sound as if they cannot be applied to disciplines other than mathematics and it seems difficult to understand what would be wrong with applying these rules to philosophy.

What is interesting from Pascal's precise and brief formulation is the fact that it is not, as usually believed, demonstration that is at the heart of this method. Nor are axioms essential to the geometrical method as it is often said. This misconception sometimes leads to an inappropriate identification of the geometrical method with the axiomatic method. However, the definitions *are* essential to the new geometrical method. Every demonstration has to set out from a definition. It is nothing more than a deduction of concepts from the concepts virtually included in the definition at the beginning. Hobbes considers a demonstration as a mere chain of definitions (OL I 252–58; De corpore 3, 20, §6), and Leibniz follows him in that.[15] Given this crucial role of definitions for the new geometrical method, it does not come as a surprise that the concept of the definition moved to the center of the long-lasting discussion of geometrical method in the seventeenth and eighteenth centuries.

While the opponents of the geometrical method had to agree that its results did indeed produce necessary conclusions, to dispute the geometrical method they instead questioned the certainty of its very beginnings, of its foundation, that is, of definitions and axioms. Both of these concepts became the subject of hot debate. Their disputed status in terms of certainty served as a bulwark of the critics of the geometrical method (Pascal 2000, 573 [no. 101]). Theologians and other defenders of Christian religion pointed to axioms as undemonstrated assumptions and to definitions as arbitrary human settings lacking the true essence of defined things on which nevertheless all the wonderful demonstrations rested, thus making the entire building of mathematics uncertain in its foundation.[16]

In principle, the problem with the axioms was easily solved by Hobbes, followed by Leibniz, by simply reducing axioms to demonstrations (OL 1 105–6; De corpore 2, 8, §25; Leibniz A II, 1 281; A VI, 2 480). Both philosophers argued that axioms are indeed assumptions, considered self-evident and therefore accepted by everybody. But as soon as anyone raised a doubt about any axiom it had to be demonstrated from definitions alone. Similar statements can be found in Spinoza.[17] Hobbes and Leibniz both showed this in an exemplary way for the famous (and in their days suddenly disputed) axiom that the part is smaller than the whole (Hoffmann 1974, 12–14; Goldenbaum 2008).

The question of definitions—whether they could be formulated at will by human beings and were thus under our control although lacking correspondence to the essences of real things, or if they were something objective albeit incomprehensible—was more problematic. This discussion continued from the seventeenth century well into Kant's days. The point of disagreement was whether definitions of things, which were not mere products of our minds, but real things, independent of our minds, could be defined by human beings at all.[18] The opponents argued that we had to empirically find the real things' properties and to build nominal definitions of collections of properties of a thing. There was no way to come up with essential definitions as we could provide in mathematics where we could construct the objects of our definitions and thus know their essence. The defenders of the use of the geometrical method beyond geometry emphasized the continuity between objects of pure mathematics (which were considered to be products of human minds) and objects of natural science (which were considered to be products of God's creation), thus allowing the use of the geometrical method within science and philosophy.[19]

DEFINITIONS IN THE FRAMEWORK OF THE NEW GEOMETRICAL METHOD

Arnauld and Nicole in their *L'art de penser* maintained the traditional distinction between nominal and real definition (Arnauld and Nicole 2011, 325–31; L'art de penser 1, 12), rooted in Aristotle's *Organon* (Anal. Post. 2, 7–10). While the first was nothing more than words by which we named things, either by convention or by custom, without knowing the essence of a thing the latter would allow us to understand whether the defined thing was real or at least possible in reality. Pascal did not accept any other than nominal (i.e., merely arbitrary definitions).[20] Likewise, he also saw axioms as a starting point without being able to secure their certainty, thereby making all our knowledge generally limited.[21] Denying our ability to come up with any *real* definition simply meant we could not know the essence of anything or, put differently, have any objective knowledge about the natural world. Therefore, we had to rely exclusively on experience to know something about the external world. We could observe and look for regular patterns.

Although it is Thomas Hobbes who is often blamed for the doctrine that definitions are arbitrary, he in fact rose to Pascal's challenge. We have seen already that Hobbes successfully addressed the problem of the axioms by showing they could be demonstrated if needed. But he also developed a new general approach to the other problem—that of real definitions. The way he does this sheds quite some light on how the new geometrical method of early modern time was indeed new—he connects the issue of definitions with Galileo's new science of mechanics.[22] Hobbes considered geometrical figures as produced by mechanical motion. A circle is produced by the motion of one endpoint of a straight line around the other endpoint. Thus Hobbes first introduced this new mechanical approach to definitions systematically into philosophy and demanded *causal definitions* (or genetic definitions) in philosophy in order to produce *necessary* conclusions about reality. A definition that includes the mechanical cause of the thing to be defined can serve to deduce all the properties of the thing (OL 1 71–73; De corpore 1, 6, §13).[23] To wit, such causal definitions provide the opportunity to deduce *any possible* property of the circle, even of properties of which we are not yet aware.

But while Hobbes and Pascal both claimed that definitions were arbitrary, Hobbes accepted nominal (i.e., arbitrary) definitions only for "absolute knowledge," that is, for factual knowledge. Such absolute knowledge, according to Hobbes, is provided only through experience, and does not require knowing the cause of the thing (Hobbes 1994, 35; Leviathan vii, 3). Therefore, in the case of empirical knowledge we are free to *name* observed sensations and thus can produce arbitrary definitions. But when we can produce the mechanical cause of a thing to be defined we can provide causal definitions. While we cannot draw any necessary but only probable conclusions from nominal definitions, causal definitions guarantee necessary conclusions. This has been noticed in Hobbes scholarship in recent decades (Jesseph 1999, 198–205).

Interestingly, Hobbes's innovation of causal definitions was then adopted (together with the geometrical method) by all rationalists—by Spinoza (TIE §69–71), by Leibniz (see below), and by Christian Wolff (Cassirer 1974, 2:521–25; Goldenbaum 2011b). With the exception of Spinoza, they hardly used the explicit form of the geometrical method but all their works follow this very method, that is, they begin with definitions and deduce the entire argument from them.[24] It should be noticed that all rationalists were mathematicians! Of course, Descartes and Leibniz were geniuses in mathematics whose achievements are still recognized today. However, Spinoza and Wolff well knew the most recent mathematics of their time and could follow the ongoing discussions. And the model for causal definitions is clearly the geometrical construction of figures. But neither Hobbes nor other rationalists stopped there.

The new approach to geometrical method, based on a new concept of causal definitions, was no longer Euclid's method but went far beyond his project. The most significant difference between the new geometrical method and what ancient geometers did is its extension beyond geometry. The most famous example is of course Spinoza, who wrote a metaphysics or rather an ethics according to this method. Descartes saw all his science as mere mathematics (AT 2 268). But Hobbes had already claimed that there cannot be any *science* that does not use the geometrical method or draw conclusions from causal definitions.[25] While no merely empirical discipline could ever turn into science because no result of these disciplines could aim for certainty, that is, for necessary knowledge, mathematics, optics, mechanics, and—famously—politics could become science because they all started from causal definitions. Hobbes's surprising inclusion of politics and Spinoza's treatment of ethics among strict sciences follow precisely the model of the causal definition as suggested for early modern geometry: knowing the mechanical cause of a thing, as the mechanical motion bringing about a circle, leads to certain knowledge of the effect. Knowing the mechanical causes that bring about a commonwealth, we can know the commonwealth, its rules and needs with necessity, with absolute certainty, a priori.[26]

Of course, all rationalists acknowledged the limits of their reason. They already knew the dilemma famously formulated by Einstein: "How is it possible that mathematics, being a product of human thinking independent of all experience, fits the objects of reality? Can human reason, without experience, explore the properties of real things, by mere thinking? There is a short answer to this, according to my opinion: To the extent that the propositions of mathematics relate to reality, they are not certain, and to the extent they are certain, they do not relate to reality" (Einstein 1921, 3–4). We could not know of external things without experience and experience could not provide us with the essence of things. We were thus forced to give provisional names, that is, nominal definitions, of the things we knew through experience. Nominal definitions were considered to be placeholders for the time being. Hybrids are considered possible. There can be a thing like a commonwealth, for Hobbes, Spinoza, or Wolff, which cannot be known through and through by causal definitions because the biological nature of human beings is still largely unknown. However, one can define human beings for the time being as animals with some use of reason, based on experience. Using these provisional definitions, one can then find merely theoretical explanations using nothing but known terms. In this way, one will be able not only to explain the rules of politics or human behavior but to predict other phenomena sufficiently.

Christian Wolff used this method systematically to reduce the gap between a priori knowledge and experiential knowledge. That is not only true for his experimental physics. When he wrote about methods to increase the growing of grain, he distinguished between facts we know from experience and the causes of some phenomena, which we know with certainty and have under control (reproducing grain) (Wolff 1734; Goldenbaum 2011b). Although we cannot know the essence of the plants yet, we can come to know some causal processes of the plants' growth and can predict the outcome with a high degree of certainty.

Spinoza used this geometrical method in his theoretical published work on ethics and in his work on politics and biblical hermeneutics. According to Tschirnhaus's reports to Christian Wolff, Spinoza had developed a method for finding and constantly improving definitions in empirical natural science[27] using experiments, starting with arbitrary nominal definitions and increasingly replacing parts of them with causal definitions (Goldenbaum

2011b, 29–41). Tschirnhaus, who was above all a mathematician and engineer (among other things, he invented Meissen porcelain), further developed this method of defining and redefining objects of natural science based on empirical research, as did Christian Wolff. It was their goal to improve the definitions of *real* things, not only geometrical figures, in such a way that valid conclusions could be drawn from them *necessarily*, thus extending the realm of the geometrical method far beyond geometry. In their view, we could indeed learn to know such real natural things in the same degree as God, although only to the extent to which we could *generate* them. Although we could not know natural things thoroughly, we could always try to know some of their properties in a causal way and thus *necessarily*, or a priori.

All these attempts clearly show that these rationalists used the geometrical method in the most general way to explain not only geometrical figures but as many phenomena of the real world as possible, by finding their causal definitions (using the analytical method). This is held to be true even if we can generate a thing in a way different from the way in which it was actually produced. Whenever a thing is produced and is thus possible, free of contradiction, its essence can be known. These essences (i.e., causal definitions) are connected to one another and have to be compatible, that is, they build a coherent conceptual structure of the world. Although we can only know a small number of particulars in such an a priori manner (because we can generate them), due to their absolute certainty no empirical knowledge can ever contradict them. Thus we can know some eternal and fixed structures, to which all empirical knowledge of particular things *must* cohere, allowing us to build one coherent structure of the world (although it will always remain incomplete). It is seldom noticed that exactly this position was already held by Galileo: "all these properties [of things in nature] are in effect virtually included in the definitions of all things; and ultimately, through being infinite, are perhaps but one in their essence and in the Divine mind" (Galilei 1967, 104).

ADEQUATE IDEAS AND CAUSAL DEFINITIONS

The mathematician and rationalist Descartes did not speak of causal definitions. But a kind of prehistory of causal definitions can be found in his discussion of adequate ideas with Arnauld, on the basis of Descartes' Fourth Meditation. The term "adequate ideas" is, of course, more familiar to us from Spinoza and Leibniz, as well as from Wolff. (Hobbes did not use it, perhaps because of his avoidance of traditional scholastic metaphysics in general.) Descartes uses the term cautiously. He states, "if a piece of knowledge is to be adequate it must contain absolutely all the properties which are in the thing which is the object of knowledge" (CSM 2 155; AT 7 220; Meditations, Fourth Set of Replies). Interestingly, adequate ideas have the same capacity as causal definitions, namely the capacity to virtually include all properties that belong to the cognized/defined thing.

It is in this first emergence of adequate ideas in rationalist modern philosophy that we are likewise confronted with the sensitivity of theology regarding adequate ideas. Descartes immediately adds a caveat: "Hence only God can know that he has adequate knowledge of all things. A created intellect, by contrast, though perhaps it *may in fact possess adequate knowledge* of many things, *can never know that* he has adequate knowledge unless God grants it a special revelation of the fact" (ibid., emphasis added). Why is the talk about adequate ideas immediately turning to theology? Because having adequate knowledge of things makes us like God—knowing things as well as He does in His omniscience. Descartes was as aware as anyone of the theological concerns regarding Galileo. Therefore, in spite of his enthusiastic statements about the certainty of deduction and intuition (both of which are available to us) in his early writings, especially in the *Rules*, he has to backpedal and grant that God could have made the world in a way that would be completely incomprehensible to us, in opposition even to what we hold to be mathematically necessary.[28] Of course this position caused headaches for Leibniz and other rationalists.

This is not the only reason adequate ideas are, from their first appearance in Descartes' discussion with Arnauld, an extremely sensitive topic in terms of theology. This discussion about the Fourth Meditation is titled "De vero et falso" and deals with the question of how error arises—although the perfect being, God does not deceive us. Descartes does not ascribe the reason for our error to God or to human reason. Rather, Descartes considers it to be an *easy* thing to know many things adequately as long as our *vis cognoscendi* is adequate to the thing to be known, "and this can easily occur" (ibid.). Descartes finds the cause of error, as we all know, in our will. This is where Descartes struggles to argue in favor of the free choice of our will. And this topic of free will is obviously tainted by deep theological concerns.

What is important for my point is rather Descartes' first concern. In order to know that we have the power to cognize things adequately and to know "that God put nothing in the thing beyond what it [our mind] is aware of"—we would have to know everything. Thus our power of knowing would have to equal the infinite cognizing power of God, which is clearly impossible. Descartes cautiously concludes that we do not need adequate ideas to conceive the *real* distinction between two subjects, here mind and body. And he suggests that we may be content with "complete" ideas that would give us *all the properties of a thing* (i.e., as much as an adequate idea) anyway without claiming its adequacy. By this distinction of adequate and complete ideas, the first owned by God alone and the second available to us, Descartes guarantees a limit to what can be known by human beings. In his *Rules* and then in *Discours*, Descartes did attribute to human beings a capability of knowing things with certainty by relying on intuition and deduction, or the geometrical method.

Interestingly, Descartes' cautious distinction between adequate and complete ideas was not upheld by his followers. For Spinoza it is precisely *our* adequate ideas that provide for our sharing of God's intellect, allowing for certainty of our knowledge (E II, 37–40s2) and overcoming our lack of freedom (E IV). Adequate ideas will even make us immortal (E V 38–42s). Spinoza defines "adequate idea" as "an idea which, insofar as it is considered in itself, without relation to an object, has all the properties, or intrinsic denominations of a true idea" (E II, d4; C 447). Thus he explicitly denies correspondence of an idea with an external object as a criterion for adequacy and thereby denies the traditional understanding of adequacy in Aristotelian scholastics as correspondence of idea and ideatum.[29] For Spinoza, having an adequate idea is to provide the proximate cause of the thing to be known, that is, the idea that causes an idea, or to define a thing by its cause, if considered under the attribute of extension.

In contrast to Descartes, Spinoza holds that we can have such adequate ideas and produce more of them when following the geometrical method and working to obtain increasingly causal definitions (or at least partially causal definitions), using nominal definitions as mere placeholders. Of course, being finite, we can never come even close to God's intellect. God knows everything adequately and moreover intuitively; however, we can get to know some important adequate ideas, which may then provide a general structure to lead our empirical research in a safe way. That is, because "the fixed and eternal things" (TIE 101; C 41) are so closely connected to the particular things, their knowledge will help us to get a more coherent knowledge of the latter. Thus Spinoza allows human beings to have adequate ideas and even sees these ideas as divine knowledge, which we share with God, clearly deviating from the cautious position of Descartes. When Spinoza discusses inadequate ideas and explains error he even ironically uses the example of free will as an exemplary inadequate idea (E 2 p35sch).

Interestingly, and seldom noticed, this rationalist position is very close to that of Galileo, who claims (just as the theologians complained in the earlier quoted trial file): "I say that as to the truth of the knowledge which is given by mathematical proofs, this is the *same* that Divine wisdom *recognizes*" (Galilei 1967, 103; emphasis added). Of course, Galileo admits a difference between divine and human knowledge—a difference consisting in God's thoroughgoing *intuitive* knowledge in contrast to human *discursive* knowledge. But still, he vindicates "a few" intuitive insights to human beings too.[30]

Even Leibniz, the committed Christian philosopher, accepted that we have the capability to have adequate ideas. And he also agreed that they are the same in us as in God, to the extent that we have them, because they are *necessarily* true. As Spinoza does, Leibniz connects them with genetic or causal definitions, which necessarily provide truth. It is interesting that in Wolffianism, when it comes to German translations, the term "idea adaequata" is bluntly translated as "complete idea" [vollständiger Begriff] (Sittenlehre 1745), thereby ignoring Descartes' careful distinction between complete ideas available to human beings and adequate ideas available to God. However, while all rationalists agree that human beings can have a certain number of necessary demonstrations (i.e., a priori knowledge equaling divine knowledge, the latter claim not being shared by Hobbes), this view is moderated by their awareness that such a priori knowledge is very limited in human beings and has to be supplemented by experience.

Leibniz on Causal Definitions, Adequate Ideas, and Necessitarianism

Given the theological concerns with the geometrical method and adequate ideas, it is rather surprising how close the Christian philosopher Leibniz's positions on these topics came to those of Spinoza and Hobbes, especially in light of the cautious attitude of Descartes about religiously sensitive issues. In his well-known *Meditations on Knowledge, Truth and Ideas* from the period of the mature Leibniz, the German philosopher introduces a full-fledged schema of different types of ideas. He first aligns with Descartes to distinguish between obscure and clear ideas but then further splits the clear ideas into clear and confused and clear and distinct, thereby adding a new type: clear and confused ideas (L 291–95; A 6, 4, N. 139, 585–86).[31] Further, Leibniz divides distinct ideas into inadequate and adequate ideas, and the latter again into symbolic or intuitive ideas. An idea is adequate if everything that has gone into a distinct knowledge of the thing is also known distinctly "or if the analysis has been done to the end." Thus, when we can provide distinct knowledge of all partial concepts of a concept we can know it adequately. Leibniz cautiously adds that he does not know whether we have any perfect example of adequate ideas within human knowledge but the knowledge of numbers would come close to it.

Adequate ideas—from Descartes via Spinoza to Leibniz—are those that provide a complete knowledge of all the properties of their subject, independent of any knowledge of correspondence. How does Leibniz relate adequate ideas to causal or genetic definitions? He explains this very systematically in a text he did not publish, *On Synthesis and Analysis* (L 229–34; A 6, 4, N. 129) (the title refers to the two aspects central to the geometrical method, as mentioned above). He begins with the traditional distinction of nominal and real definitions as still taught in the *Logic of Port-Royal*, but then emphasizes one particular kind of real definition, which displays the reality of things to us, namely causal or genetic definitions. Starting with nominal definitions, the collection of names of properties of a thing known by experience, he defines them as *distinct* concepts because it is necessary to *distinguish* and name the single properties of the subject to come up with nominal definitions. Confused ideas though, for which we cannot give single properties although we somehow recognize a thing in its entirety, do not allow yet for any definition. They may be made more (and more) distinct though by analysis, that is, by further distinguishing their parts.

In contrast to such nominal definitions, being a mere listing of properties or, rather, their names, Leibniz then defines real definitions as including the *possibility* of the defined thing, or freedom from contradiction. His example is—surprise!—the definition of a circle; specifically, Euclid's definition of a circle as produced by the motion of a straight line in a plane around one of its unmoved endpoints. This definition, clearly a causal definition as introduced by Hobbes, is for Leibniz *a real definition in an exemplary way* because it *displays* the demanded possibility of its subject. Leibniz does not even mention any other type of real definitions. He then somewhat laconically concludes: "Hence it is useful to have definitions involving the generation of a thing, or if this is impossible, at least its constitution, that is a method by which the thing appears to be producible or at least possible" (L 230–31; A 6, 4, N. 129, 541). Just as Hobbes and Spinoza did, Leibniz here extends the scope of causal definitions by way of any construction of a thing even if its actual cause might have been another one. Any construction of a thing that *can* generate it provides a clear guarantee of its possibility.

In the same work, Leibniz provides an explicit statement on the relation between an adequate idea and a genetic or causal definition: such genetic definitions *are* adequate ideas because they immediately display the possibility of the defined thing, that is, without an experiment or test or observation, as well as without the need to show the possibility of something else in advance. Such an adequate idea is given whenever the thing can be analyzed into its simple primitive concepts, which is precisely the case in geometrical causal definitions. "Obviously, we cannot build a secure demonstration on any concept unless we know that this concept is possible. . . . This is an a priori reason why possibility is a requisite in a real definition" (L 231; A 6, 4, N. 129, 542).

It is somewhat ironic that Leibniz uses this opportunity to criticize Thomas Hobbes for having claimed (as indeed he did) that all definitions are arbitrary and nominal. Leibniz knew full well that Hobbes also provided the other type of causal definitions, as mentioned earlier, which are not arbitrary. Leibniz continues that we "cannot combine notions arbitrarily, but the concepts we form out of them must be possible . . . Furthermore, although names are arbitrary, once they are adopted, their consequences are necessary, and certain truths arise which are real even though they depend on characters which have been imposed" (ibid.).

Leibniz emphasizes the necessity of the consequences as they follow from adequate ideas—that is, from causal definitions. These adequate ideas or genetic definitions are further praised for their special capacity: "From such ideas or definitions, then, there can be demonstrated all truths with the exception of identical propositions, which by their very nature are evidently indemonstrable and can truly be called axioms" (ibid.). *True* axioms are exclusively identical propositions. How close Leibniz is to Hobbes here can be seen in the following sentence in which he makes the unusual claim that even common axioms can actually be demonstrated—which Leibniz had in fact learned from Hobbes very early in his philosophical career (Goldenbaum 2008, 53–94). His critical statement against arbitrary definitions may thus have been directed at those critics of the geometrical method who questioned it for its uncertain starting point—the axioms.

As is well known, Leibniz makes another bold claim, not so horrifying to mathematicians but to theologians. He says that a reason can be given for each truth "for the connection of the predicate with the subject is either evident in itself as in identities, or can be explained by an analysis of the terms.

This is the only, and the highest, criterion of truth in abstract things, that is, things which do not depend on experience—that it must either be an identity or be reducible to identities" (L 232; A 6, 4, N. 129, 543).[32] From here, Leibniz states, the elements of eternal truths can be deduced and a method provided for everything if they are only cognized *as demonstratively as in geometry*. Of course, God cognizes *everything* in this way, that is, a priori and "sub specie aeternitatis," because He does not need any experience. While He knows everything adequately and intuitively, we can grasp hardly anything in this way and have to rely on experience. Notwithstanding, Leibniz then recommends the development of empirical sciences that combine a priori knowledge with experiment in mixed sciences, which are supposed to enrich human knowledge. There is no question that Leibniz walks thereby precisely in the paths of Galileo, Hobbes, and Spinoza, being much less cautious in terms of theology than Descartes.

AVOIDING THE NECESSITARIANISM OF THE GEOMETRICAL METHOD

So, what is so problematic about the geometrical method? According to *this* new geometrical method, which epistemologically goes far beyond Euclid, we, as humans, can have a priori knowledge, considered by rationalists (with the exception of Hobbes) to be divine knowledge, although of a very small number of things. Because we can deduce every property from genetic definitions, the degree of certainty of our knowledge of these things will be no less than that of God's knowledge, although He, of course, knows everything intuitively while we know it for the most part by the hard work of demonstrations. What is more challenging even, is that the converse is true as well: according to the *new* geometrical method, God's capacity of knowing things functions in the same way as that of our knowing. It is because He *constructed/created* all things in the universe that He knows them all a priori. It is only our finiteness and our limited capacity for intuition that hinders us from knowing *everything* a priori like the master geometrician God. As a result, the difference in knowledge between us and God would not be ontological but merely a difference in degree. That is precisely what Galileo had claimed (who is often quoted but not as often fully understood); namely that the book of nature is written in mathematical signs (1960, 183–84). Taking the knowledge of everything through causal definitions (that is, adequate ideas) into the scope of divine knowledge opens an avenue for the endless extension of human knowledge far beyond geometry.

But this avenue, *potentially* leading to necessary knowledge about everything, seemed to lead into strict determinism, thus threatening free will. This can be seen in the cases of Hobbes and Spinoza, who both were straight determinists. In contrast, it was precisely the recognition of this threat of determinism that led Henry More to his rejection of Descartes and even more of Cartesianism.[33] Leibniz, embracing the geometrical method, was fully aware of his dangerous intellectual neighbors (heretics and determinists), and worked hard to secure his metaphysics against strict determinism in order to distinguish his metaphysical and epistemological project from theirs. He had been working on this since he first studied Hobbes and Spinoza in Mainz between 1670 and 1672. The result is his well-known distinction of necessitating versus inclining at the end of the heading of paragraph 13 of the *Discourse on Metaphysics*. But, notwithstanding his obvious rejection of Hobbes's and Spinoza's strict determinism, Leibniz clearly shares the new geometrical method, as a *philosophical* method, with the infamous philosophers. Moreover, it is this new method based on the genetic or causal definition that provides the basis of Leibniz's logic of containment (Di Bella 2005, 80–95).

While only God can have a priori knowledge of the complete notions of individuals, we can at least have a priori knowledge of *abstracta*, although we have to rely on empirical knowledge when it comes to individuals (L 331–38; A II, 2, N. 14). This distinction, closely related to the distinction between necessary and contingent truths, gave Leibniz sufficient confidence to present at least the headings of his *Discourse on Metaphysics* to Arnauld in 1686, with the long section 13 being especially provocative in respect to free will. By this time, Leibniz had already worked out his new metaphysics (based on the problematic *new* geometrical method), which would make modern science compatible with Christian dogmatics and free will with (softened) determinism.

Finally, in spite of Leibniz's strong emphasis on the different ontological status of specific/abstract truths and contingent truths (he held since the *Confession of the Philosopher*) and then on the logical distinction of concrete and abstract things (since 1676) both aiming to secure contingency and to block strict determinism, he always maintains the containment theory. But this view (that the predicate of a true proposition must be included in its subject) clearly retains a general similarity between the two kinds of concepts because both—specific (or full) concepts of abstract things as much as complete concepts of individuals—must include all their predicates and can be known a priori by Him who generated them. This view is the core of the geometrical method! It was this theory that would lead to paragraph 13 of the *Discourse of Metaphysics*, according to which the complete concept of any individual was known by God and would include every single event that would ever happen to us.

At first glance, the mere claim that we cannot know individual things by a priori knowledge as God does, but only by observation and empirical research, does not sound at all new or promising and seems simply to confirm God's omniscience and the limits of our reason. Leibniz's conception is more subtle, which can already be felt by the vehemence of the theologians' protestations against him.[34] According to Leibniz, even if human beings cannot know individuals a priori but only through empirical study or by history, God does know the concepts of individual substances a priori. Moreover, God chose them as belonging to the best of all possible series of things when He created this world. Because of that choice, led by God's intellect, there cannot be any contradiction among the things of one series or one world. What is crucial here is that Leibniz's approach to contingent things assures us—from the very beginning—of the coherence of all phenomena of this world that will ever occur to us in our experience. This is so even if *we* do not yet see it. Because there is nothing arbitrary in God's creation—*nihil sine ratione*—we can take for granted that there is a universal coherence of the world in spite of our own limited approach. It is this view that deviates from Luther and the Protestant way of thinking, wherein which such an intelligibility of the world to humans is bluntly denied. According to that view, human reason has been corrupt since the fall and thus must fail to understand. Moreover, we cannot even know whether God would have wanted to create a coherent world. God is hidden from us and we can know about Him only through faith and revelation.

Leibniz is often said to be an optimist. The true optimism that can indeed be ascribed to the rationalist philosopher lies less in his belief that this is the best of all possible worlds than in the comprehensibility of the world based on the comprehensibility of God, thanks to the *new* geometrical method, based on genetic definitions. Moreover, this method not only enables us to have a priori knowledge in mathematics and other fields of merely conceptual knowledge, but provides us with a new approach to empirical research to obtain contingent truths. For Leibniz, learning the many predicates of individuals through experience does not mean simply gathering and collecting data, and watching out for common patterns from which to abstract rules. Rather, our gathered data are supposed to fit into a larger theoretical framework, known by God and—partially—by us.

This framework includes of course those full specific notions of abstract things, which we as humans are able to know a priori because their number of predicates is finite. Because these eternal abstract truths can never contradict any predicate of a complete notion, they can provide a strong framework for our empirical work, which is available to our finite knowledge. When we come to learn about new facts by experience and by history, we can expect these single historical facts to fit into the theoretical framework like the pieces of an unfinished puzzle, and build a more complete notion of an individual and its action.

Of course, this infinite process of learning can never be conclusive because it is infinite. Nevertheless, our expectation (based on the conviction of a theoretical framework that is known by God a priori and thus exists) together with the available specific notions of abstract things we have at hand a priori, provides powerful tools. It is as if we had an unfinished map, a compass, and a watch, which, with our general framework of terrestrial geography, can guide an expedition into an unknown area. Such equipment can help us to recognize coherence and causal interconnectedness in the otherwise confusingly rich abundance of single facts of empirically obtained knowledge. Therefore, Leibniz's (Spinoza's and Hobbes's) approach to empirical research is completely different from any empiricist approach to nature or history, the latter being a mere collecting of facts while looking for patterns or similarities to abstract from them, and thus to find rules.

While we distinguish between natural science as hardcore science (such as physics, chemistry, biology or, increasingly, medicine) on the one hand and humanities and social sciences on the other, Leibniz (as well as Hobbes and Spinoza) instead distinguishes demonstrative knowledge (using specific notions and dealing with abstract things such as geometrical figures) from the empirical sciences (relying on empirical knowledge *in addition* to a priori knowledge). Thus for Leibniz, human history and the humanities are not really different from natural sciences in their searching for factual and contingent truths connected through a theoretical framework of a priori eternal truths available to us through the geometrical method. He was confident that empirical knowledge can be turned into science through a theoretical framework of a priori knowledge. This stands in sharp contrast to Locke.

Leibniz is following the path of Hobbes and Spinoza, and he will be in turn followed by Wolff and Tschirnhaus.

CONCLUSION

The trouble with the geometrical method in the seventeenth and eighteenth centuries was neither its ponderous way of thinking nor its lack of success. Rather it was the turmoil about human haughtiness and the threat that its determinism would destroy free will, the arbitrary choice of the will of God as well as that of human beings. The correspondence of Leibniz and Clarke exemplifies the different approaches to God's free will. According to Leibniz, nothing can happen without a sufficient reason, which proves the existence of a God, who in His perfection could not have chosen an arbitrarily functioning world. Clarke (and Newton), on the other hand, count any act of will on God's part *as* a sufficient reason (Leibniz-Clarke, 11; 2nd Reply, #1).

Two things caused deep anxiety and anger regarding this method: (1) the attempt to extend the geometrical method to nature, to humans, and to society (taking mathematization of nature for granted), thus providing human beings with a God-like a priori knowledge beyond mathematics; and (2) the threat of determinism. These threats forced theologians and Christian philosophers to reject rationalism and the geometrical method altogether. In sharp contrast to rationalism, Locke would even deny the possibility of natural science because we could not have any real definitions beyond mathematics and morals: "This *way* of getting and *improving our Knowledge in Substances only by Experience* and History, which is all the weakness of our Faculties in this State of *Mediocrity*, which we are in this World, can attain to, makes me suspect, that natural Philosophy is not capable of being made a Science. We are able, I imagine, to reach very little general Knowledge concerning the Species of Bodies, and their several Properties" (Locke 1975, 645; Essay 4, 12, 10). Kant would declare that there would "never be a Newton for a blade of grass" (Kant 2000, 268–71), pointing us instead to design theory in biology admitting causal explanations alone for mathematics and mechanics, or applied mathematics.

Thus, the opposition between the two philosophical camps of rationalism and empiricism was not the result of different approaches to experience as is often claimed. Rather, it was their different and opposing stances toward the geometrical method and the mathematization of nature. This method was in no way external to rationalist philosophy. As much as rationalist philosophers differ in their philosophical systems, they all agree that human beings can arrive at a priori knowledge (through deducing from definitions), independent of experience, and that this knowledge is somehow "divine," that is, as certain as God's knowledge. In contrast, empiricists and theologians are eager to deny such a possibility, and therefore must rely exclusively on knowledge by experience. To be sure, empiricists do not trust experience any more than rationalists do. Rather they deny that we are capable of any better knowledge (except within mathematics). Thus it is the different approach to the *new* geometrical method that provides the explanation for the two schools of early modern philosophy. While rationalists see the mathematization of nature and the geometrical method as avenues to comprehend God's creation, sharing a priori knowledge with Him, the empiricists ally with the Christian belief (emphasized more strongly by Protestants) that human reason is corrupted due to the fall and that God as well as the essences of the created things are hidden from us. The geometrical method of the early modern period was much more than a way of demonstration. It was a new epistemological approach to true knowledge of the external world, based on the mathematization of nature, completely different from any traditional, empirical approach of natural philosophy.

ABBREVIATIONS

A	Leibniz, G. W. 1924–ongoing. *Sämtliche Schriften und Briefe* (cited by Roman number for series, Arabic number for volume, and Arabic number for page. Complete pieces are referred to by "N." with Arabic number)
AT	Descartes, R. 1996. *Œuvres de Descartes*
C	Curley, E., ed. and trans. 1985. *Spinoza, The Collected Works of Spinoza*
CSM	Descartes, R. 1985–88. *The Philosophical Writings of Descartes*
GP	Leibniz, G. W. 1875–90. *Die philosophischen Schriften*
KAA	Kant, I. 1900–ongoing. *Gesammelte Schriften*
L	Leibniz, G. W. 1969. *Philosophical Papers and Letters*
LEIBNIZ-CLARKE	Leibniz, G. W., and S. Clarke. 2000. *Correspondence*
OL	Hobbes, T. 1839. *Opera Philosophica quae latine scripsit omnia*
SITTENLEHRE	Spinoza, B. 1744
TIE	Spinoza, B. Tractatus de Intellectus Emendatione

NOTES

1. The literature about the geometrical method in early modern philosophy is to a large extent focused on Spinoza's *Ethics*. See though Cassirer 1974, 1:136–44, 512–17, and 2:48–61, 86–102; Schüling 1969 (who confuses geometrical method with axiomatic method); see also Hecht 1991. On Spinoza's use of the geometrical method, see Hubbeling 1964, 1977; Curley 1986a; and De Dijn 1986. See also Curley 1986b; Klever 1986; Matheron, 1986; and Goldenbaum 1991.

2. Due to the common confusion of axiomatic and geometrical method, Wolters even sees Spinoza's *Ethica more geometrico demonstrata* as exemplary for the degeneration of the axiomatic method into a mere external tool of presentation. See Wolters 1980, 7.

3. This was the view of Hegel and the German romantics which has been canonized in the influential German history of philosophy shaped by Hegel's view, as for example in Windelband: "The deep motion of a god-filled mind is expressed in the driest form, and the subtle religiosity appears in the stiff armor of fixed chains of conclusions" (Windelband 1919, 212; my translation).

4. Breger calls the concept of motion an essential driving engine for the conceptual transition of mathematics of the seventeenth century, especially in the development of the concept of function. He continues: "The concept of motion has not only paved the way to the problem of rectification, to the introduction of the transcendent, and as a tool to investigate limit processes; it also contributed to legitimizing the continuum (and thereby eventually infinitesimal methods). The concept of motion makes it implausible that the continuum could have gaps. The mechanical thinking makes the geometrical lines appear as homogenous and all points on them as equally justified: the limitation to points and lines which can be constructed in this or that way, no longer appears as a necessary condition of exactness but as an unnatural limitation, which was to be overcome through a new boundary line between mathematics and mechanics" (Breger 1991, 45; my translation).

5. "To clarify this fact one has to reflect on the modern form of geometry Spinoza had in mind. In fact, it is not the Euclidean but Cartesian geometry that is the systematic model for him. In analytic geometry, the number refers to space, i.e., a mere mode of 'thinking' refers to a mode of 'extension' in such a way that a gapless, one-to-one correspondence happens between both. Every dependence between figures in space is mirrored in a dependence between quantities in numbers: thus here one and the same connection is expressed in

two different forms" (Cassirer 1974, 117–18; my translation); "En fait, l'idée du cercle est une image, resçue par' l'esprit, une peinture faite à l'imitation d'un modèle externe; l'idée cartésienne du cercle est un concept né de l'activité proprement intellectuelle, de la force native de l'esprit. Le cercle et son idée appartiennent à deux orders different" (Brunschvicg 1904, 771).

6. "But because the almighty God gave the ability to us human beings to perfectly conceive the numbers and quantities, and did not keep anything for Himself, we easily fall into the awkward thought that we could also be the master of all the other objects of our knowledge (cognoscibilia), and could have the sufficient reasons under our control" (Löscher 1735, 119; my translation).

7. "I only say this [. . .] that the author deduces the stubbornness [of Pharaoh in Exodus 7, 13 and following] from the nexus or the fatal connection of all things, and in this way ascribes it to God according to his preestablished harmony. This nexus is the soul of the whole system of the mechanical philosophy" (Lange 1735, 25; my translation). This is directed against the author of the *Wertheim Bible*, Johann Lorenz Schmidt, who had produced a Wolffian translation of the Pentateuch; cf. Goldenbaum 2004, 236.

8. "Yes, one should accept, so to speak, only genetic demonstrations, or those that are taken from the generation of the subject, so that one will know in advance perfectly, how the subject came about. . . . These demonstrations are considered the only ones that provide science, a true knowledge: from this it follows that all other knowledge, proved in other ways, is opinion only, and cannot be trusted" (Löscher 1735, 126; my translation).

9. "Accordingly, we would have only nominal definitions of God and many other objects from where one could not even see whether the thing is possible [rem esse possibilem]" (Löscher 1735, 129; my translation).

10. "It is an obvious pedantry, if one plants oneself with one's mathematical method in other disciplines so broadly; the most rude thing, though, is doing such a thing with the Holy Scriptures and (N.B.) in theology" (Lange 1735, 2, §4; my translation).

11. "One takes such a reason not only to be a reason, or puts it as such in the intellect. It is supposed to be in reality as well, and not in any different way. 'Sufficient,' in this philosophy, does not mean something what we can be content with, knowing it according to its constitution. Nay, it means something so strong, perfect, and adequate, that it is sufficient everywhere, and nothing more can be asked without lacking reason" (Löscher 1735, 119; my translation).

12. "6. Asserirsi e dichiararsi male qualche uguaglianza, nel comprendere le cose geometriche, tra l'intelletto umano e divino" (Dok. 20; Car. 387r–393r).

13. Galileo makes Salvati refer to "our Academician," obviously Galileo himself, "who had thought much upon this subject and according to his custom had demonstrated everything by geometrical methods so that one might fairly call this a new science. For, although some of his conclusions had been reached by others . . . they had not been proven in a rigid manner from fundamental principles. Now, since I wish to convince you by demonstrative reasoning rather than to persuade you by mere probabilities, I shall suppose that you are familiar with present-day mechanics so far as is needed in our discussion" (Galilei 1954, 6).

14. Thus it became the strategy of Rüdiger, Hoffmann, Crusius, Löscher, and Lange to emphasize the fundamental difference between mathematical knowledge and scientific knowledge of natural things whereby mechanical theory counted as applied mathematics. Whereas mathematics dealt with figures and numbers—produced by humans and thus arbitrarily—natural science, as well as metaphysics and theology, dealt with God's creation and thus with natural things (see Löscher 1735–42, 128–29; 1742, 78). Therefore, only mathematical concepts could be known by us in their very essence, whereas the essences of God's creatures remained hidden to us. We could know them only by observation, experience (equated with sense perception), induction, and abstraction. This approach well explains the enthusiasm of German Pietism for Locke.

15. The manuscripts of Leibniz's *Elementa Iuris naturalis* can be seen as an exercise in demonstrating by chains of definitions (A VI, 1, N. 12). See also Leibniz's letters to Chapelain from the first half of 1670, in A VI, 1, N. 24.

16. There are various attempts to avoid the rationalist geometrical method by modernizing scholastic philosophy through empirical research, adopting elements of modern science while attacking the geometrical method, that is, the central role of causal definitions, if used beyond geometry. Thus the *L'Essai de logique*, published by Abbé Edme Mariotte, a gifted experimenter (coauthored by Roberval) appears as a turn against Galileo's and Descartes' mathematization of nature. Our knowledge of nature is restricted to observation and experiment (cf. Roux 2011, 63–67), a clearly empiricist move against disciples of Descartes. Mariotte considers even the principle of inertia as the result of experience (ibid., 104). On Roberval and Mersenne, see also Fouke (2003, 75–76).

17. "What I want to emphasize about this passage is that in it Spinoza shows himself to be willing, when one of his fundamental assumptions is questioned, to provide further argument for this assumption. He does not regard his axioms as argument-stoppers, principles so fundamental that they neither require nor can be given any further argument. Instead, he offers to demonstrate his axi-

oms by appealing to his definitions. It is interesting that in the final version of the Ethics all four of these axioms are removed from the list of fundamental assumptions. Three become propositions. One becomes a step in a demonstration" (Curley 1986b, 157); see also Klever (1986).

18. In full agreement with Lutheran theologians and with Crusius, Kant argues that philosophy cannot begin with definitions because its objects do not depend on human minds as the objects of mathematics do [CrR B740–763]. He has to ignore however, to make this argument, that Wolff (and other rationalists) are fully aware of this problem and indeed make the production of good definitions of real/natural things a task that has to precede any demonstration. In case no sufficiently clear and distinct definition can be found, a nominal definition can serve as a placeholder for the time being from which hypothetical knowledge can be deduced as long as no contradictions emerge. This is seen, to some extent, by Engfer (1982, 56).

19. Thus, the young Kant sharply distinguishes between mathematical bodies and natural bodies whereby the former do not have any internal force while the latter in fact do own such a force: The latter "has a power in itself, through itself to enlarge the force which was awakened in it by an external cause of its motion, thus that it can include grades of force which did not originate from external cause of motion and which are larger than it. Therefore they cannot be measured by the same measure as the Cartesian [mechanical] force and have another estimation" (KAA 1 140 §115).

20. "On ne reconnaît en géométrie que les seules definitions que les logiciens appellant défintions de nom, c'est-à-dire que les seules impositions de nom aux choses qu'on a clairement designées en termes parfaitement connus; et je ne parle que de celles-là seulement. . . . D'où il paraît que les definitions sont très libres, et qu'elles ne sont jamais sujettes à être contredites; car il n'y a rien de plus permis que de donner à une chose qu'on a clairement désignée un nom tel qu'on voudra. Il faut seulement prendre garde qu'on n'abuse de la liberté qu'on a d'imposer des noms, en donnant le meme à deux choses différentes" (Pascal 2000, 156).

21. "Nous connaissons la vérité non seulement par la raison mais encore par le cœur. C'est de cette dernière sorte que nous connaissons les premiers principes et c'est en vain que le raisonnement, qui n'y a point de part, essaie de les combattre" (Pascal 2000, 573).

22. The "mechanical" definition by mechanical motion to produce a geometrical object occurred accidentally in ancient mathematics, not in any systematic,

conscious way though. See Breger 1991, and on Hobbes and Roberval see Jesseph 1999, 117–25; 1996, 86–92.

23. "Hobbes does not think anymore of motion as an inner quality and constitution of bodies but as a mere mathematical relation, which we can construe on our own and therefore conceive. With this one step, the transition from Bacon to Galileo is accomplished. The analysis of natural objects does not end in abstract 'entities' but in laws of the mechanism, being nothing else but the concrete expressions of the laws of geometry" (Cassirer 1974, 2:47–48).

24. While Leibniz hardly used this method explicitly (cf. though his treatise in favor of the election of the Polish king, in Leibniz [1924–ongoing, 3–98]; A IV, 1, N. 1), he completely agreed with Hobbes about demonstrations as mere chains of definitions (Leibniz A VI, 1, N. 12; see also Leibniz A II, 1, N. 24, 153).

25. "In his quatuor partibus continetur quicquid in philosophia naturali, demonstratio proprie dicta explicari potest. Nam si phaenomenωn naturalium speciatim causa reddenda sit, puta quales sint motus, et virtutes corporum cœlestium, et partium ipsorum, ea ratio ex dictis scientiae partibus petendea est, aut omnino ratio non erit, sed conjectura incerta" (OL 1 62–65; De corpore i, 6, §6). "Scientia intelligitur de theorematum, id est, de veritate consequentiarum. Quando vero de veritate facti agitur, non proprie scientia, sed simpliciter cognitio dicitur. Itaque scientiae a quidem, qua scimus propositum aliquod theorema esse verum, est cognitio a causis, sive a generatione subjecti per rectam ratiocinnationem derivate" (OL, 2:92; De homine ii, 10, §4).

26. "And as the art of well building is derived from principles of reason, observed by industrious men that had long studied the nature of materials and the divers effects of figure and proportion, long after mankind began (though poorly) to build, so, long time after men have begun to constitute commonwealths, imperfect and apt to relapse into disorder, there may principles of reason be found out by industrious meditation, to make their constitution . . . everlasting. And such are those which I have in this discourse set forth" (Hobbes 1994, 220; Leviathan xxx, 5).

27. I take it to be an understatement even when Curley states: "I am not persuaded that Spinoza was such a radical anti-empiricist" (1986b, 156). In addition to Spinoza's own desire to develop a theory of experimentation (TIE 102–3; C 42), we also have evidence from Tschirnhaus via Wolff that Spinoza experimented himself (cf. Corr 1972, 323–34).

28. "Pour les veritez eternelles, je dis derechef que sunt tantum veræ aut possibiles, quia Deus illas veras aut possibiles cognoscit, non autem contra versa

à Deo cognosci quasi independenter ab illo sint verae" (AT, 1: 145, 149–50; Descartes to Mersenne, April 15, 1630 and May 6, 1630).

29. "Seule la géométrie cartésienne permet de rapporter la vérité à l'autonomie de l'intelligence; les propriétés d'une courbe se déduisent en effet de la définition analytique de cette courbe, c'est-à-dire d'une équation abstraite, sans recours à la considération directe de la figure. Seule elle permet d'interpréter la notion spinoziste de la convenance. La convenance n'implique plus l'antériorité de l'objet par rapport au sujet, mais la correspondence du sujet qui comprend et de l'objet qui est endendu, le parallélisme de deux orders d'existence qui ne suffisent à eux-mêmes, qui n'interfèrent jamais" (Brunschvicg 1904, 772). Therefore, *adequatio* in Spinoza is understood completely differently from scholastic tradition, well known by Spinoza according to Brunschvicg.

30. "The Divine intellect, by a simple apprehension of the circle's essence, knows without time-consuming reasoning all the infinity of its properties. Next, all these properties are in effect virtually included in the definitions of all things; and ultimately, through being infinite, are perhaps but one in their essence and in the Divine mind. Nor is all the above entirely unknown to the human mind either, but it is clouded with deep and thick mists, which become partly dispersed and clarified when we master some conclusions and get them so firmly established and so readily in our possession *that we can run over them very rapidly*" (Galilei 1967, 103–4; emphasis added). This is very similar to Descartes' understanding of intuition as not exclusively instantaneous but also as a *quick running through*: "necesse est illas iteratâ cogitatione percurrere, donec à primâ ad vltimam tam celeriter transierim, vt fere nullas memoriae partes relinquendo rem totam simul videar intueri" (AT 10 409 [Reg. XI]).

31. I have discussed this momentous innovation of Leibniz elsewhere; see Goldenbaum 2011a.

32. Loemker translates "justify" instead of "giving a reason," which sounds to me more like Hume than Leibniz.

33. "Sed si ullubi magnopere culpandus sit nobilissimus Philosophus, ob illud potissimùm eum reprehendum censeo, quòd Mathematico suo Genio ac Mechanico in Phenomenis Naturae explicandis nimium quantum indulferit. Eam tamen interim agnosco summorum Ingeniorum felicitatem, ut vel vitia eorum & errores aliquam virtutis speciem habeant atque fructum. Et profectò mihi planè incredibile videtur, nisi ingentem illam spem concepisset demonstrandi Omnia ferè Mundi Phaenomena ex necessariis Mechanicae legibus, eum

unquam tot tantàque tentare voluisse, aut tentata potuisse perficere" (More 1711, 58).

34. The following argument of a student of the influential Lutheran (Pietist) theologian Budde addresses only one although central point of criticism: "Nam tunc Deus mundum non eligit, quia optimus est, sed optimus est, quia eligit. . . . Hinc quidquid Deus elegit . . . non est optimum moraliter per se, sed ob Dei electionem" (Budde 1712, 72, §5).

REFERENCES

Arnauld, A., and P. Nicole. 2011. *La logique ou l'art de penser.* Ed. D. Descotes. Paris: Champion.

Aubrey, J. 1898. *Brief Lives, Chiefly of Contemporaries, set down by John Aubrey, between the Years 1669 & 1696.* Ed. A. Clark. 2 vols. Oxford: Clarendon Press.

Breger, H. 1991. "Der mechanizistische Denkstil in der Mathematik des 17.Jahrhunderts." In *Gottfried Wilhelm Leibniz im Philosophischen Diskurs über Geometrie und Erfahrung*, ed. H. Hecht, 15–46. Berlin: Akademie Verlag.

———. 2008. "Leibniz's Calculation with Compendia." In *Infinitesimal Differences. Controversies between Leibniz and his Contemporaries*, ed. U. Goldenbaum and D. Jesseph, 185–98. Berlin: De Gruyter.

Brunschvicg, L. 1904. "La revolution cartesienne et la notion spinoziste de la substance." *Revue de Métaphysique* 12: 755–98.

Budde, J. F. 1712. *Q.D.B. V. Doctrinae orthodoxae de origine mali contra recentiorum quorundam hypotheses modesta assertio.* Jena: Mullerus [resp. Georg Christian Knoerr].

Cassirer, E. 1946. "Galileo's Platonism." In *Studies and Essays in the History of Science and Learning, Offered in Homage to George Sarton*, ed. M. F. Ashley Montagu, 277–97. New York: Henry Schuman.

———. 1974. *Das Erkenntnisproblem in der Philosophie und Wissenschaft der neueren Zeit.* 4 vols. Darmstadt: Wissenschaftliche Buchgesellschaft.

Corr, C. A. 1972. "Christian Wolff's Treatment of Scientific Discovery." *Journal of the History of Philosophy* 10: 323–34.

Curley, E., ed. and trans. 1985. *Spinoza, The Collected Works of Spinoza.* Princeton, N.J.: Princeton University Press.

Curley, E. 1986a. *Behind the Geometrical Method.* Princeton, N.J.: Princeton University Press.

———. 1986b. "Spinoza's Geometric Method." In *Central Theme: Spinoza's Epis-*

temology, vol. 2 of Studia Spinozana, ed. E. Curley, W. Klever, and F. Mignini, 152–69. Alling: Walther & Walther.

Descartes, R. 1985–88. *The Philosophical Writings of Descartes*. Trans. J. Cottingham, R. Stoothoff, and D. Murdoch. 3 vols. Cambridge: Cambridge University Press.

Descartes, R. 1996. *Œuvres de Descartes*. Ed. C. Adam and P. Tannery. 11 vols. Paris: Vrin.

De Dijn, H. 1986. "Conceptions of Philosophical Method in Spinoza: Logica and Mos Geometricus." *The Review of Metaphysics* 40, no. 1 (September): 55–78.

Di Bella, S. 2005. *The Science of the Individual: Leibniz's Ontology of Individual Substance*. Dordrecht: Springer.

Einstein, A. 1921. *Geometrie und Erfahrung. Erweiterte Fassung des Festvortrages gehalten an der Preussischen Akademie der Wissenschaften zu Berlin am 27 January 1921*. Berlin: Springer.

Engfer, J. 1982. *Philosophie als Analysis. Studien zur Entwicklung philosophischer Analysiskonzeptionen unter dem Einfluß mathematischer Methodenmodelle im 17. und frühen 18. Jahrhundert* (=Forschungen und Materialien zur deutschen Aufklärung, edited by Norbert Hinske). Stuttgart-Bad Cannstatt: Frommann-Holzboog.

Fouke, D. 2003. "Pascal's Physics." In *The Cambridge Companion to Pascal*, ed. N. Hammond, 75–101. Cambridge: Cambridge University Press.

Gabbey, A. 1982. "Philosophia Cartesiana Triumphata: Henry More (1646–1671)." In *Problems of Cartesianism*, ed. T. M. Lennon, J. M. Nicolas, and J. W. Davis, 171–250. Montreal: McGill-Queen's University Press.

———. 1995. "Spinoza's Natural Science and Methodology." In *The Cambridge Companion to Spinoza*, ed. D. Garrett, 142–91. Cambridge: Cambridge University Press.

Galilei, G. 1907. *Opere. Edizione Nazionale*. Vol. 19. Ed. A. Favaro. Firenze: Barbèra.

———. 1954. *Dialogues Concerning Two New Sciences*. Trans. H. Crew and A. de Salvio, with an introduction by A. Favaro. New York: Dover Publications.

———. 1960. "The Assayer." In *The Controversy on the Comets of 1618*, ed. S. Drake and C. D. O'Malley, 151–356. Philadelphia: University of Pennsylvania Press.

———. 1967. *Dialogue Concerning the Two Chief World Systems*. Trans. S. Drake. 2nd ed. Berkeley: University of California Press.

Goldenbaum, U. 1991. "Daß die Phänomene mit der Vernunft Übereinstimmen sollen. Spinozas Versuch einer Vermittlung von geometrischer Theorie und experimenteller Erfahrung." In *Leibniz im philosophischen Diskurs über Geometrie und Erfahrung. Studien zur Ausarbeitung des Erfahrungsbegriffes in der neuzeitlichen Philosophie*, ed. H. Hecht, 86–104. Berlin: Akademie Verlag.

———. 2004. *Appell an das Publikum. Die öffentliche Debatte in der deutschen Aufklärung 1697–1786. Sieben Fallstudien*. Berlin: Akademie Verlag.

———. 2008. "Vera Indivisibilia in Leibniz's Early Philosophy of Mind." In *Infinitesimal Differences: Controversies between Leibniz and his Contemporaries*, ed. U. Goldenbaum and D. Jesseph, 53–95. Berlin: De Gruyter.

———. 2011a. "Die Karriere der Idea clara et confusa von Leibniz über Baumgarten zu Mendelssohn—Von der Epistemologie über Theologie zur Kunst." In *Berichte und Abhandlungen. Sonderband 11: Pluralität der Perspektiven und Einheit der Wahrheit im Werk von G. W. Leibniz. Beiträge zu seinem philosophischen, theologischen und politischen Denken*, ed. F. Beiderbeck and S. Waldhoff, 265–83. Berlin: Akademie Verlag.

———. 2011b. "Spinoza—ein toter Hund? Nicht für Christian Wolff." *Zeitschrift für Ideengeschichte* 5, no. 1: 29–41.

———. 2016. "How Theological Concerns Favor Empiricism over Rationalism." In *Leibniz Experimental Philosophy*, ed. A. Pelletier, 37–63. Stuttgart: Steiner.

———. Forthcoming. "How Kant Was Never a Wolffian, or Estimating Forces to Enforce *Influxus Physicus*." In *Leibniz and Kant*, ed. B. Look. Oxford: Oxford University Press.

Goldenbaum, U., and D. Jesseph, eds. 2008. *Infinitesimal Differences: Controversies between Leibniz and His Contemporaries*. Berlin: De Gruyter.

Hecht, H. 1991. *Gottfried Wilhelm Leibniz im Philosophischen Diskurs über Geometrie und Erfahrung. Studien zur Ausarbeitung des Erfahrungsbegriffes in der neuzeitlichen Philosophie*. Berlin: Akademie Verlag.

Hobbes, T. 1839. *Opera Philosophica quae latine scripsit omnia*. Ed. G. Molesworth. London: Bohn.

———. 1994. *Leviathan, with Selected Variants from the Latin Edition of 1668*. Ed. E. Curley. Indianapolis, Ind.: Hackett.

Hoffmann, J. E. 1974. *Leibniz in Paris: His Growth to Mathematical Maturity*. Cambridge: Cambridge University Press.

Hubbeling, H. G. 1964. *Spinoza's Methodology*. Assen: Van Gorcum.

———. 1977. "The Development of Spinoza's Axiomatic (Geometric) Method: The Reconstructed Geometric Proof of the Second Letter of Spinoza's Correspondence and Its Relation to Earlier and Later Versions." *Revue international de philosophie* 31: 53–68.

Jesseph, D. 1996. "Hobbes and the Method of Natural Science." In *The Cambridge*

Companion to Thomas Hobbes, ed. T. Sorell, 86–106. Cambridge: Cambridge University Press.

———. 1999. *Squaring the Circle: The War between Hobbes and Wallis*. Chicago: Chicago University Press.

Kant, I. 1900–ongoing. *Gesammelte Schriften*. Berlin: De Gruyter.

———. 1998. *Critique of Pure Reason*. Ed. and trans. P. Geyer and A. W. Wood. Cambridge: Cambridge University Press.

———. 2000. *Critique of the Power of Judgment*. Ed. P. Guyer. Trans. P. Guyer and E. Mathews. Cambridge: Cambridge University Press.

Klever, W. 1986. "Axioms in Spinoza's Science and Philosophy of Science." In *Studia Spinozana 2, Central Theme: Spinoza's Epistemology*, ed. E. Curley, W. Klever, and F. Mignini, 171–95. Alling: Walther & Walther:

Lange, J. 1735. *Der philosophische Religionsspötter*. Halle.

Leibniz, G. W. 1875–90. *Die philosophischen Schriften*. Ed. C. Immanuel Gerhardt. 7 vols. Berlin: Weidmann.

———. 1924–ongoing. *Sämtliche Schriften und Briefe*. Berlin: Akademie Verlag.

———. 1969. *Philosophical Papers and Letters*. Ed. and trans. L. E. Loemker. Dordrecht: Reidel.

———, and S. Clarke. 2000. *Correspondence*. Ed. R. Ariew. Indianapolis, Ind.: Hackett.

Locke, J. 1975. *An Essay Concerning Human Understanding*. Ed. P. H. Nidditch. Oxford: Oxford University Press.

Löscher, V. E. 1735. "Quo ruitis?" In *Frühaufgelesene Früchte der Theologischen Sammlung von Alten und Neuen, worinnen nur die neuesten Bücher, Kirchen=Begebenheiten, u.s.f. vorkommen*. Leipzig: Braun.

Matheron, A. 1986. "Spinoza and Euclidean Arithmetic: The Example of the Fourth Proportional." In *Spinoza and the Sciences*, ed. M. Greene and D. Nails and trans. D. Lachterman, 125–50. Dordrecht: Reidel.

More, H. 1711. *Epistola H. Mori ad V.C., quae Apologiam complectitur pro Cartesio, quaeque Introductionis loco esse poterit ad universam Philosophiam Cartesianam*. 4th ed. London.

Pascal, B. 2000. *Œuvres completes*. Ed. M. le Guern. 2 vols., vol. 2. Paris: Gallimard.

Roux, S. 2011. *L'Essai de logique de Mariotte. Archéologie de logique de Mariotte*. Paris: Garnier.

Schüling, H. 1969. *Die Geschichte der axiomatischen Methode im 16. und beginnenden 17. Jahrhundert*. Hildesheim: Olms.

Spinoza, B. 1744. *B.v.S. Sittenlehre widerleget von dem berühmten Weltweisen unserer Zeit Herrn Christian Wolff.* Frankfurt: Varrentrapp.

Windelband, W. 1919. *Geschichte der neueren Philosophie.* Leipzig: Breitkopf & Härtel.

Wolff, C. 1734. *A Discovery of the True Cause of the Wonderful Multiplication of Corn; With Some General Remarks upon the Nature of Trees and Plants.* London: J. Roberts.

Wolters, G. 1980. *Basis und Deduktion.* Berlin: De Gruyter.

12

PHILOSOPHICAL GEOMETERS AND GEOMETRICAL PHILOSOPHERS

CHRISTOPHER SMEENK

> Since an exact science of [colors] seems to be one of the most difficult things that Philosophy is in need of, I hope to show—as it were, by my example—how valuable mathematics is in natural Philosophy. I therefore urge geometers to investigate nature more rigorously, and those devoted to natural science to learn geometry first. Hence the former shall not entirely spend their time in speculations of no value to human life, nor shall the latter, while working assiduously with an absurd method, perpetually fail to reach their goal. But truly with the help of philosophical geometers and geometrical philosophers, instead of the conjectures and probabilities that are blazoned about everywhere, we shall finally achieve a science of nature supported by the highest evidence.
>
> —ISAAC NEWTON, *Optical Papers*, 1672

IT IS COMMON TO REGARD NEWTON as the apotheosis of mathematized natural philosophy in the seventeenth century. For example, the *Principia Mathematica* is the culmination of Dijksterhuis's grand narrative of mechanization (1961), marking the transition to a thorough mathematization of science. Accounts like this reflect Newton's transformative contributions to natural philosophy and the central role of mathematics in his achievements. Newton frequently characterized his methodology, as in the epigraph to this chapter, as distinctive and capable of achieving greater evidential support than that of his contemporaries, due to its mathematical character. Those guilty of blazoning about mere conjectures are presumably mechanical philosophers, including those working in the Cartesian tradition as well as members of the Royal Society such as Hooke and Boyle. The remark cannot be dismissed as merely reflecting the brashness of youth, the overly dogmatic stance of a twenty-seven-year-old that became more moderate with age. The emphasis on the role of mathematics in achieving certainty in natural philosophy, as well as the contrast with the errant ways of others, recurs with variations in methodological remarks throughout Newton's career.

Newton's pronouncements reflect a striking position regarding the role of mathematics in natural philosophy. We can give an initial characterization of his position by considering two questions central to seventeenth-century debates about the applicability of mathematics. First, how are we to understand the distinctive universality and necessity of mathematical reasoning? One common way to preserve the demonstrative character of mathematics was to restrict its domain, as far as possible, to pure abstractions. The subject matter of mathematics is then taken to be abstracted from the changeable natural world, consisting of quantity and magnitude themselves rather than the objects bearing quantifiable properties. Yet restricting the domain in this way makes it difficult to see how mathematics relates to natural phenomena. How could the book of nature be written in a language of pure abstractions? Second, what is the proper role of mathematical reasoning in natural philosophy? Many followed Aristotle in consigning mathematics to a subordinate role. On this view, mathematical demonstrations do not contribute to scientific knowledge because they do not proceed from causes. The demand for such demonstrations was difficult to satisfy in mathematics, especially for those who rejected formal causality. Mathematics could not fulfill the main aim of natural philosophy, namely to provide demonstrations reflecting the essences of things and nature's causal order. A related concern was also pressing for the mechanical philosophers: a merely mathematical demonstration fails to provide an intelligible mechanical explanation. For advocates of this line of thought, the very title of Newton's masterpiece, *Philosophiae Naturalis Principia Mathematica*, would have been extremely perplexing: how could natural philosophy be based on *mathematical* principles?

Newton's "mathematico-physical" approach, as Halley characterized the *Principia* in his ode, reflected an alternative line of thought in seventeenth-century debates on the status of mathematics. In particular, Thomas Hobbes and Isaac Barrow, Newton's predecessor as Lucasian chair, both held that the demonstrative character of mathematics does not require a restriction to quantity regarded abstractly. Hobbes's materialist mathematics left no place for a distinction between "pure" and "applied" mathematics: real bodies and their properties were the proper subject matter of mathematics. Hobbes further argued that demonstrations proceeding from the properties of real bodies satisfied demands for knowledge based on causes. For Hobbes the demonstrative character of mathematics, far from conflicting with the inclusion of physical concepts regarding body and motion in the definitions, actually resulted from it. There was then no reason to regard mathematics as inherently subordinate, unable to advance the aims of natural philosophy. Like Hobbes, Barrow offered a defense of the scientific status of mathematics, but with geometric demonstrations satisfying a kind of formal causality. He also collapsed the distinction between "pure" geometry and physics. Barrow regarded geometrical objects as generated through motion, leaving no gap between space and continuous magnitude, studied geometrically, and motions in real space, studied in physics. Newton developed a position similar to that of Barrow and Hobbes, although I will not here trace their influence in detail.

My aim is to articulate Newton's position regarding the mathematization of nature. This is challenging because Newton, unlike Hobbes and Barrow, never stated a systematic philosophy of mathematics. Yet it is crucial to articulate Newton's position given his enormously creative and influential contributions to mathematics and natural philosophy. On my reading, Newton regards the traditional response to the first question, which takes mathematical concepts to apply to *abstract* rather than *material* entities, to be deeply mistaken. Newton holds instead that "rational mechanics" offers as exact a description of material objects as the description of abstract allegedly provided by mathematics. Yet he does not take this exactness to be directly revealed in experience. An exact description underlies experience, but the underlying quantitative description can only be reconstructed from observations within an appropriate framework.

Regarding the second question, the aim of natural philosophy is taken to be the articulation of the appropriate framework for uncovering fundamental quantites and regularities. Newton develops an account of force in the *Principia* that provides such a framework, in that it underwrites theorems relating properties of observed motions to properties of a given force law (and vice versa). Making a convincing case in favor of the force of gravity to his contemporaries required overcoming opposition to this new approach to natural philosophy, which placed forces characterized quantitatively at the center of investigation. It is also not clear how to bring to bear the traditional demand that a demonstration must proceed from causes. Mathematics provides an inferential framework for understanding the forces relevant to phenomena, and for extracting an exact quantitative description of motion from inexact experience. While it is certainly appropriate to ask whether this approach has succeeded in discovering the true forces of nature from complex phenomena, it is less clear that there is a valid remnant of the Aristotelian demand for causes.

This chapter will develop and defend this reading of Newton's position as follows. The next section focuses on Newton's discussions of the nature and status of geometry in the preface to the *Principia* and related texts, in which he argues that mechanics and geometry do not differ, as was traditionally assumed, in terms of their objects of study. Rather than treating geometry as a self-sufficient inquiry focused on distinctive, abstract entities, Newton characterized it as relying on rational mechanics for an account of the generation of geometrical objects, and differing primarily in its more restricted scope. The intelligibility of the objects of study for both geometry and rational mechanics depends on understanding how the object is generated; geometry describes objects as generated from a restricted set of allowed constructions, whereas rational mechanics studies real motions produced by forces. The next section relates Newton's views on geometry to the aims of natural philosophy and how he pursued these in the *Principia*. Natural philosophy led Newton to extend geometry in two different senses: first, by rejecting Descartes' restrictions to particular types of curves, and second, by including reasoning using first and last ratios to allow treatment of instantaneous quantities. This extension was required to deal with the complexity of real phenomena. Newton's sophisticated methodology, briefly described here, aimed to extract the underlying forces from a study of phenomena. Following these discussions of the nature of geometry and Newton's reformulation of the aims of natural philosophy such that mathematical principles play a central role, the final section turns to Newton's views on the certainty of mathematics.

GEOMETRY AND MECHANICS

The *Principia* begins with a discussion of the relationship between geometry and mechanics:

> *Geometry* does not teach how to describe these straight lines and circles, but postulates such a description. For *geometry* postulates that a beginner has

> learned to describe lines and circles exactly before he has reached the threshold of *geometry*, and then it teaches how problems are solved by these operations. To describe straight lines and to describe circles are problems, but not problems in *geometry*. *Geometry* postulates the solution of these problems from mechanics and teaches the use of the problems thus solved. And *geometry* can boast that with so few principles obtained from other fields, it can do so much. Therefore, *geometry* is founded on mechanical practice and is nothing other than that part of *universal mechanics* which reduces the art of measuring to exact proportions and demonstrations. But since the manual arts are applied especially to making bodies move, *geometry* is commonly used in reference to magnitude, and mechanics in reference to motion. In this sense, *rational mechanics* will be the science, expressed in exact propositions and demonstrations, of the motions that result from any forces whatever and of the forces that are required for any motions whatever. [. . .] Since we are concerned with natural philosophy [. . .] we concentrate on aspects of gravity, levity, elastic forces, resistance of fluids, and forces of this sort, whether attractive or repulsive. And therefore our present work sets forth mathematical principles of natural philosophy. (*Principia*, 381–82)[1]

There are several intriguing claims in this passage, which Newton explored at greater length in unpublished manuscripts.[2] Although epistemology of geometry is not the central focus of an extended treatise, these texts suffice to elucidate aspects of his position and how it contrasts with views of his contemporaries.

Geometry is founded on mechanical practice in the sense that it turns to mechanics for the construction or generation of the objects used in geometrical reasoning. Newton adopts a kinematic conception of geometry, in which an object such as a curve is understood in terms of how it can be generated by motion. One can impose restrictions on the permissible ways of generating curves and other geometric objects, such as using only a compass and straightedge. But, as Newton emphasizes at the start of the passage, the repertoire of constructions is not the subject matter of, nor is it fixed by, geometry itself: mechanics rather than geometry determines the permissible constructions. This repertoire includes far more than straightedge and compass constructions. Newton's rejection of constraints on the methods used to generate a curve is more explicit in the *Geometria*: "We are free to describe them [plane figures] by moving rulers around, using optical rays, taut threads, compasses, the angle given in a circumference, points separately ascertained, the unfettered motion of a careful hand, or finally any mechanical means whatever. Geometry makes the unique demand that they are described exactly" (MP 7:289). Newton's only constraint on the generation of curves is that their construction must be "exact."

Newton's emphasis on "exactness" counters a common view that the subject matter of mechanics cannot be described with sufficient precision to be studied geometrically. He rejects the "common belief" that "nothing could possibly be mechanical and at the same time exact" as a "stupid one" (MP 7 289). The common view mistakenly treats the flaws of particular instances of manual generation of curves as a general failing of mechanics. The appropriate contrast, Newton argues in the *Geometria*, concerns the different aims of mechanics and geometry: mechanics concerns the form and generation of continuous magnitudes, whereas geometry is the science of measurement of such quantities. What Newton means by the exactness characteristic of both can be discerned in his formulation of a new set of postulates for geometry; the third postulate allows a quite general construction, which "has a kinship with mechanical description by moving rulers." The third postulate states: "To draw any line on which there shall always fall a point which is given according to a precise rule by drawing from points through points lines congruent to given ones" (MP 7 389). The demand for exactness is reflected in the requirement that there is a precise rule for generating the curve. Newton goes on to discuss how this postulate licenses the construction of a wide variety of curves.

Newton implicitly rejects the criteria of intelligibility proposed in Descartes' *Géométrie*, one of the texts that had inspired his early work in mathematics.[3] Descartes' criteria was formulated in terms of the means of generating a curve as well as its algebraic representation, with the scope of geometry limited to curves generated via specific generalizations of compass and straightedge constructions, or, expressed algebraically, to curves represented by closed polynomials. (Descartes hoped to prove that these two characterizations were equivalent, but failed [Bos 2001].) This excluded "mechanical" or "transcendental" curves from consideration, for example the logarithmic spiral and cycloid. Newton rejected both grounds Descartes gave for regarding such curves as falling outside the scope of geometry, and studied the properties of nonmechanical curves in his early mathematical research in the 1660s as well as the *Geometria*. Newton discovered how to represent this more general class of curves algebraically via an infinite series expansion (Guicciardini 2009, chap. 7). We will see the significance of the resulting broader scope for geometry in the next section.

Newton regards "rational mechanics" as the exact science of the generation of the motions of real bodies, due to the combination of inertia and forces acting upon them. Shifting from "manual powers" to "forces" does not alter the idea that the intelligibility of a given curve depends on understanding how it is generated, given an appropriate conception of force. One of Newton's achievements was the recognition of how much was required in clarifying the conceptions of inertia and force, in order to undertake the project of determining the true motion of bodies.[4] Unlike his contemporaries, he argued that providing an account of the dynamics governing motion could not be adequately founded on geometrical relationships among bodies alone, but required in addition an appeal to "absolute" structures (namely, intervals of spatial distance and temporal duration, and a way of identifying locations over time). These structures underwrite a contrast between inertial (moving in a straight line at uniform velocity) and noninertial motion. With this contrast in place, it is in principle possible to consider a physical trajectory as generated by the net force acting on a body in much the same manner as the geometry student manually producing a curve. And, importantly, it is possible to draw inferences in the opposite direction as well: that is, given the trajectory, to determine the net force that would produce the required motion.

Geometry and rational mechanics are thus both exact sciences with a common subject matter: geometry measures the properties of objects whose generation is described by mechanics. It seems more apt to call this physicalizing geometry rather than mathematizing the study of motion. Our sensible experience of the geometrical properties of objects and their trajectories may be vague or inexact. The trajectories of real bodies are still suitable objects of study for mechanics and geometry, even though they are only accessible via a combination of observation and calculation. The tension between the exactness of geometry and mechanics and the character of sensations is resolved by taking geometry to apply directly to physical objects, whose geometrical properties are immanent in sensation rather than directly apparent. Since the properties are not ascribed to abstract mathematical entities with a distinctive ontological status, there is no place for a worry to arise regarding how mathematical entities can stand in relation to, or represent, physical objects.

Newton's account of the nature of geometry stands in stark contrast with the Cartesian tradition, as well as the traditional Aristotelian account of pure and mixed mathematics. Newton's position was not unprecedented, and it is particularly useful to compare Newton's position with that of

Barrow.[5] Barrow's *Geometrical Lectures* (2006) highlighted the utility of treating curves in terms of generating motions rather than as a collection of points. Several of Barrow's predecessors solved problems such as that of finding tangents and areas of a curve based on such a kinematic conception, but Barrow treated the generation of curves by motion as the appropriate foundation for geometry rather than simply a useful heuristic. Barrow argued further that the definition of curves in terms of their generating motions satisfied traditional demands for causal arguments, and regarded geometry as the most fundamental branch of mathematics because it deals directly with magnitudes generated by motion.[6]

Newton was almost certainly influenced by Barrow's defense of the scientific status of mathematics in his earlier *Mathematical Lectures*. Barrow held that mathematical demonstrations satisfy a version of formal causality.[7] He also addressed a problem facing any empiricist epistemology of geometry, namely how sense experience relates to geometric reasoning. Barrow characterized the contribution of sensation as limited to establishing that geometrical postulates reflect a real possibility. We can see that a straight line is a real possibility by considering an actual line, and further recognizing that there is no obstacle to making it straighter. Sensation does not provide an inductive base for geometrical arguments, but instead establishes that geometrical postulates are not empty or vacuous. Against the view that geometrical objects are merely mental entities, Barrow asserts that "all imaginable Geometrical Figures are really inherent in every Particle of Matter," even if they are inaccessible to the senses (Barrow 1734, 76), just as the statue is in a block of marble, waiting to be uncovered by the sculptor's chisel. Barrow further rejected the traditional distinction between pure and mixed mathematics. It was common, following Aristotle, to regard pure mathematics as restricted to the study of quantities abstracted from material objects, whereas mixed mathematics applied to the mathematical properties imperfectly instantiated by sensible objects. Since for Barrow the instantiation of geometrical properties in sensible objects underwrites their intelligibility, this contrast makes no sense. Barrow collapsed this distinction by taking geometry to apply directly to material objects, even though their exact geometrical properties are not immediately revealed in sensation.

The *Principia*'s preface echoes this position, but Barrow's effort to ground mathematics solely on geometry, as he conceived it, is incompatible with Newton's mathematical practice. By the time Barrow presented his *Geometrical Lectures* in 1668, Newton had made many strikingly innovative discoveries in what he called the fluxional analysis of curves. Newton's first treatise, "To Resolve Problems by Motion" (1666), began like Barrow with a kinematic conception of curves. Newton appeals to the continuity of motion generating a curve to justify the use of limiting procedures. Yet Newton's development of the calculus depended on combining a kinematic conception with ideas from algebra foreign to Barrow's approach. In particular, Newton discovered a generalization of the binomial theorem to noninteger exponents, which allowed him to treat curves such as the logarithmic spiral using an infinite series expansion. This result was inspired by Wallis's *Arithmetica Infinitorum*, and Newton's mature 1671 treatise develops fluxional analysis based on a kinematic approach to curves used in concert with algebraic techniques.[8]

The importance of these algebraic techniques is largely hidden from view in the *Principia*, which Newton wrote in a synthetic, geometric style. Newton's wide range of mathematical techniques peeks through the chinks in the armor of synthetic geometry. The types of problems Newton handled earlier using fluxional analysis are treated in the *Principia* based on a geometrical treatment of limits (described briefly later in this chapter). Newton had several reasons for adopting a geometrical style.[9] For many of the problems in the *Principia*, geometrical methods may have been the most efficient and direct calculational tool. But it is also the case that Newton's views regarding mathematics shifted, as he developed admiration for Greek geometry based on studying Pappus and Apollonius in the 1670s. While I do not have the space to explore the issue fully here, it is clear that the *Principia*'s proofs required combining geometrical reasoning (suitably extended) with a variety of algebraic techniques. Insights from fluxional analysis are crucial in a number of places, for example the proofs in section 8 and in various cases where Newton states the area under a curve without specifying his method for finding it. Book 2 includes a terse summary of fluxional analysis in Lemma 2, which Newton needed to treat the problem of an object falling through a resisting medium. In sum, Newton extended the scope of "geometry" in the *Principia* to handle a broad class of curves, leading to a much richer conception of geometry than that allowed by either Barrow or Descartes.

Following his remarks on the relationship between mechanics and geometry, Newton turns to giving a positive characterization of the basic problem of natural philosophy—"to discover the forces of nature from the phenomena of motions and then to demonstrate the other phenomena from these forces" (*Principia*, 382). Pursuing this project is possible because the trajectories generated by the interplay of inertia and the forces of nature can be determined exactly. The *Principia* offers an understanding of physical trajectories as generated by a combination of inertia and forces of nature that is analogous to that provided by a kinematical conception of geometrical curves as generated by a moving point or figure. Given this conception of rational mechanics as the project of discovering the generating forces for the trajectories of real objects, a number of questions about "forces of nature" are pushed to one side. For the inferential connection between the forces of nature and phenomena of motion, going in both directions, can be fully specified given a mathematical characterization of the force, without an account of its underlying source. (Although Newton in various places acknowledges the interest of determining the seat or cause of the forces of nature, an answer to such questions would not alter the *Principia*'s project. Hence the causal question, which Barrow responded to by defending a version of formal causality, is left aside.) Next I will turn to study a few aspects of the argumentative structure of the *Principia* that reflect Newton's response to the basic problem of philosophy. His response reflects the challenge of *accessibility* of the true motions: inferring the true motions and the forces responsible for them from the complex apparent motions of bodies we experience is a laborious and difficult, yet far from hopeless, undertaking.

HANDLING COMPLEXITY IN THE *PRINCIPIA*

The *Principia* provides a framework for drawing inferences from observed trajectories to the underlying causes of motion, namely inertia and forces. This is analogous to analytical problems in geometry, in which a curve is given and the mechanism for generating the curve is to be found. In the first steps toward the *Principia*, taken in the manuscript *De Motu*, Newton treated the planetary trajectories, in effect, as given curves from which he inferred an inverse-square force law. If Newton had left it at that, the argument would have established the inverse-square force law for gravity in much the same way as earlier results had been established in Galilean-Huygensian mechanics: namely, that an inverse-square law produces trajectories with several striking features in common with observed motions. Even though the real motions are far too complicated for Kepler's laws to hold exactly, the account would be explanatory in much the same way as Galileo's treatment of projectile motion, despite its failure to account for air resistance.

It is a common mistake, however, to read the *Principia* as giving nothing more than a more elaborate version of this argument. Newton recognized the limitations of this initial argument and developed a sophisticated approach in the *Principia* to overcome them.[10] There is a disanalogy between the analytic problem in geometry and the problem in natural philosophy: a trajectory is not "given" to the natural philosopher as the starting point of investigation, as it is to the geometer. The further steps Newton took in the *Principia* were driven in part by the challenge of reaching conclusions regarding real motions despite their enormous complexity. Rather than lowering the standard of success to require only qualitative agreement with the phenomena, Newton proved a number of results that allowed him to assess the impact of removing various idealizations.

From De Motu *to the* Principia

Christopher Wren offered Edmond Halley and Robert Hooke the reward of "forty-shilling book" for a proof that elliptical planetary trajectories follow from a force varying as the inverse square of the distance from the sun. Neither of them was up to the challenge, and Halley posed the problem to Newton on a visit to Cambridge in 1684.[11] The brief manuscript Newton composed in response would have been sufficient to secure Newton a place in the history of mechanics (Gandt 1995). The *De Motu* achieved a unification of the Galilean-Huygensian theory of uniformly accelerated motion with Kepler's treatment of planetary motion, based on Newton's innovative treatment of force. But this was only the first step on the road to the *Principia*.

The central result of the *De Motu* brings together a generalization of Galileo's treatment of free fall with Kepler's area law, recognized as a general feature of motion under a central force (i.e., a centripetal force whose magnitude depends only on distance to the force center). Following Galileo, the distance traveled by a body starting at rest, undergoing uniform acceleration, is proportional to the square of the elapsed time. Newton realized that this proportionality holds for *finite* elapsed times in the case of uniform acceleration, but it is also valid *instantaneously* for arbitrary centripetal forces. A generalization of Galileo's result then provides Newton with a precise quantitative measure of a trajectory's deviation from straight, inertial motion at each point of the orbit given the magnitude of the force, which is allowed to vary: the deviation produced by any centripetal force is proportional to the square of the elapsed time, "at the very beginning of its motion."[12]

Newton next established, as his first theorem, that what we now know as Kepler's area law holds for any central force: elapsed time of motion along the trajectory can be represented geometrically by the area swept out by a radius vector from the force center.

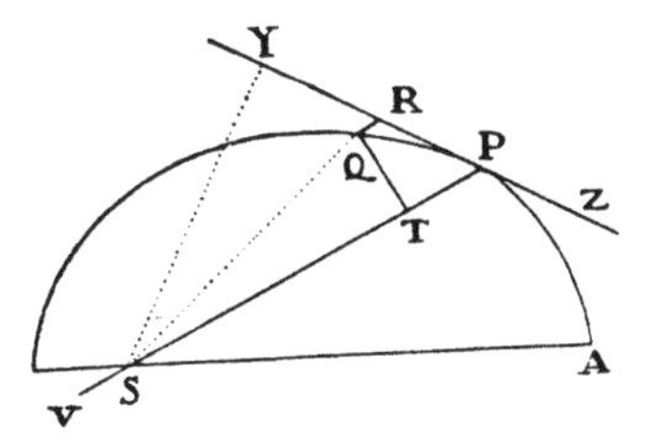

Figure 1. Figure from proposition 6 in the *Principia* (and theorem 3 of the *De Motu*).

Combining these two results leads to a general expression relating the geometrical properties of the trajectory to the magnitude of the force and the law describing its variation. The deviation produced by the force acting at point P is represented by a line segment QR, directed toward the force center (in Newton's figure, 3.1). This displacement is proportional to the product of the force F acting on the body with the square of the time elapsed, $QR \propto F \times t^2$, as shown by the generalization of Galileo's law. From Kepler's area law, $t \propto SP \times QT$, it follows that $F \propto QR/(SP^2 \times QT^2)$.

This theorem also illustrates how Newton extended classical geometry to treat instantaneous quantities, without explicit appeal to fluxional analysis. The result holds instantaneously, in the limit as the point Q approaches the point P. The continuity of the generation of the curve guarantees, according to Newton, that a limiting value for the ratio of evanescent quantities exists, and it is given by the ratio of finite quantities. Enlarging the scope of geometry to include line segments of varying length allowed Newton to handle instantaneous quantities and the limiting behavior of their ratios. Newton used theorem 3 by establishing connections between the "evanescent" figure $QRPT$ and finite quantities characterizing the trajectory, such as the radius of a circle or the *latus rectum* of an ellipse. This leads to an expression for the force law entirely in terms of finite quantities unblemished by evanescent quantities that vanish in the limit as $Q \rightarrow P$. Consideration of the figure $QRPT$ thus allows Newton to handle the differential properties of the curve geometrically.

With these results, Newton generalized Huygens's earlier treatment of uniform circular motion to arbitrary curvilinear trajectories, and this opened the way for considering a variety of forces sufficient for motion along different plane curves. In particular Newton could directly address Wren's question. Halley and his colleagues would have naturally wondered whether an inverse-square force sufficient for perfectly circular orbits would have to be supplemented by a secondary cause to account for elliptical motion. Newton's results show convincingly that a simple inverse-square force alone is sufficient for Keplerian motion. But more important, Newton concluded that planets held in their orbits by an inverse-square force directed at the sun move "exactly as Kepler supposed." Kepler's area law has special standing as a direct consequence of any central force, and Kepler's first and third laws hold for an inverse-square force, specifically. According to Kepler's first law, the planets follow elliptical trajectories. An inverse-square force directed at the focus is sufficient to produce this motion if, in addition, the second law holds with respect to the focus of the ellipse (that is, the radius vector from the focus sweeps out equal areas in equal times). Thus, insofar as Kepler's laws hold *exactly* for each planet, one can infer an inverse-square force between the sun and each of the planets. Kepler's third law is a specific instance of a general result linking periodic times to the exponent in the force law (theorem 2). Furthermore, the ratio of the radii to the periods (a^3/P^2) is the same for all of the planets, leading to the conclusion that a *single* inverse-square force directed at the sun suffices.

The *Principia* grew out of a number of questions provoked by *De Motu*.[13] One line of thought forced Newton to abandon the simple picture of the planets following stable elliptical orbits, and the argument just summarized connecting these motions with a force law. The motion of the planets was apparently due to an inverse-square force directed at the sun, and similarly the motion of Jupiter's moons was due to an inverse-square force directed at Jupiter. How are these forces related? Such questions may have led Newton to apply what became the *Principia*'s third law to combine the distinct forces due to the planets and the sun. Then the planets can no longer be described as responding to a single inverse-square force directed at a fixed sun, as there are other forces to take into account: the attraction of the planets on the sun, causing it to orbit a common center of gravity rather than remaining fixed, along with the interactions among the planets. Newton eloquently noted the consequences:

> By reason of the deviation of the Sun from the center of gravity, the centripetal force does not always tend to that immobile center, and hence the planets neither move exactly in ellipses nor revolve twice in the same orbit. There are as many orbits of a planet as it has revolutions, as in the motion of the Moon, and the orbit of any one planet depends on the combined motion of all the planets, not to mention the action of all these on each other. But to

> consider simultaneously all these causes of motion and to define these motions by exact laws admitting of easy calculation exceeds, if I am not mistaken, the force of any human mind. (Hall and Hall 1962, 280)

Newton recognized the multifaceted challenge to reasoning from complex phenomena to the underlying forces, and book 1 of the *Principia* lays the groundwork for Newton's response in book 3.

The Motion of Bodies

How is one to proceed in the face of curves that no human mind can comprehend, as Newton so pointedly remarked? The results in book 1 of the *Principia* support a novel response to the challenges posed by the complexity of real phenomena.[14] Newton's strategy proceeds in stages, beginning with a heavily idealized and tractable description of motion, and adding complexity as demanded by observations. I will emphasize three aspects of the *Principia* that are needed for this approach to be viable: first, the generality of Newton's treatment of force and trajectories; second, his handling of limiting relations to insure the durability of measurements; and third, the controlled way in which he handles idealizations.

The generality of Newton's approach allows him to establish relationships between trajectories and force laws, rather than simply giving a collection of striking consequences for particular cases. Several theorems in book 1 hold for *arbitrary* central forces.[15] The force law itself is the unknown quantity to be determined based on the phenomena, so this level of generality is crucial. It leads to a much stronger inference, spanning a variety of alternative force laws rather than restricting consideration to a single force law (as in the *De Motu*). The level of generality in the treatment of force is matched by the broad scope of curves Newton allows as possible trajectories. The collection of curves deemed intelligible by Descartes did not include a possible representation of the actual trajectory traced by Mars (idealized as a point mass), as a result of the complexity Newton acknowledged in the previous extract quote. This problem would arise even for the motion of a pendulum bob, idealized again as a point mass, in a Huygensian cycloidal pendulum. Newton also proved (in Lemma 28) that there is no solution in terms of a Cartesian geometric curve to the so-called Kepler problem, which arises in finding the position of a planet as a function of time on a Keplerian ellipse. The solution required the use of what Newton called "geometrically irrational curves," which cannot be defined in terms of an equation with a finite number of terms. Hence celestial mechanics forces a conception of geometry broad enough to regard such curves as legitimate.

Proposition 1.45 illustrates the importance of both senses of generality; it shows how apsidal motion depends on the exponent of the power law of the underlying force. The apsides are the points of maximum and minimum distance from one focus of an elliptical orbit; in the case of apsidal precession, a body does not form a closed orbit and the apsides shift slightly with each revolution, by an amount given by the apsidal angle. The first corollary of 1.45 states that for nearly circular orbits if the apsidal angle θ is given by $n = (\theta/\pi)^2$, then the force is given by $f \propto r^{n-3}$. Based on this result, specific phenomena can be taken as measuring the parameters appearing in the force law, in the sense that there are lawlike connections between the phenomena and parameters, within some delimited domain.[16] Establishing such connections requires quantifying over a range of different force laws, which would not be possible without Newton's level of generality of force laws and trajectories. Furthermore, the relationship between the phenomena and the parameter of the force law is robust, in the sense that if the apsidal angle is *approximately* θ, the force law is approximately $f \propto r^{n-3}$. With this proposition in hand, Newton can infer that the force law is approximately inverse square despite the complexity of the planetary orbits—it suffices to have established that they are very nearly elliptical with stable apsides.

This general treatment of force and trajectories makes it possible to regard observed motions as "measuring" parameters of the force law. This is obviously a heavily theory-dependent account of measurement. How stable are these results as one adds complexity, or changes the theory? Section 10 of the *Principia* provides an example of how Newton responds to this concern, by considering the relation between earlier results regarding constrained motion, in particular Huygens's treatment of pendulums, and his own approach.[17]

Several lines of argument in the *Principia* depend on using pendulums as measuring devices, and the case for universal gravity in particular relies on measurements of the local strength of surface gravity and its variation with latitude. Yet the claim that appropriately designed pendulums measure surface gravity depends, in Huygens's account, on a Galilean conception of gravity, which treats gravity as uniform acceleration directed along parallel lines. Rather than simply making an argument that Huygens's results are a useful approximation to universal gravity, Newton proves theorems that specify a precise limiting relationship (although Newton does not explicitly state it in these terms): Galilean gravity holds in the asymptotic limit of universal gravity at or below the surface of a uniformly dense spheroid as the ratio of any vertical distance of interest d to the radius of the earth R goes to zero, $d/R \rightarrow 0$.[18] This ensures that the relationships between physical quantities asserted in Huygens's theory retain approximate validity. At or below the surface of a uniformly dense spheroid, Huygens's results still describe lawlike relationships among quantities mirroring the lawlike relationships within Newton's theory. This claim is restricted to a specific domain, but it is the domain for which Newton needed to establish that pendulums could be used as measuring devices.

The results in section 10 thus allow Newton to argue that Huygens had successfully measured the time of free fall, not because of his adoption of Galilean gravity as the underlying theory, but in spite of it. In the domain Huygens considered, the physical relationships asserted by the superseded theory of Galilean gravity approximately match those of Newton's own theory. Hence the measurements of surface gravity—and, indeed, the entire body of evidence in favor of Galilean-Huygensian mechanics—could be subsumed within Newton's theory. Evidence consisting of theory-dependent measurements can be durable enough to survive transitions in underlying theory.

Turning to the third point, Newton notes that the first ten sections of book 1 concern "bodies attracted toward an immovable center, such as, however, hardly exists in the natural world" (1999, 561). Since gravity is a mutual interaction, it is unphysical to regard one body as fixed—instead, a system of interacting bodies will orbit a common center of gravity. It is also unphysical to treat these interacting bodies as if they are point masses, rather than having finite extent. The force the earth exerts on a nearby body is composed out of the attractive forces on each part of the body from each part of the earth. Newton realized in composing the *Principia* that the resultant force may not have a simple form even if the component forces do. Newton's predecessors often acknowledged similar limitations, then argued that the theory nonetheless provides an approximate description of the world. Newton's response is strikingly different, as he considers in detail the consequences of relaxing both of these simplifying assumptions.

Newton obtained limited results for systems of interacting bodies in section 11. A number of results regarding bodies orbiting a fixed force center carry over directly to the case of two interacting bodies. In the three-body case for an inverse-square force, Newton's results consisted of a series of corollaries describing various effects in a three-body system (such as the earth-moon-sun system) in qualitative terms. Although Newton regarded these as "imperfect," they indicate various consequences of treating the mutual interactions among all three bodies. Sections 12 and 13 consider bodies of finite extent, and Newton established a remarkable feature of an inverse-square force law. For if we idealize the earth as a spherical body whose density varies with the radius, the total force on a nearby body will be an inverse-square force directed at the center of the earth with the absolute measure of the force given by the total mass. He further proved the famous result that two spherical bodies interacting via an inverse-square force can be treated as mass points interacting with forces depending on their total masses directed at their respective centers.

These three sections reveal a quite sophisticated approach to handling idealizations (Smith 2001; 2002b). Rather than treating his initial account as a good approximation despite its unphysical idealizations, Newton developed the mathematics needed to assess the effects of removing the idealizations. The initial account is itself exact, given a precise quantitative treatment, yet acknowledged to differ from the true trajectory. Newton's results, while incomplete, could be used to characterize qualitatively the departures from the initial idealized treatment due to many-body interactions and the finite extent of real bodies. These results made it possible to identify the kind of contrasts between theoretical calculations and observation one would expect to see as a result of specific idealizations, and to assess whether removing a particular idealization would lead to an exact trajectory closer to the true trajectory. Many of the effects identified in these sections were already relevant to assessing the application of the theory to the solar system, treated in book 3.

Together these three aspects of book 1 support an approach to the complexity of real motions that proceeds in successive stages, dubbed the "Newtonian style" by Cohen (1980). My characterization of this style follows Smith's much richer accounts (2001; 2002b). At a given stage the physical trajectories will be treated as if they were produced by an explicit combination of forces. The resulting exact trajectory is expected to differ from the true trajectory, because the derivation, to be tractable, will include a number of presumably false idealizing assumptions. Even so, the idealizations can be used in making inferences concerning values of physical quantities as long as these inferences are robust, as in the case of proposition 1.45. This inference holds even if the physical trajectory is only approximately an elliptical orbit with stable apsides. Of course, this inference clearly depends on the basic Newtonian framework and the earlier stages of the description of the system. Newton showed, however, that in the specific case of Huygens's pendulum measurements, conclusions based on measurements in this sense can be durable through theory change.

The demand for exactness at each stage of inquiry also lends particular significance to discrepancies between the calculated trajectories and observations. Results like those in sections 11 through 13 put Newton in a position to assess whether particular systematic deviations from idealized models can be eliminated by dropping specific assumptions and developing a more complicated model. If the empirical deviations from the ideal case are of this kind, then the research program can proceed by relaxing the idealization. But it is also possible to identify systematic deviations that instead reveal deeper problems with the entire framework of book 1, exemplified by Mercury's anomalous perihelion motion. Newton's treatment of idealizations allows for observations to continue to guide research, even though the identification of the deviations in question presupposes that the simplest idealized models are approximately correct.

Arguing from the Phenomena

The body of results in book 1 provided Newton with mathematical tools to infer the underlying forces of nature from the phenomena of motion. Nature also cooperated in providing a system of bodies that could be described to a very good approximation as interacting via a single force, whose motion had long been studied and described with obsessive precision. In the opening sequence of book 3 Newton argues from observed regularities of the motion of the planets and their satellites to the properties of the force of gravity. Here is not the place to review that famous argument (see, in particular, Harper 2011), but I will briefly contrast it with the argument in the *De Motu*.

Based on the results of the *De Motu*, Newton could have claimed that idealized descriptions of bodies moving in response to gravity agree qualitatively with observed planetary motions. Newton had shown that Kepler's laws hold for bodies moving in response to an inverse-square force directed at the sun, regarded as fixed. Yet there are two shortcomings of an argument based on these results. First, the inference presupposes that the planet moves exactly on an ellipse with a force directed at the sun at one foci; the argument does not apply to *nearly* elliptical trajectories or forces directed nearly at one foci.[19] Once Newton considered the effect of mutual interactions, he realized that the planets do not follow exactly elliptical trajectories—blocking the *De Motu* argument. Second, there are challenges regarding the status of various idealizations that could not be addressed. Is it legitimate, for example, to treat Mars and the sun as point-masses in calculating their trajectories, and to neglect the gravitational attraction Mars exerts on the sun?

In the *Principia* Newton overcame both of these shortcomings. The opening sequence of book 3 gives an argument for universal gravitation that is robust, in the sense that the inferences do not require that the antecedent holds exactly in order to reach conclusions regarding the force law (as in proposition 1.45). Much of the remainder of book 3 is devoted to discussing the effects of dropping various idealizations to give a more realistic description, for example in studying the earth-moon-sun system. These are the kinds of questions that the results discussed earlier put Newton in a position to answer, albeit provisionally.

The further results in book 3 embodied a distinctive approach to problems in celestial mechanics that would set the agenda for the eighteenth century.[20] Newton treated celestial motions as consequences of the gravitational interactions among a system of bodies. The goal at any given stage of inquiry would be to provide an exact description of the motions that would result given several idealizations. This exact trajectory could then be compared with observations. At each stage the theoretical description was well controlled in the sense that the idealizations needed to derive a specific solution were explicitly identified. Any systematic discrepancies could then be used to identify physically significant features that had been initially excluded in the description. Finally, Newton's further results indicating how the behavior of the system changes as a result of dropping particular idealizations provided the basis for distinguishing troubling discrepancies from those that one could expect to be handled at the next stage of theoretical refinement. Insofar as one encounters only discrepancies of the latter sort, it is plausible to take the understanding of the physical trajectory as generated by the forces explicitly taken into account by that stage as very nearly accurate, albeit not exact. One can imagine approaching a completely exact description of the physical trajectory, with a fully specified account of the forces generating the motion, as a kind of limit. This should not be regarded, however, as approaching a closer and closer match with a preexisting description of the true trajectory. The projects of developing a more detailed theoretical account of the motion and of giving a more exact observational characterization often progressed in tandem. In many cases, without the structure provided by the theoretical skeleton, observations would remain a complex, amorphous mass.

The problem of the moon exemplifies the importance of Newton's approach. Seventeenth-century astronomers struggled to describe the moon's motion with accuracy comparable to that achieved for the planets. In order to assess whether the details of the moon's motion were compatible with gravitational theory, Newton had to first develop a more accurate description of the motion itself, and he tackled the theoretical and observational problems simultaneously. The physical trajectory only becomes intelligible via a combination of observations and a theoretical understanding of how it is generated. Newton's aim was to account for the various known inequalities in the lunar orbit as a consequence of the perturbing effect of the sun's gravity and other features of the earth-moon-sun system. He was ultimately not able to make substantial progress with regard to accuracy over the ideas of Jeremiah Horrocks, which were formed in the 1630s. Newton's approach to the problem proved to be more influential. His eighteenth-century successors were able to employ new mathematical methods that made it possible to enumerate all of the perturbations at a given level of approximation, namely as all of the terms at a given order in an analytic expansion. It was only with a more sophisticated mathematics that astronomers could fully realize the advantages of approaching the complexities of the moon's motion via a series of approximations. Although it drew on tools unavailable to Newton, this work followed Newton's approach to achieving high levels of precision by using a series of well-controlled idealizations.

MATHEMATIZATION AND CERTAINTY

Newton directly addressed the contribution mathematics makes to achieving the "highest evidence" for claims in natural philosophy in only a few places in his published work; in addition to the *Principia*'s preface, the most detailed discussion appears in query 31 of the *Opticks* (1717). Newton's main concern was to characterize his natural philosophy in terms of analysis and synthesis, as part of a defense of his methodology and its fruits in reply to continental critics. As Domski (2010) has emphasized, in this and similar passages Newton locates his methodology in a historical tradition, as a revival and extension of ancient mathematical practice. This concern is nearly orthogonal to questions regarding the epistemology of geometry that are our main focus. Guicciardini (2009) provides a thorough exegesis of query 31 and related passages, drawing on a careful assessment of analysis and synthesis in Newton's mathematical practice (309–28). Yet, as Guicciardini acknowledges, these passages are not sufficient to elucidate

Newton's philosophy of geometry. Hence, I will turn to an unpublished manuscript customarily called "De Gravitatione" (hereafter DG), rather than Newton's sparse methodological remarks. Passages in the DG suggest that Newton regarded the certainty of geometry as grounded in knowledge of the nature of space. For the metaphysical account of space to play such a role, we must have access to its structure. Here I will briefly consider the epistemology of geometry suggested in the DG, and argue that in the *Principia* geometry becomes entangled with dynamics.

Much of DG is devoted to a detailed critique of Cartesian views regarding space, body, and motion. Newton argued, in contrast to the Cartesian identification of space and body, that an adequate definition of motion had to be formulated in terms of "some motionless being such as extension alone, or space in so far as it is seen to be truly distinct from bodies" (Newton 2004, 20–21). His ensuing discussion of the ontological status of this distinct entity and its properties sheds some light on the status of geometry. The object of geometry is clearly space itself, and not, by contrast with Barrow, properties of objects. Barrow held that were "the Hand of an Angel" to polish a solid particle of matter, a perfectly spherical surface would be revealed. Sphericity is a property of a particle of matter, rather than being a mental entity or having a distinctive mode of being. Newton also rejected the latter two options, but ascribed geometrical properties to space rather than bodies: "For thus we believe all those spaces to be spherical through which any sphere ever passes, being progressively moved from moment to moment, even though a sensible trace of that sphere no longer remains there. We firmly believe that the space was spherical before the sphere occupied it, so that it could contain the sphere; and hence as there are everywhere spaces that can adequately contain any material sphere, it is clear that space is everywhere spherical" (Newton 2004, 23). Newton attributed geometrical properties to all regions of space. He further argued that bodies bear geometrical properties derivatively, as "determined quantities of extension" endowed with additional attributes, such as impenetrability. Finally, the geometrical structure ascribed to space itself, along with an independent structure relating locations at different times, is necessary, as Newton argued in a famous rebuttal of Descartes, for an adequate definition of motion.[21]

Some of Newton's remarks in DG suggest a straightforward empiricist conception of geometry. The main challenge facing such an account, as Torretti (1978) succinctly put it, is that "geometrical objects [. . .] are nowhere to be found in experience exactly as geometry conceives them" (254). Barrow and Newton alike regarded geometry as treating properties of real entities—bodies and space itself, respectively—that are not directly present in experience. How then does experience bear on the concepts and theorems of geometry? Barrow's position seems to be that experience can establish the existence of geometric properties. Any actual line is not straight, for example, but we can see that it can always be straightened, leading to a geometric line as a limit. It is unclear exactly how experience reveals that this refinement of our experience is a real possibility.[22] Newton held that space itself is insensible, so its structure must be discerned based on the geometric properties of objects. We obtain the idea of space from experience by abstracting away dispositional and sensible features of bodies, leaving us with an "exceptionally clear idea of extension," namely "the uniform and unlimited stretching out of space in length, breadth and depth" (Newton 2004, 22). In this context, the question analogous to that confronting Barrow regards how the character of sense experience justifies this "clear idea" and the six properties of space Newton goes on to elucidate. Taking bodies as the basis for abstraction is particularly problematic given the status of Newton's account of body in DG: he presented it as *sufficient* to ground our sense experience, but did not claim that it is the only such account (McGuire 2007, §6; Stein 2002). Such an uncertain account of body cannot provide firm foundations for geometric knowledge, and in any case Newton's arguments do not follow Barrow's approach in appealing to abstraction and refinement of our experience of bodies.

Newton's argument for the infinity of space, the second property he discusses at length, takes a quite distinctive approach: it depends on a simple geometric construction, carried out in the imagination. This appeal to the imagination, as a form of access to truths about the nature of space, marks a further critical response to Descartes.[23] Descartes rejected the scholastic idea that knowledge of geometry is derived from sense experience via a process of abstraction, and instead held that geometry makes claims regarding the true and immutable natures of geometric objects. The eternal and immutable natures exist, in some sense, as innate ideas in created minds. Newton clearly did not adopt this Cartesian approach fully; in a later unpublished manuscript intended as a fifth rule for philosophizing (to be added to those in the *Principia*), probably reflecting Locke's influence, Newton explicitly rejected the possibility of innate ideas (Koyré 1965, 272). Yet the argument for the infinity of space is based on the ability to imagine the following construction (cf. Domski 2012). Consider opening up a given triangle by rotating one of its sides around a vertex, with the other sides fixed, and defining a sequence of points where the line segment extending the rotated side meets a line segment extending the opposite fixed side. There is no final point of this sequence; Newton concluded that the real line defined by the sequence is longer than any finite length. The contrast with Descartes is clear, regarding both the strength of this conclusion (namely, that space is *infinite* rather than merely of *indefinite* extent, as Descartes held) and the roles of imagination and understanding. Newton further argued that we can understand the infinity of space positively rather than merely as a lack of boundaries, as illustrated by geometrical objects with finite surface areas and infinite lengths.[24] The later manuscript "Tempus et Locus"[25] includes a wide variety of other examples illustrating our ability to comprehend infinite quantities in mathematics.

It is challenging to find a coherent epistemology of geometry in the DG. Although I will not argue the point here, on my reading there are unresolved tensions among the lines of argument Newton gives regarding the nature of space.[26] These tensions can be avoided to some extent because the different properties of space Newton considers are nearly independent of one another. The infinity of space does not determine the other local features Newton identifies, such as Euclidean geometrical structure ascribed directly to space, and the relation among spaces at different times needed to define motion.

More important, however, I regard DG as a transitional text in which Newton had not fully developed the insights regarding dynamics that are crucial to the *Principia*. The accounts of body, space, and time, and the relations among them, differ substantively in the *Principia* and in later texts such as "Tempus et Locus."[27] Far from providing philosophical foundations for the account of space and time in the *Principia*, as Domski (2012) argues, DG was written prior to Newton's recognition of the importance of specifying particular measures of space and time and the full implications of the relativity of motion. There are only hints of Newton's innovative conceptions of force and mass in DG (Biener and Smeenk 2012), and developing these ideas led to the distinctive account of geometry reflected in the *Principia*'s preface and related texts discussed earlier in this chapter. The considerations of measurement and dynamics that are central to the *Principia* lead to a physical conception of geometry. On this account, the geometrical properties of space are related to sense experience only quite indirectly, by virtue of their role in a dynamical account of motion.

The *Principia* starts with the phenomena of motion and our observations of the trajectories of bodies over time. All of our descriptions of motion are implicitly made with respect to some relative space—for example, the motion of balls dropped from the dome of St. Paul's Cathedral (in one of Newton's experiments) was described relative to the building, with times determined by a pendulum clock. An initial description of the trajectory of the body could start by assigning spatial dimensions with respect to this relative space; obviously, the trajectory will generally not be the same with respect to a different relative space. In formulating the relativity of motion in corollaries 5 and 6, Newton recognized that some specific choices of relative space do not lead to any dynamical differences in the descriptions of the motion; in anachronistic language, there is an equivalence class of relative spaces that give dynamically equivalent descriptions. There are also relative spaces that do lead to dynamical differences. If the relative space used to describe these motions is itself rotating, as St. Paul's does due to the motion of the earth, the true forces arising as interactions among bodies will not match the observed accelerations (due, in this case, to the Coriolis force). In principle, once such an acceleration is discovered one could describe the motion more accurately using a different relative space (in this case, by acknowledging earth's motion), with the hope of eventually determining the true motions. At the end of the famous scholium on space and time, Newton remarks that the entire *Principia* was composed so that the true motions could be found, and that "the situation is not entirely desperate" (*Principia*, 414). A relative space that is not suitable should always be revealed, in principle, by bodies accelerating without an identifiable physical force, but isolating such an effect requires first identifying and characterizing all the other forces relevant to the bodies' motion—and doing so requires employing the full framework of the *Principia*. The results described earlier in this chapter provide the framework to carry out a study of this kind for real motions, such as the motion of the objects in the solar system. Newton's success in characterizing a single force responsible for all of these motions, despite their complexity, makes a compelling case against desperation, in the sense that the study of trajectories within the quantitative framework of the *Principia* suffices to establish the underlying dynamics.

Geometrical ideas enter into this description of motion in two different senses. First, there are the measurements of geometrical properties and congruence made with respect to a chosen relative space, which are refinements of our subjective experience of distances and times. Geometry in this sense is not tied to the structures of "absolute space" described in DG, because the relationship between the chosen relative space and absolute space is unknown. Absolute space is itself insensible, as Newton remarks (*Principia*, 414), and the connection between a given relative space and absolute space arises only at the end of inquiry, so to speak—corresponding to a stage of inquiry in which all accelerated motions have been attributed to a physical force, with no discrepancies. Despite the inaccessibility of absolute space, it plays a fundamental role in Newton's dynamical analysis of motions: the crucial distinction between accelerated and nonaccelerated motion cannot be adequately captured, as Newton famously argued in the Scholium, in terms of motion with respect to relative spaces. Second, geometry enters directly into the formulation of the dynamics, as part of the characterization of the force law. The force of gravity, for example, depends on the distances among interacting bodies. Yet the geometrical structure required to fulfill this role is evanescent, in the sense that the force law depends only on the configuration of bodies at a given instant.[28] This second sense of spatial geometry connects back to the first via a dynamical account of how spatial measurements can be performed—for example, by using rigid bodies that can be moved from one location to another in order to make assessments of congruence. The subtlety of this connection between the dynamical role of geometry and geometrical properties revealed by measurements follows from the relativity of motion. The instantaneous geometrical properties may be shared with respect to different relative spaces, yet comparisons of locations at different times, as will be required for any spatial measurements, implicitly depend on a specific choice of relative space.

CONCLUSION

In closing, let me return to the general theme of the mathematization of nature by reflecting on the sense in which Newton's natural philosophy is based on mathematical principles. I have argued that Newton rejected one objection to this idea as simply reflecting a confusion: mechanics is no less exact in describing its objects of study than geometry. This is particularly clear given his kinematic conception of curves and its close parallel with the mechanical generation of curves via forces. Yet in order to maintain that geometry studies the properties of material rather than abstract entities, Newton is forced to regard these properties as remote from direct sensory experience. The main challenge is then one of epistemic accessibility: granted that there is an underlying quantitative structure that it is the aim of natural philosophy to uncover, how can this be done? The most striking aspect of the *Principia* is the depth of Newton's insight and mathematical resourcefulness in responding to this problem. The *Principia*'s mathematical framework provided a way of reasoning from the evidence provided by observed trajectories to claims about the underlying dynamics. The use of a mathematical framework to implement research via a controlled sequence of successive approximations is Newton's most influential contribution to the mathematization of nature within physics. The object of geometry, absolute space, is pushed to lie beyond our possible sense experience, as an ideal limit that may be reached only at the end of the series of successive approximations. It is then no surprise that Kant identified the nature and status of our knowledge of geometry as a pressing foundational problem for Newtonian science.

ABBREVIATIONS

MP Whiteside, D. T. 1967–81. *The Mathematical Papers of Isaac Newton* (MP is followed by a number that represents the volume, sometimes followed by page numbers)

NOTES

1. References to the *Principia* are to Newton 1999 [1726].

2. The unpublished manuscripts include, in particular, the *Geometriae* (ca. 1692, published in MP, 7), a long treatise on geometry Newton undertook in the 1690s, whose opening discussion amplifies the themes of the preface, as well as shorter fragments. My comments here are indebted to the discussion in Guicciardini (2009, chapters 13 and 14) and Domski (2003; 2010).

3. See Domski (2003). Newton also criticizes the Cartesian, algebraic approach to the classification of curves in the Appendix of *Arithmetica Universalis*.

4. Newton takes up this question in "De Gravitatione" and the famous Scholium to the definitions (see Stein 1967; Earman 1989; Rynasiewicz 1995; DiSalle 2006).

5. The comparison between Barrow and Newton, which I return to later in the chapter, is explored in much greater depth by Dunlop (2012a; 2012b). See

Guicciardini (2009, chapter 8) for a detailed contrast between Barrow's mathematics and Newton's fluxional analysis, and Mahoney (1990) and Stewart (2000) for further discussion of Barrow's mathematics.

6. Hobbes defended a similar view, namely that the scientific status of mathematics depended on the generation of geometrical magnitudes by motion, in an extended polemic with John Wallis, described with a clear account of the philosophy of mathematics involved by Jesseph (1999). Newton studied De Corpore closely in his student years, but I am not aware of any evidence regarding his assessment of the Hobbes–Wallis exchange.

7. Barrow was responding to debates regarding the scientific character of mathematical demonstrations going back to Piccolimini's 1547 treatise (see Mancosu 1996); considered in the context of these debates, Barrow advocates an unorthodox account of formal causality.

8. Guicciardini (2009) provides a masterful overview of Newton's development of the calculus, which draws on the manuscripts and Whiteside's editorial apparatus in the MP; see also Smith (2005).

9. For discussions of the mathematical style of the *Principia* and Newton's reasons for adopting it, see Whiteside (1970); MP, 6; Mahoney (1993); and Guicciardini (2009, chapters 10–12).

10. Smeenk and Schliesser (2013) give a more thorough treatment of the structure of the *Principia* along these lines, which is particularly indebted to discussions of Newton's methodology with George Smith and his publications on the topic (2001; 2002a; 2002b).

11. There are two distinct problems: first, *given* the orbit or trajectory, find a force law sufficient to produce it, and, second, *given* the force law and initial position and velocity, determine the trajectory. It is not known precisely what problem Halley posed to Newton, but Newton's response addresses the first.

12. This result appears in the *Principia* as Lemma 10, and the proof of the Lemma makes it clear that the proportionality only holds "ultimately" (or "in the limit" as the elapsed time goes to zero).

13. The manuscript was extended in a series of revisions leading to the *Principia*; see Herivel (1965) and MP, 6. Two other questions would likely have been raised by Halley: how do these ideas apply to the moon, with its apparently non-Keplerian motion, and to comets, which move through a much wider range of distances from the sun?

14. The *Principia*'s first two books consider the motions of body generally, that is under a variety of different force laws, moving in spaces without resis-

tance (book 1) and with resistance of different kinds (book 2). Book 3, "The System of the World," gives the argument for universal gravitation and further implications of gravity.

15. Most of the results are restricted to forces varying as $f \propto rn$, for integer values of n, but several important theorems also hold for rational values of n (e.g., prop. 1.45).

16. Here I draw on the discussion of measurement given, in roughly this sense, by Harper (2011), in his defense of a Newtonian ideal of empirical success. Cf. Dunlop (2012b) for a different take on measurement, emphasizing connections with Barrow and the importance of judgments of congruence.

17. This brief discussion of section 10 summarizes the conclusions of Smeenk and Smith (2012).

18. The restriction to a uniformly dense spheroid is needed to ensure that gravity inside the spheroid varies directly with distance. This relationship is formulated with regard to a specific physical situation. The Galilean theory does not approximate Newton's theory in all cases, and the asymptotic limit is needed because in Newton's theory the case described by Galilean gravity only obtains in the limit and not in any physically realizable case.

19. Newton had proved that for a body moving on an elliptical trajectory with the force directed at the *center*, the force varies directly with the distance. Yet for elliptical orbits that are nearly circular, as in the case of the planets, the two foci and the center are not far apart. As Smith (2002a) indicates, an inference to an inverse-square force from Kepler's first law is not robust, in the sense that the conclusion does not follow if the antecedent only holds approximately.

20. See Wilson (1989); Smith (2014, 262) for more detailed discussions of Newton's contribution to celestial mechanics.

21. For discussions of this argument, see Stein (1967); Earman (1989); Rynasiewicz (1995); DiSalle (2006).

22. See Dunlop (2012a) for a sympathetic reconstruction of Barrow's appeal to practice as a response to this question.

23. See, in particular, McGuire (2007) for an assessment of Newton's debt to, and contrasts with, Descartes, as well as Domski (2012), who draws interesting parallels with Proclus's neo-Platonic philosophy of mathematics, and Stein (2002). See also Biener and Smeenk (2012) for a discussion of the account of body in DG parallel to this discussion of space.

24. Newton probably had cases such as Torricelli's trumpet in mind, which Barrow (1734) discussed in lecture 16 of his *Mathematical Lectures*.

25. Dated to the 1690s, and translated by McGuire (1978).

26. But see Domski (2012) for the most persuasive and careful attempt at a coherent account.

27. There is much more continuity in other aspects of Newton's views that are not my focus here, such as the infinity of space and its relation to God.

28. For further discussion, see, in particular, Stein (1991).

REFERENCES

Barrow, I. 1734. *The Usefulness of Mathematical Learning Explained and Demonstrated: Being Mathematical Lectures Read in the Publick Schools at the University of Cambridge.* London: Stephen Austen.

———. 2006. *The Geometrical Lectures of Isaac Barrow.* Trans. J. M. Child. New York: Dover.

Biener, Z., and C. Smeenk. 2012. "Cotes' Queries: Newton's Empiricism and Conceptions of Matter." In *Interpreting Newton: Critical Essays*, ed. A. Janiak and E. Schliesser, 105–37. Cambridge: Cambridge University Press.

Bos, H. J. M. 2001. *Redefining Geometrical Exactness: Descartes's Transformation of the Early Modern Concept of Construction.* New York: Springer Verlag.

Cohen, I. B. 1980. *The Newtonian Revolution: With Illustrations of the Transformation of Scientific Ideas.* Cambridge: Cambridge University Press.

Dijksterhuis, E. J. 1961. *The Mechanization of the World Picture.* Trans. C. Dikshoorn. Princeton, N.J.: Princeton University Press.

DiSalle, R. 2006. *Understanding Space-time: The Philosophical Development of Physics from Newton to Einstein.* Cambridge: Cambridge University Press.

Domski, M. 2003. "The Constructible and the Intelligible in Newton's Philosophy of Geometry." *Philosophy of Science* 70, no. 5: 1114–24.

———. 2010. "Newton as Historically-Minded Philosopher." In *Discourse on a New Method: Reinvigorating the Marriage of History and Philosophy of Science*, ed. M. Domski and M. Dickson, 65–89. Chicago: Open Court Pub Co.

———. 2012. "Newton and Proclus: Geometry, Imagination, and Knowing Space." *The Southern Journal of Philosophy* 50, no. 3: 389–413.

Dunlop, K. 2012a. "The Mathematical Form of Measurement and the Argument for Proposition I in Newton's *Principia*." *Synthese* 186, no. 1: 191–229.

———. 2012b. "What Geometry Postulates: Newton and Barrow on the Relationship of Mathematics to Nature." In *Interpreting Newton: Critical Essays*, ed. A Janiak and E. Schliesser, 69–101. Cambridge: Cambridge University Press.

Earman, J. S. 1989. *World Enough and Space-Time: Absolute versus Rational Theories of Space and Time.* Cambridge, Mass.: MIT Press.

Gandt, F. de. 1995. *Force and Geometry in Newton's* Principia. Trans. C. Wilson. Princeton, N.J.: Princeton University Press.

Guicciardini, N. 2009. *Isaac Newton on Mathematical Certainty and Method.* Cambridge, Mass.: MIT Press.

Hall, A. R., and M. B. Hall. 1962. *Unpublished Scientific Papers of Isaac Newton.* Cambridge: Cambridge University Press.

Harper, W. L. 2011. *Isaac Newton's Scientific Method: Turning Data into Evidence about Gravity and Cosmology.* New York: Oxford University Press.

Herivel, J. 1965. *The Background to Newton's* Principia. Oxford: Oxford University Press.

Jesseph, D. 1999. *Squaring the Circle: The War between Hobbes and Wallis.* Chicago: University of Chicago Press.

Koyré, A. 1965. *Newtonian Studies.* Cambridge, Mass.: Harvard University Press.

Mahoney, M. S. 1990. "Barrow's Mathematics: Between Ancients and Moderns." In *Before Newton: The Life and Times of Isaac Barrow*, ed. M. Feingold, 179–249. Cambridge: Cambridge University Press.

———. 1993. "Algebraic vs. Geometric Techniques in Newton's Determination of Planetary Orbits." In *Action and Reaction*, ed. P. Theerman and A. F. Seeff, 183–205. Newark: University of Delaware Press.

Mancosu, P. 1996. *Philosophy of Mathematics and Mathematical Practice in the Seventeenth Century.* Oxford: Oxford University Press.

McGuire, J. E. 1978. "Newton on Place, Time, and God: An Unpublished Source." *British Journal for the History of Science* 11: 114–29.

———. 2007. "A Dialogue with Descartes: Newton's Ontology of True and Immutable Natures." *Journal of the History of Philosophy* 45, no. 1: 103–25.

Newton, I. 1999 [1726]. *The Principia, Mathematical Principles of Natural Philosophy: A New Translation.* Trans. I. B. Cohen and A. Whitman. Berkeley: University of California Press.

———. 2004. *Isaac Newton: Philosophical Writings.* Ed. A. Janiak. Cambridge: Cambridge University Press.

Rynasiewicz, R. 1995. "By Their Properties, Causes and Effects: Newton's Scholium on Time, Space, Place and Motion–I. The Text." *Studies in History and Philosophy of Science* 26, no. 1: 133–53.

Smeenk, C., and E. Schliesser. 2013. "Newton's Principia." In *The Oxford Handbook of the History of Physics*, ed. J. Buchwald and R. Fox, 109–65. Oxford: Oxford University Press.

Smeenk, C., and G. E. Smith. 2012. "Newton on Constrained Motion." Unpublished manuscript.

Smith, G. E. 2001. "The Newtonian Style in Book II of the *Principia*." In *Isaac Newton's Natural Philosophy*, ed. J. Z. Buchwald and I. B. Cohen, 249–314. Cambridge, Mass.: MIT Press.

———. 2002a. "From the Phenomenon of the Ellipse to an Inverse-Square Force: Why Not?" In *Reading Natural Philosophy: Essays in the History and Philosophy of Science and Mathematics to Honor Howard Stein on his 70th Birthday*, ed. D. B. Malament, 31–70. Chicago: Open Court.

———. 2002b. "The Methodology of the *Principia*." In *Cambridge Companion to Newton*, ed. I. B. Cohen and G. E. Smith, 138–73. Cambridge: Cambridge University Press.

———. 2005. "Newton's Research in Mathematics." Unpublished manuscript.

———. 2014. "Closing the Loop." In *Newton and Empiricism*, ed. Z. Biener and E. Schliesser, 262–332. Oxford: Oxford University Press.

Stein, H. 1967. "Newtonian Space-time." *Texas Quarterly* 10: 174–200.

———. 1991. "On Relativity Theory and Openness of the Future." *Philosophy of Science* 58, no. 2: 147.

———. 2002. "Newton's Metaphysics." In *Cambridge Companion to Newton*, ed. I. B. Cohen and G. E. Smith, 256–307. Cambridge: Cambridge University Press.

Stewart, I. 2000. "Mathematics as Philosophy: Proclus and Barrow." *Dionysius* 18: 151–81.

Torretti, R. 1978. *Philosophy of Geometry from Riemann to Poincaré*. Dordrecht: D. Reidel.

Whiteside, D. T., ed. 1967–81. *The Mathematical Papers of Isaac Newton*. 8 vols. Cambridge: Cambridge University Press.

———. 1970. "The Mathematical Principles Underlying Newton's *Principia Mathematica*." *Journal for the History of Astronomy* 1: 116–38.

Wilson, C. 1989. "The Newtonian Achievement in Astronomy." In *Planetary Astronomy from the Renaissance to the Rise of Astrophysics*, ed. R. Taton and C. Wilson, Part A: Tycho Brahe to Newton, 1:233–74. Cambridge University Press.

CONTRIBUTORS

Roger Ariew is professor and chair of the Department of Philosophy at the University of South Florida. He is author of *Descartes and the First Cartesians* and *Descartes among the Scholastics*, and editor and translator of Descartes's *Philosophical Essays and Correspondence* and Pascal's *Pensées*. He is currently working on *Descartes' Correspondence: A Historical-Critical Edition and English Translation*, with Theo Verbeek and Erik-Jan Bos et al.

Richard T. W. Arthur is professor of philosophy at McMaster University. He is author of *G. W. Leibniz: the Labyrinth of the Continuum*; *Natural Deduction*; and *Leibniz*. He has published and presented extensively on early modern philosophy, history and philosophy of science and math, and philosophy of modern physics. He is former president of the Canadian Society for History and Philosophy of Science and is a member of the board of the North American Leibniz Society.

Lesley B. Cormack is professor of history and dean of the Faculty of Arts at the University of Alberta. She is author of *Charting an Empire: Geography at the English Universities 1580–1620* and coauthor of *A History of Science in Society: From Philosophy to Utility*. She has published essays on the history of science and is president of the Canadian Society for the History and Philosophy of Science.

Daniel Garber is Stuart Professor of Philosophy at Princeton University. He is author of *Descartes' Metaphysical Physics*; *Descartes Embodied*; and *Leibniz: Body, Substance, Monad*. He is coeditor of *The Cambridge History of Seventeenth Century Philosophy* and the author of essays on the history of philosophy and science in the early modern period. He is working on a monograph on the seventeenth-century scholastico-novator Jacobus Fontialis.

Ursula Goldenbaum is associate professor in the Department of Philosophy at Emory University. She the author of monographs on Spinoza and on public debates in the eighteenth century and coeditor of *Infinitesimal Differences: Controversies between Leibniz and His Contemporaries*. She is a member of the board of the *Journal of the History of Ideas* and president of the North American Leibniz Society.

Geoffrey Gorham is professor of philosophy at Macalester College and resident fellow of the Minnesota Center for Philosophy of Science at the University of Minnesota. He has published numerous articles on early modern philosophy and science in the *Journal of the History of Philosophy,* the *British Journal for the History of Philosophy, Studies in the History and Philosophy of Science,* and *Early Science and Medicine.*

Benjamin Hill is associate professor of philosophy and a founding member of the Rotman Institute of Philosophy at Western University. His interests include Lockean empiricism, the connections between sensory-based beliefs and scientific knowledge, and the status of medicine and medical knowledge in the seventeenth century. He is coeditor of *The Philosophy of Francisco Suárez* and *Sourcebook in the History of the Philosophy of Language.*

Dana Jalobeanu is reader in philosophy and director of the Institute of Research in Humanities at the University of Bucharest. She is coeditor of the *Journal of Early Modern Studies,* executive editor of *Society and Politics,* and co-organizer of the Bucharest–Princeton Seminar in Early Modern Philosophy. She is author of *The Art of Experimental Natural History: Francis Bacon in Context* and coeditor of *Vanishing Matter and the Laws of Nature: Descartes and Beyond.*

Douglas Jesseph is professor of philosophy at the University of South Florida. He is author of *Berkeley's Philosophy of Mathematics*; *Squaring the Circle: The War between Hobbes and Wallis*; and essays on early modern philosophy, mathematics, and methodology.

Carla Rita Palmerino is professor of history of modern philosophy at Radboud University, Nijmegen, The Netherlands, and professor of philosophy at the Open University. Her research focuses on seventeenth-century theories of matter and motion, with emphasis on debate concerning the ontological and mathematical foundation of Galileo's new science, diagrammatic reasoning, and the use of thought experiments in early modern science and philosophy.

Eileen Reeves is professor of comparative literature at Princeton University. She is author of *Painting the Heavens: Art and Science in the Age of Galileo*; *Galileo's Glassworks: The Telescope and the Mirror*; *Evening News: Optics, Astronomy, and Journalism in Early Modern Europe*; and coauthor of *On Sunspots*. She is an associate member of the Program in the History of Science and chair of the Department of Comparative Literature.

Edward Slowik is professor of philosophy at Winona State University and a resident fellow at the Minnesota Center for Philosophy of Science at the University of Minnesota. His area of research is the history and philosophy of science, with emphasis on the philosophy of space in contemporary physics and early modern philosophy. He is tauthor of *Cartesian Spacetime*.

Christopher Smeenk is associate professor in philosophy and director of the Rotman Institute of Philosophy at Western University. His research focuses on the interplay between theory and evidence in physics and includes projects on Newton's methodology, the discovery of general relativity, and recent work in early universe cosmology.

Justin E. H. Smith is University Professor of History and Philosophy of Science at the University of Paris. He is author most recently of *Nature, Human Nature, and Human Difference: Race in Early Modern Philosophy*.

Kurt Smith is professor of philosophy at Bloomsburg University of Pennsylvania. He is author of *Matter Matters: Metaphysics and Methodology in the Early Modern Period*; *The Descartes Dictionary*; and several chapters, journal articles, and encyclopedia entries on early modern theories of ideas and ideational representation. His research focuses on conceptions of the unconscious that emerge in the modern period.

C. Kenneth Waters is Canada Research Chair in Logic and Philosophy of Science. His research is on the epistemology and metaphysics of scientific practices, especially investigative practices using genetics and theoretical practices in evolutionary biology. He is coeditor of *Scientific Pluralism* (Minnesota, 2006).

INDEX OF NAMES

INDEX OF SUBJECTS